Instrument Transducers

Instrument transducers

An introduction to their performance and design

HERMANN K. P. NEUBERT

SECOND EDITION

CLARENDON PRESS · OXFORD
1975

Oxford University Press, Ely House, London W. 1

GLASGOW NEW YORK TORONTO MELBOURNE WELLINGTON

CAPE TOWN IBADAN NAIROBI DAR ES SALAAM LUSAKA ADDIS ABABA

DELHI BOMBAY CALCUTTA MADRAS KARACHI LAHORE DACCA

KUALA LUMPUR SINGAPORE HONG KONG TOKYO

ISBN 0 19 856320 5

© OXFORD UNIVERSITY PRESS 1963, 1975

FIRST EDITION 1963
SECOND EDITION 1975

TYPESET IN NORTHERN IRELAND AT THE UNIVERSITIES PRESS, BELFAST
AND PRINTED IN GREAT BRITAIN BY J. W. ARROWSMITH LTD., BRISTOL

Preface

THIS second edition of 'Instrument Transducers' maintains, in a climate of growing interest and sophistication, the emphasis on transducers as entities with respect to physical principles, performance analysis, design, and construction. Aspects of mechanical and electrical design procedures, of specific materials, and of modern construction techniques—often sadly neglected in 'user's' books on transducers—are discussed in greater detail than available elsewhere, against the background of exacting performance requirements.

The second edition has been entirely re-written; text, illustrations, and lists of references have been up-dated and the consistent International System of Units (SI) has been employed throughout. (There is a brief introduction to the SI and comprehensive conversion tables of units at the end of the book.) In particular, sections on mathematical models of bilateral transducers, on semiconductor and thin-film strain gauges, and on six-component sting balances have been included, and the chapter on force-balance transducers has been greatly extended, including a number of original contributions. In many other places new illustrations and tables have been added to match the up-dated text.

The second edition again aims at experimental physicists and electrical engineers active in the fast-expanding field of instrumentation and control engineering, which now covers not only research and development work, but also numerous branches of modern industry.

This new edition could not have been written without the support by my former colleagues at the Royal Aircraft Establishment, Farnborough, and the help offered by many friends in industry. I also wish to extend my sincere thanks to the staff of the Clarendon Press for their co-operation in matters of presentation, and to my wife for her unfailing patience.

The material of the book has been drawn from many sources, identified in the text, which cannot all be mentioned here, though permission by the Controller of H. M. Stationery Office to reproduce the following Crown Copyright illustrations is acknowledged: Figs 2.2, 2.3, 4.1.4–4.1.14, 4.2.13, 4.2.24, 4.3.8, 4.3.18–4.3.20, 4.4.21, 5.2–5.11, 5.21, 5.23, 5.25.

Contents

1. Introduction

INSTRUMENT transducers are devices which, for the purpose of measurement, turn physical input quantities into electrical output signals, their output–input and their output–time relationship being predictable to a known degree of accuracy at specified environmental conditions. The aim of this book is to introduce the reader to the appreciation, the performance, and the design of instrument transducers conforming with the above definition.

The title *Instrument transducers* [1] is to indicate the purpose of our type of transducers: measurement. It therefore excludes sender-type transducers such as loudspeakers, actuators, or vibrators. The frequently encountered concept of 'sensor' has been reserved for sensing elements employed in complex transducers where they appear, for instance, as strain sensors in pressure transducers, or as displacement sensors in force-balance transducers.

Chapter 2 gives an over-all view of instrument transducers with respect to their input and output quantities, linked by the particular transducing principle employed. It discusses the problems concerning the classification of instrument transducers. It also offers a critical assessment of electromechanical analogies and develops a unified theory of bilateral electromechanical transducers. There follows a brief introduction to feedback-type transducers.

Chapter 3 discusses the performance of instrument transducers with mechanical input elements, constituting first- and second-order vibratory systems which may be excited by static, sinusoidal, or transient input quantities. The three essential elements of these systems, inertia, compliance, and damping, are treated with respect to their concept and realization. This is followed by a discussion of pressure-sensitive elements and the effect of pressure inlet pipes on the dynamic response of instrument transducers.

Chapter 4, by far the longest of all, deals with the performance and design of electrodynamic, resistive, inductive, capacitive, and piezoelectric transducers in the light of their output characteristics. After an introduction to the basic physical principles involved the discussion goes into details of design parameters and materials peculiar to individual types of transducers, and deals with problems arising in their construction.

Chapter 5 is devoted to force-balance transducers with a feedback path between transducer output and input. It develops a unified theory of the static and dynamic performance of acceleration and pressure transducers employing either electrodynamic or electrostatic servo-actuators, and concludes with a brief survey of actual servo-transducer designs.

In recent years the sheer number and wide diversity of instrument transducers available, or described, have grown to such an extent that not even a cursory survey for the user could cover the whole field, let alone a book like this which was conceived as a useful guide for the designer. For instance, thermocouples and photoelectric cells, which have a well-known literature of their own, have been omitted in order to avoid unnecessary duplication, and so have microphones and hydrophones for use in communication. Likewise, the scope of the book has been restricted to instrument transducers with analogue electrical outputs; a great variety of analogue-to-digital converters are now commercially available [2]. Digital transducers based on frequency modulation of their mechanical elements, such as vibrating strings, have been known since 1919 and are still in demand. A vibrating-cylinder pressure transducer followed in 1959 [3]. Other digital transducer types have become known since ([4]–[9]), particularly for use in numerically controlled machine-tools.

Generally speaking, instrument transducers are better understood than 10 years ago, and there have been a number of notable additions to transducer technology. Piezoresistive sensors using silicon-based semiconductor elements, and new piezoelectric and magnetic materials have been added to the transducer designer's armory. More efficient associated electronic circuitry, such as charge amplifiers, operational amplifiers, and complex integrated-circuit configurations of minute size have been coming into general use. A continuously growing demand for numerous and highly sophisticated control systems in old and new branches of industry has maintained a steady growth rate in the transducer field and is likely to continue to do so. Although the image of the transducer as a minor 'gadget' in the noble family of instrumentation systems has perhaps gone by now, the struggle for priority with respect to effort and funds for new transducer developments is still on [10].

References

1. J. THOMSON. Instrument transducers. *J. scient. Instrum.* **34,** 217–21 (1957).
2. R. E. FISCHBACHER and A. E. S. MILLS. *Analogue-to-digital converters*, pp. 679–87. IEA Yearbook and Buyers Guide, London (1966).
3. ANON. Vibrating cylinder transducer. *Engineer, Lond.* **207,** 665 (1959).
4. M. L. KLEIN, F. K. WILLIAMS, and H. C. MORGAN. Digital Transducers. *Instrums. Automn.*, **30,** 1299–300 (1957).
5. J. D. COONEY and B. K. LEDGERWOOD. 31 numerically controlled point-to-point positioning systems. *Control Enging.* **5,** 105 (1958).
6. H. F. FINDEN. The inductosyn. In *Progress in automation 1* (ed. A. D. Booth), p. 82. Butterworth, London (1960).
7. A. E. SCHULER. Digital transducers—types and trends. *Instrum. Technol.* **16,** 41–6 (1969).

8. R. R. RICHARD. Development of an inherently digital transducer. *NASA tech. Note* **D-6694** (1972).
9. H. A. DOREY. Digital instrumentation. *IEE Conference on Digital Instrumentation,* London (1973).
10. Instrumentation in the 70s. *Instrum. Technol.* **17,** January issue (1970).

2. Classification of instrument transducers

In Chapter 1 we have defined instrument transducers, for the purpose of this book, as devices for the measurement of physical quantities by electrical means, i.ę. transducers with a calibrated relationship between the input and output quantities. They may be genuine energy converters, or they may require an auxiliary energy source. From this definition it is obvious that the classification of instrument transducers can be related either to the input quantities to be measured by the transducer, or to the output quantities determined by the transducing principle employed. With the majority of instrument transducers a clean separation between these two aspects of input and output is feasible, especially with all passive transducers, which require an auxiliary source of energy. Thus in the first two sections of this chapter we shall discuss the classification of instrument transducers with respect to their input (section 2.1) and output (section 2.2) characteristics, while in the body of the book input and output characteristics of major types of transducers will be treated in greater detail in Chapters 3 and 4, respectively.

However, some transducer types, e.g. electrodynamic generators, with close electromechanical coupling between the electrical and mechanical parameters, require a somewhat more complex approach in which the input and output characteristics must be considered together. Section 2.3.1 will deal with this aspect in terms of electromechanical analogies, while a unified theory, treating the coupling characteristics of bilateral electromechanical transducers without the use of analogies, will be given in section 2.3.2. Finally, section 2.4 will introduce force-balance-type transducers with feedback between input and output, which will be studied in detail in Chapter 5.

2.1. Primary quantities: input characteristics

To the user of instrument transducers, classification with respect to primary quantities to be measured is probably quite natural. If, for instance, he wants to measure dynamic pressure on a wind-tunnel model, he would hope to find a survey of various kinds of pressure transducers described by their essential features so as to make a suitable choice for his particular needs. Excellent surveys of instrument transducers from this point of view are given in Lion's [1] and Doebelin's [2] books; an even wider field is covered by Kohlrausch's compendium [3] on methods of practical physics. A fair number of useful books on transducers have now been published in this country and abroad ([4]–[11]), one having originated in Russia [12], and there are also comprehensive lists ([13]–[15]) of commercially available transducers.

TABLE 2.1

List of some primary (input) quantities of instrument transducers

Basic quantities	Derived quantities
Linear displacement	Length, width, thickness, position, level, erosion, wear, surface quality, strain, vibration
Angular displacement	Attitude, angle of turn, angle of incidence, angle of flow, angular vibration
Linear velocity	Speed, rate of flow, momentum, vibration
Angular velocity	Angular speed, rate of turn, (roll, pitch, yaw), angular momentum, angular vibration
Linear acceleration	Vibration, force, impact (jerk), mass, stress
Angular acceleration	Angular vibration, torque, angular impact, moment of inertia
Force	Weight, thrust, density, stress, torque, pressure, altitude, fluid velocity, sound, acceleration
Temperature	Heat flow, fluid flow, gas pressure, gas velocity, angle of flow, turbulence, sound
Light	Light flux and density, spectral distribution, length, strain, force, torque, frequency, number
Time	Frequency, number, statistical distribution

However, for the student of instrument transducers interested in their design and performance, this is not necessarily the best line of approach, since it impedes an efficient study of basic features such as coupling characteristics and dynamic response. Classification with respect to input quantities, however convenient it may be to the user, leads to repetition or, alternatively, to superficiality. It will therefore suffice to list in Table 2.1 the basic primary quantities and some of their more important derivatives. Although not comprehensive, this list will give the reader an idea of the multiplicity and variety of transducer inputs.

More important, from a designer's point of view, are the static and dynamic response characteristics of transducer input components, such as mass–spring-damping systems which are common to all instrument transducers with mechanical input and which are, by analogy, also applicable to transducers with heat, light, or other physical input qualities. These fundamental characteristics will be discussed in Chapter 3.

2.2. Secondary quantities: output characteristics

For the intimate study of instrument transducers classification with respect to the transducing principles provides the most suitable approach and has been adopted in this book. The transducing principle determines the electrical output characteristics of the transducer which may be resistance, inductance, or capacitance, or a complex combination of these.

The conversion of one form of energy into another is the subject of textbooks on physics. There are also monographs dealing with this particular aspect ([16]–[18]) which are of great interest to the designer of instrument transducers. The following short list (Table 2.2) gives, in rough order of usefulness, a survey of the more common physical effects employed in instrument transducers. To the research worker in this field

TABLE 2.2

List of some physical effects for use in instrument transducers

(a) ENERGY-CONVERSION PRINCIPLES

Electrodynamic ⎫	
Electromagnetic ⎬	(Generators)
Electrostatic ⎭	
Thermoelectric	(Thermocouples)
Piezoelectric	
Photoelectric	(Photo-emission)
Photovoltaic	(Photo-junction)
Magnetostrictive	
Electrostrictive	
Electrokinetic	(Streaming potential)
Pyroelectric	
Triboelectric	
Galvanic	(Electrochemical)

(b) ENERGY-CONTROLLING PRINCIPLES

Resistance ⎫	
Inductance ⎬	(Controlled by geometry)
Capacitance ⎭	
Thermoresistance	
Elastoresistance	(Strain gauge)
Hall effect	
Magnetoresistance	
Photoresistance	
Photo-ionization	
Radioactive absorption	
Ionic conduction	(Humidity)

the list may seem rather conventional, while the instrument designer and engineer may protest that some of the listed transducing principles are not yet suitable for application in high-class instrument transducers, especially with respect to secondary effects, such as sensitivity to environmental temperature variations.

2.3. Electromechanical coupling characteristics

2.3.1. Electromechanical analogies

Analogies between electrical and mechanical quantities are as old as our knowledge of electrical phenomena. Tension (voltage), current, and resistance are concepts of mechanical origin. There is no basic difference between the innocent intentions with which these words were coined and the objects to which the more sophisticated electromechanical analogies of our times are applied. Analogies serve the purpose of furthering a better understanding in one field of study by employing the 'language' of another field in order to 'explain' common characteristics. On the one hand, there are mechanical models of electrical phenomena and, on the other, complex mechanical systems conveniently analysed by reduction to well explored electrical circuits.

Additional to the latter aspect—and more important in the present context—is the possibility of combining mechanical and electrical elements of an electromechanical system in one purely electrical equivalent circuit which then can be analysed for its over-all coupling characteristics. However, this method—useful as it is in a wide range of applications—has serious shortcomings which will become apparent at closer inspection.

Electromechanical analogies are based on the formal similarities of the differential equations applicable to the mechanical systems under investigation and of their electrical equivalent circuits. Considering the expressions for mechanical and electrical power:

$$\text{mechanical: } \begin{cases} P - Fx/t = mx^2/t^3 & \text{(rectilinear system)} & (2.1a) \\ P = M\omega & \text{(rotary system)} & (2.1b) \end{cases}$$

$$\text{electrical: } \quad P = ie, \qquad\qquad\qquad\qquad\qquad (2.2)$$

where

P = power M = moment

F = force ω = angular velocity

x = displacement i = current

t = time e = voltage

m = mass v = velocity.

Equating the mechanical and the electrical power, we have

$$ie = Fx/t = mx^2/t^3 \qquad (2.3a)$$

and

$$ie = M\omega. \qquad (2.3b)$$

Multiplying numerator and denominator of the left-hand terms of eqns (2.3a) and (2.3b) by a factor n, the equations can be split into two each:

$$e/n = F = mx/t^2 \quad \text{and} \quad in = x/t = v, \qquad (2.4a)$$

$$e/n = M = Fx \quad \text{and} \quad in = \omega = 1/t. \qquad (2.4b)$$

The 'translation' factor n governs the relationship between the mechanical and the electrical analogies. In principle it may assume any magnitude and

TABLE 2.3

List of electromechanical analogies; rectilinear system; $n = 1$

Mechanical quantities		Electrical quantities	
Force	F	Voltage	e
Velocity	v	Current	i
Displacement	x	Charge	Q
Momentum.	p	Magnetic flux	Φ
Viscous resistance	c	Resistance	R
Mass	m	Inductance	L
Compliance	$1/k$†	Capacitance	C

† k represents the spring constant.

any dimensions leading to an infinite number of analogies. For practical purpose, however, our choice is limited.

Consider the rectilinear mechanical system first (eqn (2.4a)).

(a) With $n = 1$, we have

$$e = F = mx/t^2 \quad \text{and} \quad i = x/t = v, \qquad (2.5)$$

and the list of appropriate analogies of quantities is given in Table 2.3. Thus, the analogue differential equations in terms of 'charge' and 'displacement' are

$$\text{electrical:} \quad e = L\frac{d^2Q}{dt^2} + R\frac{dQ}{dt} + \frac{Q}{C}, \qquad (2.6a)$$

$$\text{mechanical:} \quad F = m\frac{d^2x}{dt^2} + c\frac{dx}{dt} + \frac{x}{1/k}, \qquad (2.6b)$$

or in terms of 'current' and 'velocity':

$$\text{electrical:} \quad e = L\frac{di}{dt} + Ri + \frac{1}{C}\int i \, dt \qquad (2.7a)$$

$$\text{mechanical:} \quad F = m\frac{dv}{dt} + cv + k\int v \, dt. \qquad (2.7b)$$

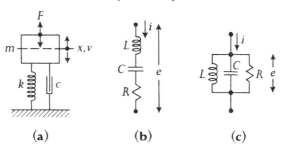

FIG. 2.1. (a) Mechanical vibratory system; (b) 'direct' electrical analogy of (a); (c) 'inverse' electrical analogy of (a).

While the electrical equations (2.6a) and (2.7a) describe a series connection with the charge or current common to all its elements, the mechanical equations (2.6b) and (2.7b) apply to a parallel configuration with the displacement or velocity as the common characteristic of the elements (Fig. 2.1(a) and (b)). This type of analogy—sometimes referred to as the 'direct' analogy—is appropriate for systems with an 'affinity' between force and voltage, such as piezoelectric transducers. Here the charge or voltage generated is directly proportional to the applied force. (See section 4.5 on piezoelectric transducers.) However, an obvious shortcoming of this analogy is the necessity to 'translate' mechanical parallel configurations into electrical series circuits.

Another workable analogy for rectilinear mechanical systems can be obtained by putting·

(b) $n = t/m$. Then we have

$$e - x/t = v \quad \text{and} \quad i = mx/t^2 = F, \tag{2.8}$$

and the corresponding list of analogies is given in Table 2.4.

The differential equations, for instance in 'voltage' terms, are

$$i - C\frac{de}{dt} + \frac{e}{R} + \frac{1}{L}\int e \, dt \tag{2.9}$$

TABLE 2.4

List of electromechanical analogies; rectilinear system; $n = t/m$

Mechanical quantities		Electrical quantities	
Force	F	Current	i
Velocity	v	Voltage	e
Displacement	x	Magnetic flux	Φ
Momentum	p	Charge	Q
Viscous resistance	c	Conductance	$1/R$
Mass	m	Capacitance	C
Compliance	$1/k$	Inductance	L

and eqn (2.7b). Both the electrical and the mechanical 'circuits' are thus parallel configurations (Fig. 2.1(a) and (c)). This 'inverse' analogy has often been used in electrodynamic generator-type transducers with an 'affinity' between velocity and voltage. However, a serious shortcoming here is the inverse frequency characteristics of the elements. The effect of mass and inductance (which are not analogue pairs) increase with frequency, while the effects of compliance and capacitance decrease. The inverse analogy thus breaks down with respect to frequency behaviour, while the direct analogy breaks down with respect to circuit configurations.

Tables 2.3 and 2.4 deal with the two most common electrical analogies of rectilinear mechanical systems. Likewise, the analogies of rotary systems can be derived from eqn (2.4b).

(c) With $n = 1$ we have

$$e = M \quad \text{and} \quad i = \omega, \tag{2.10}$$

the direct analogy.

(d) With $n = \omega/M$, the inverse analogy. Tables of analogous quantities, similar to those of Tables 2.3 and 2.4 can be compiled, if the equivalent rotational quantities are substituted for the rectilinear quantities. The circuit and frequency aspects, discussed above, also apply here.

Electromechanical analogies have a long history [19], [20] and are well documented [21], [22] and widely used. Nevertheless, the conscientious reader may still be worried by the arbitrary choice of the 'translation' factor n. In fact, an electromechanical system requires four basic dimensions, one more than the three basic dimensions necessary and sufficient for purely mechanical systems. If—as we have implied above—the n-factors are either dimensionless or have three basic mechanical dimensions only, the analogous electrical quantities are bound to emerge with wrong dimensions. But this is a characteristic of formal analogies, in contrast to identities. There is, therefore, an obvious need for a consistent theory of transducers with distinct coupling characteristics between their input and output parameters.

2.3.2. *Unified theory of bilateral electromechanical transducers*

Bilateral transducers are electromechanical energy converters which operate by reversible physical laws, such as Faraday's law of electrodynamics and the Curie's direct and reciprocal piezoelectrical effects. By virtue of their reversibility, bilateral electromechanical transducers can be employed either as 'receivers' (sensors) or 'senders' (generators) of mechanical quantities, such as dynamic forces and velocities, where a 'user' of electrical signals, or an 'originator' of electrical excitation, is thought to be positioned at the electrical transducer end. Instrument transducers, as defined in this book, are receivers; the sender characteristics of bilateral transducers are strictly not relevant, but they cannot be ignored here, since only a comprehensive generalized treatment of both receiver and sender aspects will provide a full understanding of bilateral-transducer performance.

The unified theory employs the well-established concept of electrical four-terminal, or two-port, networks combined with elementary matrix algebra; it affords generality without loss of detail and is particularly suited to bringing out common features. The treatment retains the mechanical and electrical quantities unaltered [23], [24] and derives the two basic transfer matrix equations (2.33a) and (2.33b) of a general (passive) two-port which fully describe the properties of voltage-source- and current-source-type bilateral transducers. It thus avoids the inconsistencies built into the concept of electromechanical analogies explained in section 2.3.1.

2.3.2.1. Basic two-port equations.   Consider Fig. 2.2(a) which shows an electromechanical transducer represented by a two-port, i.e. by a 'black box' which is capable of 'transducing' the mechanical quantities, force f_t and velocity v_t, into the electrical quantities, voltage e_t and current i_t, and vice versa. In the general case all quantities are complex numbers.

Since bilateral transducers are mechanical receivers (mechanical-to-electrical conversion), as well as senders (electrical-to-mechanical conversion), we shall, for convenience, choose the signs (arrows) of the quantities such that both the mechanical power P_m and the electrical power P_e are positive, if they are directed towards the two-port, as shown in Fig. 2.2(a). The signs, therefore, do not indicate the actual flow of energy through a receiver or a sender.

In SI units we have at the mechanical port,

$$f_t = \text{transducer force (N)},$$

$$v_t = \text{transducer velocity (m s}^{-1}\text{)},$$

and the mechanical power becomes

$$P_m = \tfrac{1}{2}\text{Re}(f_t v_t^*) = \tfrac{1}{2}\text{Re}(v_t f_t^*) \quad \text{(W)}, \qquad (2.11\text{a})$$

where asterisks indicate the conjugates of v_t and f_t, respectively. At the electrical port,

$$e_t = \text{transducer voltage (V)},$$

$$i_t = \text{transducer current (A)},$$

with the electrical two-port power being

$$P_e = \tfrac{1}{2}\text{Re}(i_t e_t^*) = \tfrac{1}{2}\text{Re}(e_t i_t^*) \quad \text{(W)}. \qquad (2.11\text{b})$$

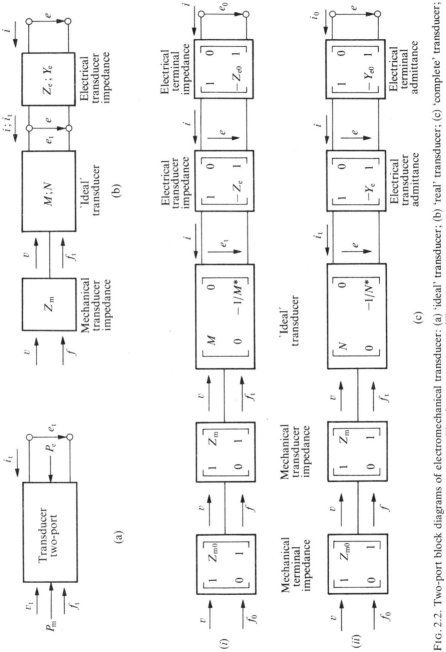

FIG. 2.2. Two-port block diagrams of electromechanical transducer: (a) 'ideal' transducer; (b) 'real' transducer; (c) 'complete' transducer; (i) voltage-source type, (ii) current-source type.

We now consider two basic modes of describing the relationship between the mechanical and the electrical quantities pertaining to an electromechanical transducer. The transducer force f_t depends on transducer velocity v_t as the first variable, and on either transducer current i_t or transducer voltage e_t as the second variable, namely,

$$f_t = f(v_t, i_t) \tag{2.12a}$$

and

$$e_t = e(v_t, i_t), \tag{2.12b}$$

or

$$f_t = f(v_t, e_t) \tag{2.13a}$$

and

$$i_t = i(v_t, e_t). \tag{2.13b}$$

2.3.2.2. Ideal transducers. In an 'ideal' transducer the terms in eqns (2.12) and (2.13) containing the mechanical and electrical impedancès or admittances are neglected and eqns (2.12a) and (2.12b), representing the transducer as an ideal voltage source, can be written generally

$$f_t = M_{fi} i_t, \tag{2.14a}$$

$$e_t = M_{ev} v_t, \tag{2.14b}$$

where M_{fi} and M_{ev} are the (complex) transducer constants related to force and current, and to voltage and velocity, respectively. Note that in a consistent system of units, such as SI units, their dimensions $(N\,A^{-1})$ and $(V\,s\,m^{-1})$ are identical and equal to $(J\,A^{-1}\,m^{-1})$. Then, from eqns (2.11a) and (2.14a), the mechanical transducer power is

$$P_m = \tfrac{1}{2}\mathrm{Re}(f_t v_t^*) - \tfrac{1}{2}\mathrm{Re}(M_{fi} i_t v_t^*) \quad (\mathrm{W}), \tag{2.15a}$$

and from eqns (2.11b) and (2.14b) the electrical power

$$P_e = \tfrac{1}{2}\mathrm{Re}(i_t e_t^*) = \tfrac{1}{2}\mathrm{Re}(i_t M_{ev} v_t^*) \quad (\mathrm{W}). \tag{2.15b}$$

With the convention of Fig. 2.2(a) energy transport requires

$$P_m = -P_e, \tag{2.16}$$

or, with eqns (2.15a) and (2.15b),

$$M_{fi} = -M_{ev}^*, \tag{2.17}$$

i.e. M_{ev} changes sign if real, but does not if imaginary. Letting

$$M_{fi} = M \quad \text{and} \quad M_{ev} = -M^*, \tag{2.18}$$

eqns (2.14a) and (2.14b) now become

$$f_t = M i_t, \tag{2.19a}$$

$$e_t = -M^* v_t. \tag{2.19b}$$

Eqns (2.19a) and (2.19b), when written in matrix form, represent the ideal voltage-source-type transducer of Fig. 2.2(a), thus

$$\begin{bmatrix} f_t \\ e_t \end{bmatrix} = \begin{bmatrix} 0 & M \\ -M^* & 0 \end{bmatrix} \begin{bmatrix} v_t \\ i_t \end{bmatrix}. \tag{2.20}$$

If, however, the transducer voltage e_t is chosen as the second independent variable, i.e. if the transducer is seen as a current source, we have with eqns (2.13a) and (2.13b) generally

$$f_t = N_{fe} e_t, \tag{2.21a}$$

$$i_t = N_{iv} v_t, \tag{2.21b}$$

where N_{fe} and N_{iv} are the (complex) transducer constants related to force and voltage, and to current and velocity, respectively. Both their dimensions (N V^{-1}) and (A s m^{-1}) are identical to $(\text{J V}^{-1} \text{m}^{-1})$.

From an argument similar to that concerning M_{fi} and M_{ev} (eqns (2.14)–(2.18)) we can write for N_{fe} and N_{iv},

$$N_{fe} = N \quad \text{and} \quad N_{iv} = -N^*. \tag{2.22}$$

Eqns (2.21a) and (2.21b) then become

$$f_t = N e_t, \tag{2.23a}$$

$$i_t = -N^* v_t, \tag{2.23b}$$

or, in matrix form,

$$\begin{bmatrix} f_t \\ i_t \end{bmatrix} = \begin{bmatrix} 0 & N \\ -N^* & 0 \end{bmatrix} \begin{bmatrix} v_t \\ e_t \end{bmatrix}. \tag{2.24}$$

2.3.2.3. *Real transducers*

(*a*) *Impedance and hybrid matrices.* We now wish to extend our study to 'real' transducers. In Fig. 2.2(b) the first box on the left represents the mechanical impedance Z_m of the transducer, while the last box may stand for an electrical impedance Z_e, or an electrical admittance Y_e. Note that $v_t = v$, since Z_m is always an impedance, but $i_t = i$ only if Z_e is an impedance; in case of an admittance Y_e, $e_t = e$.

The two-port equations (2.19a) and (2.19b) of the ideal transducer can now be extended to represent the real transducer of Fig. 2.2(b),

$$f = Z_m v + M i, \tag{2.25a}$$

$$e = -M^* v + Z_e i, \tag{2.25b}$$

and the equivalent matrix form of eqns (2.25a) and (2.25b),

$$\begin{bmatrix} f \\ e \end{bmatrix} = \begin{bmatrix} Z_m & M \\ -M^* & Z_e \end{bmatrix} \begin{bmatrix} v \\ i \end{bmatrix}, \tag{2.26}$$

is known as an 'impedance' matrix equation.

According to the rules of matrix algebra, eqn (2.26) could be converted to apply also to a current-source transducer model (but retaining the M-type transducer constant), thus

$$\begin{bmatrix} f \\ i \end{bmatrix} = \begin{bmatrix} Z_m + M^2/Z_e & M/Z_e \\ M^*/Z_e & 1/Z_e \end{bmatrix} \begin{bmatrix} v \\ e \end{bmatrix}, \tag{2.27}$$

where $M^2 = MM^*$. A simpler form derives from eqn (2.24), extended to represent a 'real' current-source-type transducer (N-form),

$$\begin{bmatrix} f \\ i \end{bmatrix} = \begin{bmatrix} Z_m & N \\ -N^* & Y_e \end{bmatrix} \begin{bmatrix} v \\ e \end{bmatrix}, \tag{2.28}$$

where $Y_e = 1/Z_e$. Eqn (2.28) is known as a 'hybrid' matrix equation, since it comprises both impedance and admittance elements. (The impedance matrix equation equivalent to the hybrid matrix equation (2.28) would also assume a more complex form.)

From the foregoing we conclude that electromechanical transducers may be divided into two groups, the choice of which depends on whether an impedance or a hybrid matrix representation gives simpler expressions (see section 2.3.2.4). In the former case the ideal transducer is seen as a voltage source (M-form), in the latter as a current source (N-form).

(b) *Transfer matrices.* In addition to the impedance and hybrid forms of the transducer equations discussed in the previous section, the so-called 'transfer' matrix equations of the general two-port theory are of particular interest in transducer analysis, since they relate the parameters at the mechanical port directly to those at the electrical port, and vice versa.

The transfer matrix equation of a voltage-source-type transducer is conveniently obtained by conversion from its impedance matrix eqn (2.26), namely,

$$\begin{bmatrix} f \\ v \end{bmatrix} = \begin{bmatrix} M + Z_m Z_e/M^* & -Z_m/M^* \\ Z_e/M^* & -1/M^* \end{bmatrix} \begin{bmatrix} i \\ e \end{bmatrix}, \tag{2.29}$$

and that of a current-source-type transducer by conversion from its hybrid matrix equation (2.28)

$$\begin{bmatrix} f \\ v \end{bmatrix} = \begin{bmatrix} N + Z_m Y_e/N^* & -Z_m/N^* \\ Y_e/N^* & -1/N^* \end{bmatrix} \begin{bmatrix} e \\ i \end{bmatrix}. \tag{2.30}$$

The advantages of combining two-port transducer representation with matrix algebra becomes particularly apparent if the transfer matrices of eqns (2.29) and (2.30) are split, according to Fig. 2.2(b), into their constituent transfer matrices. From eqn (2.29),

$$\begin{bmatrix} f \\ v \end{bmatrix} = \begin{bmatrix} 1 & Z_m \\ 0 & 1 \end{bmatrix} \begin{bmatrix} M & 0 \\ 0 & -1/M^* \end{bmatrix} \begin{bmatrix} 1 & 0 \\ -Z_e & 1 \end{bmatrix} \begin{bmatrix} i \\ e \end{bmatrix}, \tag{2.31a}$$

and from eqn (2.30),

$$\begin{bmatrix} f \\ v \end{bmatrix} = \begin{bmatrix} 1 & Z_m \\ 0 & 1 \end{bmatrix} \begin{bmatrix} N & 0 \\ 0 & -1/N^* \end{bmatrix} \begin{bmatrix} 1 & 0 \\ -Y_e & 1 \end{bmatrix} \begin{bmatrix} e \\ i \end{bmatrix}, \tag{2.31b}$$

where the centre factors represent the transfer matrices of the ideal transducers, flanked by the transfer matrices of the mechanical impedance and of the electrical impedance or admittance, respectively, of the real transducer. The constituent transfer matrices are said to be in 'cascade', or 'tandem'; their sequences are not interchangeable.

The close analogy between circuit and matrix representation, and its potentialities in system analysis and synthesis, may be illustrated further by Fig. 2.2(c), where two-ports representing a mechanical terminal impedance Z_{m0} and an electrical terminal impedance Z_{e0}, or a terminal admittance Y_{e0}, have been added to the block diagram of Fig. 2.2(b). These outer blocks may stand for sources or loads, depending on how the transducer is operated. In the receiver, f_0 is the mechanical input, or excitation, force. In a sender, e_0 is the electrical input, or driving, voltage, or i_0 the electrical input, or driving, current, as the case may be.

Eqn (2.31a) now expands (see Fig. 2.2(c)(i)) to

$$\begin{bmatrix} f_0 \\ v \end{bmatrix} = \begin{bmatrix} 1 & Z_{m0} \\ 0 & 1 \end{bmatrix} \begin{bmatrix} 1 & Z_m \\ 0 & 1 \end{bmatrix} \begin{bmatrix} M & 0 \\ 0 & -1/M^* \end{bmatrix} \begin{bmatrix} 1 & 0 \\ -Z_e & 1 \end{bmatrix} \begin{bmatrix} 1 & 0 \\ -Z_{e0} & 1 \end{bmatrix} \begin{bmatrix} i \\ e \end{bmatrix}, \tag{2.32a}$$

and eqn (2.31b) (see Fig. 2.2(c)(ii)) to

$$\begin{bmatrix} f_0 \\ v \end{bmatrix} = \begin{bmatrix} 1 & Z_{m0} \\ 0 & 1 \end{bmatrix} \begin{bmatrix} 1 & Z_m \\ 0 & 1 \end{bmatrix} \begin{bmatrix} N & 0 \\ 0 & -1/N^* \end{bmatrix} \begin{bmatrix} 1 & 0 \\ -Y_e & 1 \end{bmatrix} \begin{bmatrix} 1 & 0 \\ -Y_{e0} & 1 \end{bmatrix} \begin{bmatrix} e \\ i \end{bmatrix}. \tag{2.32b}$$

Matrix multiplication of eqn (2.32a) then yields the complete transfer matrix equation of the energized and loaded voltage-source-type transducer,

$$\begin{bmatrix} f_0 \\ v \end{bmatrix} = \frac{1}{M^*} \begin{bmatrix} M^2 + (Z_m + Z_{m0})(Z_e + Z_{e0}) & -(Z_m + Z_{m0}) \\ Z_e + Z_{e0} & -1 \end{bmatrix} \begin{bmatrix} i \\ e_0 \end{bmatrix} \tag{2.33a}$$

and eqn (2.32b) that of the current-source type,

$$\begin{bmatrix} f_0 \\ v \end{bmatrix} = \frac{1}{N^*} \begin{bmatrix} N^2 + (Z_m + Z_{m0})(Y_e + Y_{e0}) & -(Z_m + Z_{m0}) \\ Y_e + Y_{e0} & -1 \end{bmatrix} \begin{bmatrix} e \\ i_0 \end{bmatrix}. \tag{2.33b}$$

The transfer matrix equations (2.33a) and (2.33b) fully describe the performance properties of bilateral electromechanical transducers of the voltage-source and the current-source type, respectively; they are identical with eqns (2.29) and (2.30), except for the additional source and load terms.

2.3.2.4. Generalized performance analysis of bilateral electromechanical transducers. We now want to derive general expressions† for the mechanical and electrical input impedances of electromechanical receivers and senders, and for their transfer functions. For this purpose we shall need only the general transfer matrix equations (2.33a) and (2.33b).

(a) *Mechanical input impedance of receivers.* For the voltage-source-type receiver which is terminated by an electrical load, or indicator, impedance Z_{e0}, the mechanical input impedance is obtained at electrical short-circuit conditions, $e_0 = 0$. For the current-source-type receiver which is terminated by an electrical admittance Y_{e0}, electrical open-circuit conditions apply.

(i) *Voltage-source-type receivers.* From eqn (2.33a) for $e_0 = 0$,

$$f_0 = \frac{M^2 + (Z_m + Z_{m0})(Z_e + Z_{e0})}{M^*} i,$$
(2.34a)

$$v = \frac{Z_e + Z_{e0}}{M^*} i,$$
(2.34b)

and the mechanical input impedance of a voltage-source-type receiver therefore becomes

$$Z_{mi} = (f_0/v)_{e_0=0} = \frac{M^2 + (Z_m + Z_{m0})(Z_e + Z_{e0})}{Z_e + Z_{e0}}$$

$$= Z_m + Z_{m0} + M^2/(Z_e + Z_{e0}) \quad (\text{N s m}^{-1}).$$
(2.35)

(ii) *Current-source-type receivers.* Similarly, from eqn (2.33b) for $i_0 = 0$,

$$Z_{mi} = (f_0/v)_{i_0=0} = Z_m + Z_{m0} + N^2/(Y_e + Y_{e0}) \quad (\text{N s m}^{-1}).$$
(2.36)

(b) *Electrical input impedance of senders.* Since the electromechanical senders of both the voltage-source and the current-source types are terminated by a mechanical impedance Z_{m0}, the electrical input impedances are obtained in both cases at mechanical 'short-circuit' conditions, $f_0 = 0$.

(i) *Voltage-source-type senders.* From eqn (2.33a) for $f_0 = 0$,

$$0 = \frac{M^2 + (Z_m + Z_{m0})(Z_e + Z_{e0})}{M^*} i - \frac{Z_e + Z_{e0}}{M^*} e_0.$$
(2.37)

Hence, the electrical input impedance of a voltage-source-type sender becomes

$$Z_{ei} = (e_0/i)_{f_0=0} = Z_e + Z_{e0} + M^2/(Z_m + Z_{m0}) \quad (\Omega).$$
(2.38)

---

† For the specific performance analysis of individual transducer types see the appropriate sections of Chapter 4.

(*ii*) *Current-source-type senders.*   Similarly, from eqn (2.33b) for $f_0 = 0$, the electrical input admittance is

$$Y_{ei} = (i_0/e)_{f_0=0} = Y_e + Y_{e0} + N^2/(Z_m + Z_{m0}) \quad (\Omega^{-1}). \tag{2.39}$$

(*c*) *Transfer functions of receivers.*   The transfer function of an electromechanical transducer is the (complex) ratio of the appropriate output quantity (electrical in a receiver, mechanical in a sender) to the related input quantity (mechanical in a receiver, electrical in a sender). It therefore constitutes the 'frequency response' of a receiver or a sender, as the case may be.

The electromechanical receiver operates either under electrical short-circuit conditions (*M*-form, terminal impedance Z_{e0}), or under electrical open-circuit conditions (*N*-form, terminal admittance Y_{e0}). In the former case, the electrical excitation voltage e_0 vanishes and the appropriate electrical output quantity is the current i; in the latter case i_0 vanishes and e is required. Since Z_{e0} and Y_{e0} represent indicator impedance and indicator admittance, respectively, the indicated output voltage is $e_i = -Z_{e0}i$, and the indicated output current $i_i = Y_e e$.

(*i*) *Voltage-source-type receivers.*   From eqn (2.33a) for $e_0 = 0$,

$$f_0 = \frac{M^2 + (Z_m + Z_{m0})(Z_e + Z_{e0})}{M^*} i, \tag{2.40}$$

and the transfer function G_r of the voltage-source-type receiver, is therefore

$$G_r = (-i/f_0)_{e_0=0} = -M^*/\{M^2 + (Z_m + Z_{m0})(Z_e + Z_{e0})\} \quad (\text{A N}^{-1}). \tag{2.41}$$

The negative sign in $(-i)$ stems from our conventions of energy flow towards the transducer (Fig. 2.2(a)). The indicated phase reversal may, or may not, occur in a particular transducer; this will depend on the sign of all complex quantities involved, and not only on whether M is real or imaginary.

(*ii*) *Current-source-type receivers.*   Similarly, from eqn (2.33b) for $i_0 = 0$,

$$G_r = (e/f_0)_{i_0=0} = N^*/\{N^2 + (Z_m + Z_{m0})(Y_e + Y_{e0})\} \quad (\text{V N}^{-1}). \tag{2.42}$$

(*d*) *Transfer function of senders.*   The output quantity of the electromechanical sender is the (negative) velocity $(-v)$ which is the result of either a driving voltage e_0 (*M*-form) or a driving current i_0 (*N*-form). In the sender, the mechanical excitation force f_0 vanishes and the transfer function is then obtained from eqns (2.33a) and (2.33b) for mechanical short-circuit conditions, $f_0 = 0$, which apply to both voltage and current excitation.

(i) *Voltage-source-type senders.* From eqn (2.33a) for $f_0 = 0$,

$$0 = \frac{M^2 + (Z_m + Z_{m0})(Z_e + Z_{e0})}{M^*} i - \frac{Z_e + Z_{e0}}{M^*} e_0, \qquad (2.43a)$$

$$v = \frac{Z_e + Z_{e0}}{M^*} i - \frac{1}{M^*} e_0, \qquad (2.43b)$$

and, after some rearrangements,

$$G_s = (-v/e_0)_{f_0=0} = M/\{M^2 + (Z_m + Z_{m0})(Z_e + Z_{e0})\} \quad (\text{m V}^{-1}\text{s}^{-1}). \quad (2.44)$$

Except, normally, for a different phase relationship, the transfer function of eqn (2.44) for the voltage-source-type sender is identical with eqn (2.41) of the voltage-source-type receiver, as would be expected of bilateral transducers with reversible transducing laws.

(ii) *Current-source-type senders.* Similarly, from eqn (2.33b) for $f_0 = 0$,

$$G_s = (-v/i_0)_{f_0=0} = -N/\{N^2 + (Z_m + Z_{m0})(Y_e + Y_{e0})\} \quad (\text{m A}^{-1}\text{s}^{-1}). \quad (2.45)$$

Again, eqn (2.45) is identical with eqn (2.42) for the receiver, except for a different phase relationship.

In both the voltage-source and the current-source-type senders the 'indicated' output force f_i exerted by the transducer on its mechanical load impedance Z_{m0} is given by $f_i = -Z_{m0}v$.

2.3.2.5. The transducer constants. In order to obtain the performance characteristics of individual electromechanical transducers, their transducer constants M or N must be known. The derivation of these constants will be given in the appropriate sections of Chapter 4 dealing with individual transducer types. These derivations will also indicate whether it is more convenient to consider a transducer as a voltage source (M-form) or as a current source (N-form). In fact, it will be shown that electrodynamic- and electrostatic-type transducers are best treated as voltage sources, and electromagnetic- and piezoelectric-type transducers as current sources. Fig. 2.3 gives a preview of two-port configurations and characteristic transfer equations of these four transducer types. Detailed discussion must be deferred until Chapter 4, but the figure already clearly indicates the simple and unified forms of the expressions which can be arranged in the 'symmetrical' pattern shown.

Network theory employing the concept of two-ports represented by matrix equations is indeed a powerful tool in dynamic analysis. Above it has been applied—without the use of analogies—to simple electromechanical transducers, but it may be mentioned here that a generalized network theory, based on concepts similar to those employed in this section, has

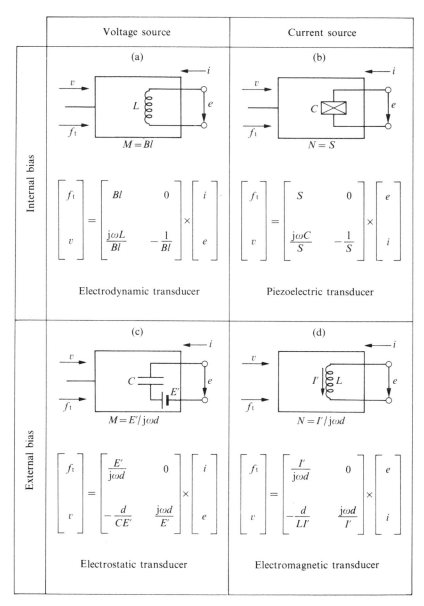

F IG. 2.3. Characteristic transfer equations of four electromechanical transducers.

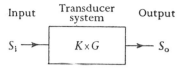

FIG. 2.4. Straightforward ('open-loop') transducer system.

recently been derived which is capable of solving problems of non-equilibrium thermodynamics involving coupling and feedback effects in non-linear and unsteady systems [25].

2.4. Feedback systems

Consider a straightforward ('open-loop') measuring system, shown in Fig. 2.4. Its input S_i and output S_o are functions of time and frequency. The system characteristics are represented by a two-port which has a complex transfer function G and a gain K. The output is the product of the complex quantities S_i and G, and of K,

$$S_o = KGS_i. \tag{2.46}$$

The majority of the measuring systems described in this book follow the simple relationship of eqn (2.46), where K represents the nominal sensitivity of the transducer or transducer system, while G introduces the time- or frequency-dependent law(s) which modify the sensitivity.†

Now consider Fig. 2.5. Here a portion of the output S_o is returned to the input terminals via a 'feedback loop' with a complex transfer function H. The sum of S_i and the returned portion of S_o appears at the input terminals of the forward path,

$$\varepsilon = S_i + HS_o, \tag{2.47}$$

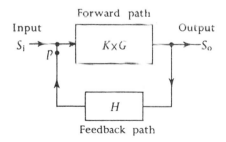

FIG. 2.5. Feedback ('closed-loop') transducer system.

† In some transducers this nominal sensitivity occurs in the pass-band above or below the cut-off frequency; in others it is the mid-frequency sensitivity between two cut-offs.

the output signal S_0 being

$$S_0 = \varepsilon KG. \tag{2.48}$$

The 'closed-loop' transfer function of the feedback system, i.e. the ratio of the output signal to the input signal, is

$$\frac{S_0}{S_i} = \frac{\varepsilon KG}{\varepsilon - HS_0} = \frac{\varepsilon KG}{\varepsilon - \varepsilon HKG} = \frac{KG}{1 - HKG}. \tag{2.49}$$

If the feedback loop is cut at P the so-called 'open-loop' transfer function of the feedback system at this point is HKG, and it is seen from eqn (2.49) that the closed-loop transfer function can be derived by dividing the 'forward' transfer function KG of the system by 1 minus the open-loop transfer function.

Eqn (2.49) is the fundamental relationship of any feedback system. With 'negative' feedback, i.e. if the feedback loop is arranged such that the feedback signal is opposing the input signal, H assumes a negative sign, and we have

$$\frac{S_0}{S_i} = \frac{KG}{1 + HKG}. \tag{2.50}$$

Transducers with negative feedback between output and input will be discussed in Chapter 5, where they are known as 'force-balance' transducers, their input forces being balanced by opposing electromechanical forces. Negative-feedback systems are widely employed in electronic amplifiers, where they actually originated. They offer great advantages with respect to reduced distortion, improved signal-to-noise ratio, and other desirable electronic characteristics, such as impedance transformation.

A large portion of books and papers on feedback systems is usually occupied with discussions of system stability to self-oscillation, and of suitable methods of 'stabilization'. Standard methods of analysis and synthesis applicable to servo-mechanisms (control systems) and electronic amplifiers have been developed and the interested reader is invited to consult the appropriate textbooks on the subject [26]–[29].

References

1. K. S. Lion. *Instrumentation in scientific research; electrical input transducers.* McGraw-Hill, New York (1959).
2. E. O. Doebelin. *Measurement systems; application and design.* McGraw-Hill, New York (1966).
3. F. Kohlrausch. *Praktische Physik*, Vols. 1–3. B. G. Teubner, Stuttgart (1968).
4. A. Goldmann and H. Bley. *Eigenschaften und Anwendungen von Messwertumformern*, Vols. 1 and 2. Franckh'sche Verlagsbuchhandlung, Stuttgart (1963).
5. P. K. Stein. *Measurement engineering*, Vols. 1 and 2. Stein Engineering Services, Phoenix, Arizona (1964).

6. H. F. GRAVE. *Elektrische Messung nichtelektrischer Grössen.* Akademische Verlagsanstalt, Frankfurt/Main (1965).
7. A. F. GILES. *Electronic sensing devices.* Newnes, London (1966).
8. C. ROHRBACH. *Handbuch für elektrisches Messen mechanischer Grössen.* V.D.I. Verlag, Düsseldorf (1967).
9. T. G. BECKWITH and N. L. BUCK. *Mechanical measurements.* Addison-Wesley, Reading, Massachusetts (1969).
10. H. N. NORTON. *Handbook of transducers for electronic measuring systems.* Prentice-Hall, Englewood Cliffs, New Jersey (1969).
11. F. J. OLIVER. *Practical instrumentation transducers.* Hayden Books, New York (1971).
12. L. A. OSTROVSKIJ. *Elektrische Messtechnik* (transl. from Russian). V.E.B. Verlag, Berlin (1969).
13. L. STARKE, Messwertaufnehmer; Bausteine elektronischer Messketten. *Elektronik*, 181–4; 371–9 (1961).
14. W. R. MACDONALD. Pressure transducers available in the U.K. *I.E.A. Year Book and Buyers' Guide 1966,* pp. 581–93. Morgan Brothers, London (1966).
15. INSTRUMENT SOCIETY OF AMERICA. *ISA transducer compendium,* Part 1: Pressure, flow, level (1969); Part 2: Sound, force, torque, motion, dimension (1970); Part 3: Temperature, chemical composition, physical properties, humidity, moisture, radiation (1971). Instrument Society of America, Pittsburgh, Pennsylvania.
16. C. F. HIX and R. P. ALLEY. *Physical laws and their effects.* Wiley, New York (1958).
17. A. C. BEER. *Galvanometric effects in semiconductors.* Academic Press, New York (1963).
18. D. W. G. BALLENTYNE and D. R. LOVETT. *A dictionary of named effects and laws in chemistry, physics and mathematics.* Chapman & Hall, London (1970).
19. W. HÄHNLE. Die Darstellung elektro-mechanischer Gebilde durch rein elektrische Schaltbilder. *Wiss. Veröff. Siemens Werken* **11**, 1 (1932).
20. F. A. FIRESTONE. New analogy between mechanical and electrical systems. *J. acoust. Soc. Am.* **4**, 249 (1933).
21. W. P. MASON. *Electro-mechanical transducers and wave filters.* Van Nostrand, New York (1942).
22. H. F. OLSON. *Dynamic analogies.* Van Nostrand, New York (1958).
23. L. CREMER and H. HECKL. *Körperschall.* Springer, Berlin (1967).
24. H. K. P. NEUBERT. Bilateral electro-mechanical transducers; a unified theory. R.A.E. Technical Report No. TR 68248 (1968).
25. G. OSTER, A. PERELSON, and A. KATCHALSKY. Network thermodynamics. *Nature, Lond.* **234**, 393–9 (1971).
26. H. W. BODE. *Network analysis and feedback amplifier design.* Van Nostrand, New York (1946).
27. C. J. SAVANT. *Basic control system engineering.* McGraw-Hill, New York (1958).
28. J. C. TRUXAL (ed.). *Control engineering handbook.* McGraw-Hill, New York (1964).
29. J. T. TOU. *Modern control theory.* McGraw-Hill, New York (1964).

3. Mechanical input characteristics

3.1. Static and dynamic response

THE primary signal applied to, and measured by, the transducer may be of a steady-state (static or quasi-static) or of a dynamic (cyclic or transient) nature. These two groups of signals are significant not only if the input is purely mechanical—such as acceleration, force, or pressure— but also, for instance, with temperature, light, and radiation input quantities. In static measurements the major concern lies with the desired linear relationship between input and output quantities, while in the case of cyclic or transient signals the dynamic response of the transducer has also to be considered. This leads to the two major subjects of this section: linearity and dynamic response.

3.11. Linearity

The input–output relationship of measuring instruments without imperfections such as hysteresis, creep, etc. is generally given by the equation

$$y = (a_0 + a_1 x + a_2 x^2 + a_3 x^3 + \ldots + a_n x^n)x, \tag{3.1}$$

where y is the output quantity and x the input quantity. The coefficients a_0, a_1, ..., a_n are the calibration factors, and it is seen from eqn (3.1) that in general the calibration curve is based on a linear relationship $(a_0 x)$ with superimposed higher orders of the input quantity x. Fig. 3.1(a)–(d) illustrates the general shapes of calibration curves. After the ideal linear curve (a), the symmetrical curve (c) with only odd orders of x-components is the most useful for instrument work, since it provides a symmetrical and usually fairly wide quasi-linear range about the origin. In a practical design symmetry in spring deflection facilitates symmetry in calibration, while biased springs and other non-symmetrical features tend to produce calibration curves of type (b) or (d). The over-all linearity of a transducer, of course, is also affected by non-linearities in the output characteristics of the electrical sensing elements, and it will be shown in Chapter 4 that a symmetrical arrangement of two sensing elements, operating in push–pull fashion, improve linearity considerably by eliminating even–order components in the characteristics of the electrical sensing elements. Good temperature compensation is often an extra bonus.

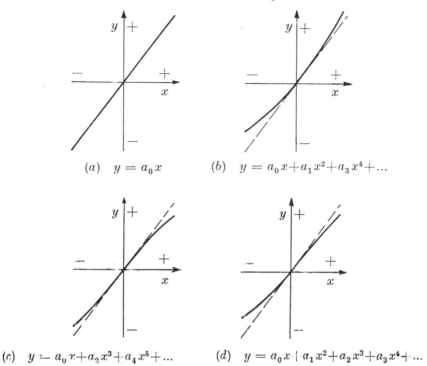

(a) $y = a_0 x$

(b) $y = a_0 x + a_1 x^2 + a_3 x^4 + \dots$

(c) $y = a_0 x + a_2 x^3 + a_4 x^5 + \dots$

(d) $y = a_0 x + a_1 x^2 + a_2 x^3 + a_3 x^4 + \dots$

FIG. 3.1. Types of calibration curves; (a) strictly linear; (b) non-linear with even-order components of x; (c) non-linear with odd-order components of x; (d) non-linear with even- and odd-order components of x.

Up to this point we have assumed that the input element of a transducer has no defects other than non-linearity, generally expressed by eqn (3.1). But in order to understand, and possibly to control, the properties of transducers we must also consider deviations from this ideal 'elastic' behaviour. The three additional imperfections, in order of practical importance, are: (a) mechanical hysteresis, (b) viscous flow or creep, and (c) elastic after-effect.

(a) *Mechanical hysteresis.* The concept of hysteresis has to cover a multitude of sins in instrument performance, such as bearing friction, backlash, loose screws, fractured or corroded components, and plain dirt. These faults follow their own unpredictable 'laws' and do not concern us here.

Mechanical hysteresis in transducer components, such as springs or pressure capsules, is a manifestation of the imperfect response of the microscopic crystal grains, integrated over the macroscopic dimensions of the strained transducer element. It is analogous to similar phenomena in magnetic and dielectric materials. When load-cycling, for instance, a steel spring, it is observed that with increasing maximum stress increasing numbers of distorted, and partly dislocated, crystals do not return to their original shapes and positions after the load has been released. The magnitude of the residual deformation depends on the maximum stress applied, but it is independent of time (frequency). The load–deformation diagram (Fig. 3.2) is a closed loop, the area of which can be shown to represent the energy dissipated into heat.

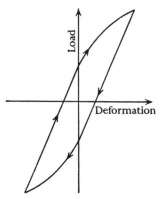

FIG. 3.2. Mechanical hysteresis, schematic diagram.

A mathematical treatment of hysteresis in general has been attempted at various stages of experimental evidence available to the investigator, though generally only with limited success. The earliest results of consequence were submitted by Lord Rayleigh [1] in 1887. They deal with magnetic hysteresis in fields of low magnitude. Other investigators [2]–[5] have added to the understanding of mechanical hysteresis in metals and other materials. For small cyclic deformations the most convenient representation of hysteresis or—as it is often called in acoustics and vibration engineering—of 'internal' or structural damping is by way of a complex Young's modulus $\bar{E}$;

$$\bar{E} = E(1+\mathrm{j}\alpha), \tag{3.2}$$

where E is the in-phase or elastic component of Young's modulus, and $\mathrm{j}\alpha E$ the contribution by hysteresis which is in quadrature with the elastic component. We notice that the hysteresis component is assumed independent of frequency. Mathematical analyses of vibratory systems with structural, as compared with viscous, damping are given in references [6] and [7].

(b) *Viscous flow or creep.*   The main difference between hysteresis and creep is the fact that the latter depends on time (Fig. 3.3). The effect is

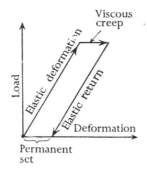

FIG. 3.3. Viscous flow or creep, schematic diagram.

due to viscous flow in the material. Its magnitude increases with increasing load and temperature. Materials with low melting-points may show quite large creep values even at room temperature, while in the best materials at high values of stress and temperature creep may become the limiting design factor [8]. Mathematical expressions have been derived to fit certain types of creep curves for certain materials, revealing a basic logarithmic creep law. The interested reader should consult the monographs on this subject [9]–[13].

(c) *Elastic after-effect.* This effect was first noticed by W. Weber in 1825 on galvanometer suspensions. When the load was applied and kept constant there was still movement (Fig. 3.4) similar to creep, which decreased with time, but in contrast to viscous flow as described above, there was also a similar relaxation towards the original position when the load was removed, i.e. there was virtually no residual deformation. The physics of 'after-effects' is complex [14] and perhaps of little practical use to the designer. It involves a range of concepts, such as stress-relaxation, creep recovery, and internal damping.

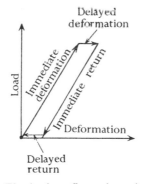

FIG. 3.4. Elastic after-effect, schematic diagram.

3.12. *Dynamic response*

If the input signal $s(t)$ to a mechanical system is a function of time, then its output $z(t)$ is also a function of time, and we can write

$$b_n z^{(n)} + b_{n-1} z^{(n-1)} + \ldots + b_2 \ddot{z} + b_1 \dot{z} + b_0 z = b_0 s(t). \tag{3.3}$$

Both sides of eqn (3.3) have similar dimensions (e.g. force) if $z(t)$ and $s(t)$ are of the same dimensions (e.g. displacement). In a linear system $b_0, \ldots, b_n$ are constant coefficients and $\dot{z}, \ddot{z}, \ldots, z^{(n)}$ are the time derivatives of z.

(a) *Sinusoidal input.* For sinusoidal input $s(t) = S \exp(j\omega t)$ the well-known solution is

$$z(t) = Z e^{j(\omega t + \phi)}, \tag{3.4}$$

and eqn (3.3) then becomes

$$(b_n/b_0)(j\omega)^n Z e^{j\phi} e^{j\omega t} + \ldots + (b_1/b_0)(j\omega) Z e^{j\phi} e^{j\omega t} + Z e^{j\phi} e^{j\omega t} = S e^{j\omega t}, \tag{3.5}$$

which may be written

$$(Z/S)e^{j\phi} = 1/\{T_n^n(j\omega)^n + \ldots + T_2^2(j\omega)^2 + T_1 j\omega + 1\}$$
$$= F(j\omega). \tag{3.6}$$

$F(j\omega)$ of eqn (3.6) is the complex frequency response of the system and $T_n, \ldots, T_1$ are coefficients with the dimension of time ('time-constants'). Then the output $z(t)$ is given by

$$z(t) = F(j\omega) S e^{j\omega t}, \tag{3.7}$$

where the real part

$$\text{Re}(z(t)) = |F(j\omega)| \, S \, \sin(\omega t + \phi). \tag{3.8}$$

The relative amplitude response is $|F(j\omega)|$, where $|F(j\omega)|_{\omega \to 0} = 1$. The phase angle ϕ between the input signal $s(t)$ and the output $z(t)$ is

$$\phi = \tan^{-1} \left\{ \frac{\text{Im}(F(j\omega))}{\text{Re}(F(j\omega))} \right\}. \tag{3.9}$$

These equations describe completely the dynamic response of the mechanical components of an instrument transducer to sinusoidal excitation, provided its calibration curve is linear.

(b) *Step-function input.* A step-function input may be represented by an infinite number of sinusoidal waves (Fourier's theorem) and can be written in the form

$$s(t) = \frac{S}{2\pi j} \int_{-\infty}^{+\infty} \frac{e^{j\omega t}}{\omega} \, d\omega. \tag{3.10}$$

The contribution to the input of an individual frequency is

$$ds(t) = \frac{S}{2\pi j} \cdot \frac{e^{j\omega t}}{\omega} \, d\omega, \tag{3.11}$$

and the corresponding output is

$$dz(t) = \frac{S}{2\pi j} F(j\omega) \frac{e^{j\omega t}}{\omega} \, d\omega. \tag{3.12}$$

Integrating eqn (3.12) over all frequencies from $-\infty$ to $+\infty$ we have

$$z(t) = \frac{S}{2\pi j} \int_{-\infty}^{+\infty} \frac{F(j\omega)}{\omega} e^{j\omega t} \, d\omega. \tag{3.13}$$

From eqn (3.13) the transient response $T(t)$ to a unit step function becomes

$$T(t) = \frac{z(t)}{S} = \frac{1}{2\pi j} \int_{-\infty}^{+\infty} \frac{F(j\omega)}{\omega} e^{j\omega t} \, d\omega. \tag{3.14}$$

Eqn (3.14) defines the transient response of our system to a unit step function in terms of the complex frequency response $F(j\omega)$. This relationship is valid for any linear system.

(c) *Generalized relation between frequency and transient response of linear systems.* The integral of eqn (3.14) may be solved by Laplace transformation ([15]-[17]), yielding

$$f(p) = \mathscr{L}\{f(t)\} = \int_{0}^{+\infty} f(t)e^{-pt} \, dt \tag{3.15a}$$

and

$$f(t) = \mathscr{L}^{-1}\{f(p)\} = \frac{1}{2\pi j} \int_{-j\infty}^{+j\infty} f(p)e^{pt} \, dp, \tag{3.15b}$$

where p is the Laplace operator. $\mathscr{L}f(t)$ represents the Laplace transformation of $f(t)$, and $\mathscr{L}^{-1}$ the inverse transformation. Writing eqn (3.14) in general form, i.e. substituting p for $j\omega$, we have

$$T(t) = \frac{1}{2\pi j} \int_{-j\infty}^{+j\infty} \frac{F(p)}{p} e^{pt} \, dp, \tag{3.14a}$$

which is identical with the inverse transformation formula of eqn (3.15b). Thus

$$\mathscr{L}\{T(t)\} = F(p)/p, \qquad T(t) = \mathscr{L}^{-1}\{F(p)/p\}, \tag{3.16a}$$

or

$$F(p) = \frac{\mathscr{L}\{T(t)\}}{1/p}. \tag{3.16b}$$

Eqn (3.16a) enables us, by way of the inverse Laplace transformation, to compute the transient response to a unit step function $T(t)$ of a linear system, if its sinusoidal frequency response $F(p)$ is known. Vice versa, by way of eqn (3.16b), $F(p)$ is obtained from the direct Laplace transformation of $T(t)$. A large number of transformation pairs have been tabulated ([18]–[20]) and there are several monographs available to the interested reader ([21]–[24]).

Summarizing, we have for a linear system (eqn (3.3)) the complex frequency response to sinusoidal excitation (eqns (3.4), (3.5), and (3.6)):

$$F(j\omega) = 1/\{1 + T_1(j\omega) + T_2^2(j\omega)^2 + \ldots + T_n^n(j\omega)^n\}, \tag{3.17}$$

T_1, T_2,..., T_n being time-constants derived from the system parameters. Then the amplitude response (eqn (3.8)) is

$$|F(j\omega)| = \{(1 - T_2^2\omega^2 + T_4^4\omega^4 - \ldots + \ldots)^2 +$$
$$+ (T_1\omega - T_3^3\omega^3 + T_5^5\omega^5 - \ldots + \ldots)^2\}^{-\frac{1}{2}}, \tag{3.18}$$

and the phase response (eqn (3.9))

$$\tan\phi = -\frac{T_1\omega - T_3^3\omega^3 + T_5^5\omega^5 - \ldots + \ldots}{1 - T_2^2\omega^2 + T_4^4\omega^4 - \ldots + \ldots}. \tag{3.19}$$

The transient response to a unit step function $T(t)$ is derived from eqns (3.16a) and (3.17) as the inverse Laplace transformation,

$$T(t) = \mathscr{L}^{-1}\left\{\frac{F(j\omega)}{j\omega}\right\}. \tag{3.20}$$

3.2. Dynamic response of basic transducer systems

3.21. Sinusoidal excitation

(a) *Systems of one degree of freedom of first order.* For the mechanical system of Fig. 3.5(a), consisting of stiffness k (N m^{-1}) and velocity (viscous) damping c (N s m^{-1}), the differential equation is

$$c\dot{z} + kz = ks(t) \tag{3.21a}$$

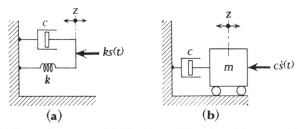

FIG. 3.5. First-order system with (a) spring and damping, (b) mass and damping.

or
$$T_1\dot{z}+z = s(t), \qquad (3.21b)$$

where the time-constant $T_1 = c/k$. The alternative first-order system, of Fig. 3.5(b), with mass m (kg) and damping c may be treated in a similar way, but since
$$m\ddot{z}+c\dot{z} = c\dot{s}(t), \qquad (3.22)$$

the time-constant $T_1 = m/c$. The complex frequency response (eqn (3.17)) at sinusoidal excitation is
$$F(j\omega) = 1/(1+j\omega T_1), \qquad (3.23)$$

which becomes in terms of amplitude and phase (eqns (3.18) and (3.19))
$$|F(j\omega)| = \{1+(\omega T_1)^2\}^{\frac{1}{2}} \qquad (3.24)$$
and
$$\phi = -\tan^{-1}(\omega T_1). \qquad (3.25)$$

Eqns (3.24) and (3.25) have been plotted in Fig. 3.6(a) and (b) as dash–dotted lines for various values of the dimensionless parameter ωT_1.

(b) *Systems of one degree of freedom of second order.* The mechanical system of Fig. 3.7, consisting of mass m, spring stiffness k, and viscous damping c,† has the differential equation
$$m\ddot{z}+c\dot{z}+kz = ks(t) \qquad (3.26)$$
or
$$T_2^2\ddot{z}+T_1\dot{z}+z = s(t), \qquad (3.27)$$

where $T_1 = c/k$ and $T_2^2 = m/k$. The complex frequency response (eqn (3.17)) for sinusoidal excitation is
$$F(j\omega) = 1/(1+j\omega T_1-\omega^2 T_2^2), \qquad (3.28)$$

which in terms of amplitude and phase (eqns (3.18) and (3.19)) becomes
$$|F(j\omega)| = \{(1-\omega^2 T_2^2)^2+\omega^2 T_1^2\}^{-\frac{1}{2}} \qquad (3.29)$$
and
$$\phi = -\tan^{-1}\{\omega T_1/(1-\omega_2 T_2^2)\} \qquad (3.30)$$

Rearranging eqns (3.29) and (3.30), and introducing
$$\omega_0^2 = \frac{1}{T_2^2} = \frac{k}{m} \quad \text{(undamped circular natural frequency } \omega_0) \qquad (3.31a)$$
and
$$h = \frac{T_1}{2T_2} = \frac{c}{2(mk)^{\frac{1}{2}}} \quad \begin{array}{l}\text{(damping ration } h \text{ being unity at 'critical'} \\ \text{damping; see section 3.33),}\end{array} \qquad (3.31b)$$

---

† For the performance of vibratory systems with structural damping see references [4], [6], [7]. For combined viscous and Coulomb (friction) damping see [70], [71].

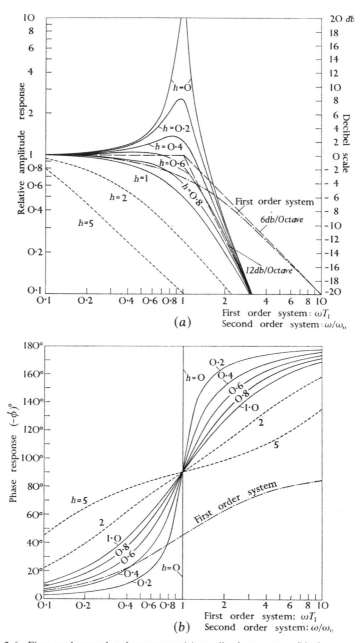

FIG. 3.6. First- and second-order systems; (a) amplitude response, (b) phase response.

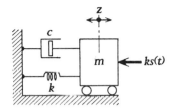

FIG. 3.7. Second-order system with mass, spring, and damping.

we have the complex frequency response (eqn (3.28)) for sinusoidal excitation

$$F(j\omega) = \frac{1-(\omega/\omega_0)^2-j2h(\omega/\omega_0)}{\{1-(\omega/\omega_0)^2\}^2+4h^2(\omega/\omega_0)^2},$$ (3.32)

which, in terms of amplitude and phase (eqns (3.18) and (3.19)), becomes

$$|F(j\omega)| = [\{1-(\omega/\omega_0)^2\}^2+4h^2(\omega/\omega_0)^2]^{-\frac{1}{2}}$$ (3.33)

and

$$\phi = -\tan^{-1}\left\{\frac{2h(\omega/\omega_0)}{1-(\omega/\omega_0)^2}\right\}.$$ (3.34)

Eqns (3.33) and (3.34) have been plotted in Fig. 3.6(a) and (b) for varying values of ω/ω_0, the damping ration h being a parameter. The curves are well known and need no explanation. The response to sinusoidal excitation of some vibratory systems of higher than second order will be discussed in Chapter 4, section 4.1.2.6 (secondary resonances).

3.2.2. Transient excitation

If the input term in eqn (3.3) is a unit step function it was shown in section 3.1.2 that the displacement response can be computed from eqn (3.16a) in conjunction with the complex frequency response of eqn (3.17).

(a) *Systems of one degree of freedom of first order.* In case of a first-order system of Fig. 3.5 we have

$$\mathcal{L}\{T(t)\} = \frac{F(p)}{p} - \frac{1}{p(1+pT_1)}$$

$$= \frac{1}{T_1}\frac{1}{p\{(1/T_1)+p\}}.$$ (3.35)

The inverse Laplace transformation of eqn (3.35), taken from any table of transform pairs ([18]–[20]), is

$$T(t) = \frac{1}{T_1}T_1\{1-\exp(-t/T_1)\} = 1-\exp(-t/T_1).$$ (3.36)

TABLE 3.1

Transient response of second-order vibratory systems

INPUT			A	B	C	D	E	F
DISPLACEMENT	VELOCITY	ACCELERATION						
			Displacement	Velocity	Acceleration	Jerk		
			Excursion	Displacement	Velocity	Acceleration	Jerk	
				Excursion	Displacement	Velocity	Acceleration	Jerk
					Excursion	Displacement	Velocity	Acceleration

This equation, when plotted against t/T_1, gives the well-known exponential curve.

(b) *Systems of one degree of freedom of second order.* The second-order system may be treated in a similar fashion. We have

$$\mathscr{L}\{T(t)\} = \frac{1}{p(1+pT_1+p^2T_2^2)},$$ (3.37)

which yields, for value of $h < 1$,

$$T(t) = 1 - \frac{e^{-h\omega t}}{(1-h^2)^{\frac{1}{2}}}\cos\{(1-h^2)^{\frac{1}{2}}\omega_0 t - \sin^{-1}h\}.$$ (3.38)

The unit step function is not, of course, the only type of transient input which may occur. Its time derivative is the 'unit impulse', or Dirac, function of infinite magnitude lasting an infinitesimal time, the product of both being unity. On the other hand, we may have to deal with an input which increases linearly with time, the so-called 'unit ramp' function, or its time integral, the 'unit parabolic' function. Also, besides the displacement response, the velocity or the acceleration response of the system may be of interest, and, in exceptional cases, the time derivative of acceleration ('jerk') or the time integral of displacement ('excursion') may be wanted. Table 3.1 has been compiled to assist in locating the appropriate response with respect to a given type of transient exictation. The general character of input and response are shown pictorially. For instance, unit step-function input of displacement (first field of third row) leads in column C to the appropriate displacement response, while column D shows the velocity response, and column E the acceleration response. (The schematic curves at the top of Table 3.1 are those applicable to low damping only.) Accurate response curves have been computed for damping ratios $0 < h < 5$ and are plotted in Fig. 3.8(a)–(f) against the dimensionless 'time' $\omega_0 t$, the damping ratio h being a parameter. The response equations, also given in Fig. 3.8(a)–(f), are either time derivatives or time integrals of their neighbours, depending on whether one moves in Table 3.1 from A to F, or from F to A. Generalized response curves for a great variety of transient inputs (e.g. triangular, rectangular, half-sinusoidal, etc.) can be found in the literature [25]–[26].

3.3. Components of seismic systems

The performance of seismic systems discussed in sections 3.1 and 3.2 of this chapter assumes ideal system components, i.e. constant seismic mass, linear springs, and damping strictly proportional to velocity. The components of practical instrument transducers, however, usually deviate appreciably from these ideal characteristics.

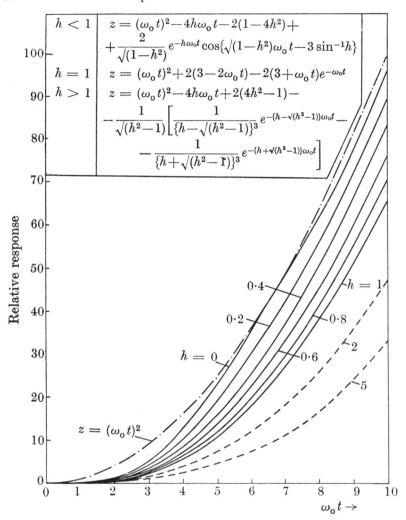

FIG. 3.8(a) Transient response of second-order system Type A (for correlation see Table 3.1).

3.3.1. Seismic mass

(a) *Rectilinear systems.* The seismic mass of a rectilinear system, i.e. a vibratory system with the seismic mass moving on a straight line, is measured in $kg(\equiv N\,s^2\,m^{-1})$. It usually consists of a conveniently shaped block of metal of high density, so that its size, and thus the size of the transducer, may be as small as possible. Brass is the material most commonly used (specific gravity 8·5), but, especially in miniature acceleration transducers, a so-called 'heavy alloy' [72] consisting of about 90 per

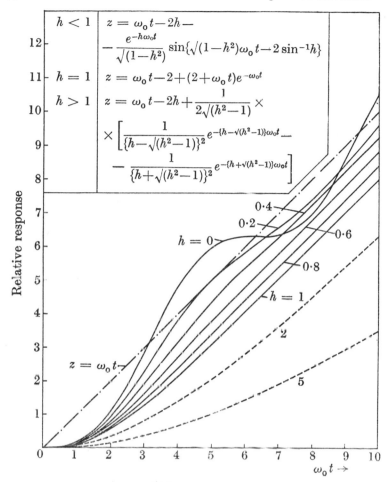

The figure inset contains the following equations:

$h < 1$
$$z = \omega_0 t - 2h - \frac{e^{-h\omega_0 t}}{\sqrt{(1-h^2)}} \sin\{\sqrt{(1-h^2)}\omega_0 t - 2\sin^{-1}h\}$$

$h = 1$
$$z = \omega_0 t - 2 + (2 + \omega_0 t)e^{-\omega_0 t}$$

$h > 1$
$$z = \omega_0 t - 2h + \frac{1}{2\sqrt{(h^2-1)}} \times$$
$$\times \left[\frac{1}{\{h - \sqrt{(h^2-1)}\}^2} e^{-\{h - \sqrt{(h^2-1)}\}\omega_0 t} - \frac{1}{\{h + \sqrt{(h^2-1)}\}^2} e^{-\{h + \sqrt{(h^2-1)}\}\omega_0 t} \right]$$

FIG. 3.8(b). Type B (see Table 3.1).

cent tungsten alloyed with nickel and copper (specific gravity 16·3–17) is of great advantage, since it produces about twice the mass at a given size. The alloy can be machined by conventional means.

If the transducer is filled with a damping liquid (see section 3.3.3) the effective seismic mass depends on the density ratio of the seismic mass to the mass of the liquid in which the seismic mass is immersed. This ratio may lead even to a negative effective mass, for instance, if the mass is hollow, i.e. buoyant with respect to the liquid. If such a 'negative' mass is rigidly coupled to the moving parts of a pressure transducer, errors due to acceleration could be compensated, as shown in Fig. 3.9, but effective sealing at low values of friction has proved difficult.

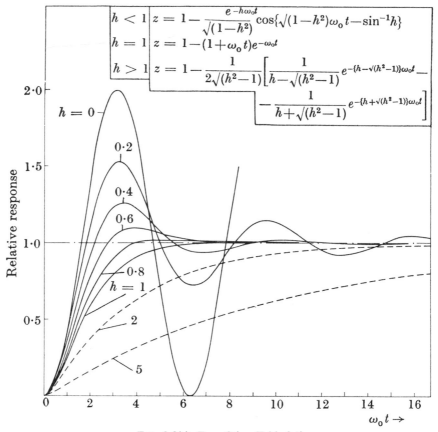

FIG. 3.8(c). Type C (see Table 3.1).

Besides this 'static' effect of the liquid on the seismic mass there is also a dynamic effect. Considering the schematic arrangement of an acceleration transducer with liquid damping (Fig. 3.10, [27]), it is obvious that the total kinetic energy of the system is the sum of the kinetic energy of the mass M,

$$\tfrac{1}{2}M(\dot{z})^2, \tag{3.39a}$$

and the kinetic energy of the damping liquid returning in the annular gap around the mass M. If m is the liquid mass in the return path, and A_1/A_2 the ratio of the piston area of the seismic mass and of the return path, then the kinetic energy of the liquid is

$$\tfrac{1}{2}m\left(\frac{A_1}{A_2}\dot{z}\right)^2, \tag{3.39b}$$

and the total energy

$$\frac{1}{2}\left\{M+\left(\frac{A_1}{A_2}\right)^2 m\right\}\dot{z}^2. \tag{3.40}$$

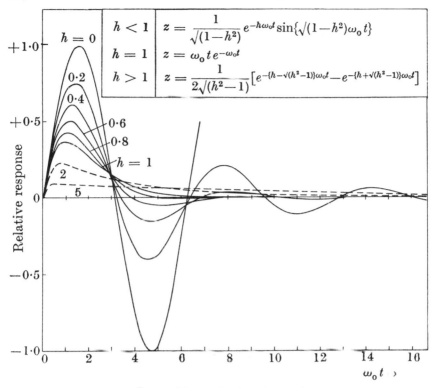

The figure box contains:

$h < 1$	$z = \dfrac{1}{\sqrt{(1-h^2)}}\,e^{-h\omega_o t}\sin\{\sqrt{(1-h^2)}\omega_o t\}$
$h = 1$	$z = \omega_o t\,e^{-\omega_o t}$
$h > 1$	$z = \dfrac{1}{2\sqrt{(h^2-1)}}\left[e^{-\{h-\sqrt{(h^2-1)}\}\omega_o t} - e^{-\{h+\sqrt{(h^2-1)}\}\omega_o t}\right]$

FIG. 3.8(d). Type D (see Table 3.1).

Eqn (3.40) shows that the seismic mass M of the transducer is increased by $(A_1/A_2)^2 m$. This contribution, which results in a decrease in natural frequency, depends greatly on the area ratio A_1/A_2, and it can be shown that if the return path is unduly restricted (in order to obtain high values of damping) the velocity, and thus the kinetic energy of the returning liquid, can be very high, resulting in an appreciably lower natural frequency than would be computed from the transducer mass only. Also, since the density, and thus the mass, of the participating liquid changes with temperature, the effective seismic mass of a liquid-damped transducer depends slightly on temperature. This applies to both the static and the dynamic contribution of the damping liquid to the effective mass.

(b) *Angular systems.* Newton's second law of mechanics for angular movement can be written

$$I\ddot{\alpha} = T, \tag{3.41}$$

where I = moment of inertia $(\text{kg m}^2 \equiv \text{N m s}^2)$,
$\ddot{\alpha}$ = angular acceleration (rad s^{-2}),
T = torque (N m).

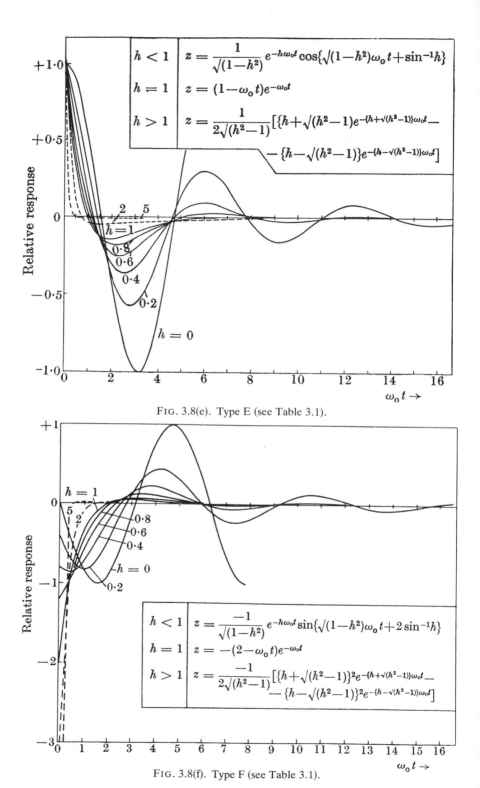

FIG. 3.8(e). Type E (see Table 3.1).

FIG. 3.8(f). Type F (see Table 3.1).

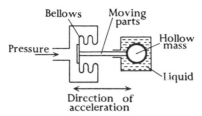

FIG. 3.9. Pressure transducer compensated for acceleration error by means of a buoyant ('negative') mass.

Tables for the computation of moments of inertia of simple solid shapes may be found in any handbook of basic engineering, or, as in case of more complex configurations, can be obtained experimentally by suspending the rotary mass as a bifilar pendulum [28]. The ideal rotary mass should have radical symmetry in order to avoid the effect of angular velocity about axes normal to the transducer axis [29]. Therefore, if the rotary mass consists of individual arms carrying masses at their ends, three equally spaced identical arms are the minimum requirement. Bartype masses should be avoided.

An angular acceleration transducer employing the rotary inertia of a liquid will be described in section 4.2 dealing with unbonded strain-gauge-type transducers. Liquid-damped angular acceleration transducers generally are subject to the same effects of fluid-density variations as rectilinear types; with large fluid volumes additional problems may arise with subsidiary flow patterns in the liquid.

3.3.2. Springs

(a) *Spring design.* An instrument spring is more often than not the heart of the electromechanical transducer and thus warrants the greatest care and consideration in design and manufacture. Frequently, pressure-sensing elements, such as capsules or diaphragms, are simultaneously the

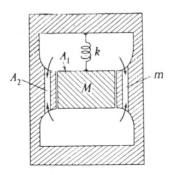

FIG. 3.10. Acceleration transducer with liquid damping, indicating flow of damping liquid.

BASIC LINEAR DEFLECTION SPRINGS

Shape of spring and mode of loading	Deflection (m)	Maximum Stress (N m^{-2})	Spring rate (N m^{-1})
Cantilever 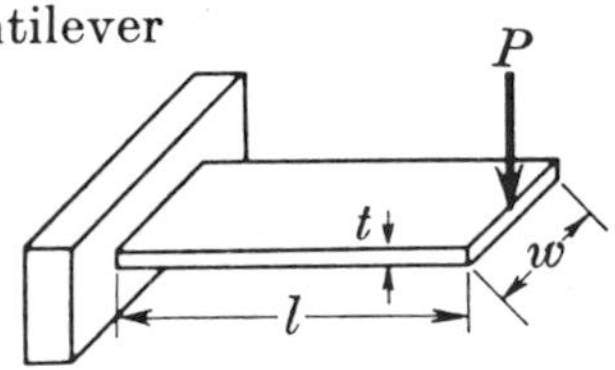	$f = \dfrac{4Pl^3}{Ewt^3} = \dfrac{2\sigma l^2}{3Et}$	$\sigma = \dfrac{6Pl}{wt^2} = \dfrac{3Etf}{2l^2}$	$\dfrac{P}{f} = \dfrac{Ewt^3}{4l^3}$
Beam 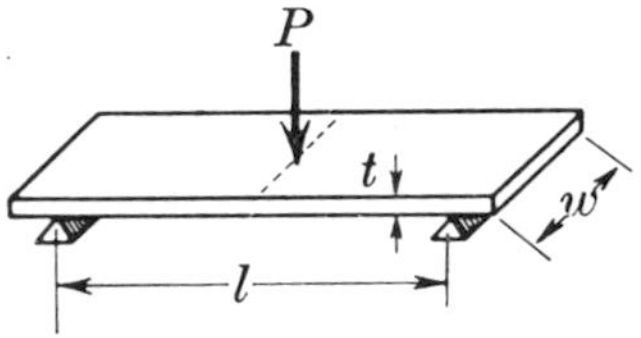	$f = \dfrac{Pl^3}{4Ewt^3} = \dfrac{\sigma l^2}{6Et}$	$\sigma = \dfrac{3Pl}{2wt^2} = \dfrac{6Etf}{l^2}$	$\dfrac{P}{f} = \dfrac{4Ewt^3}{l^3}$
Helical 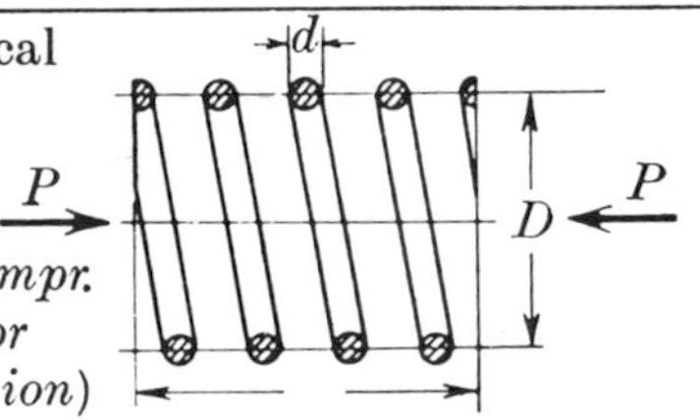 (Compr. or tension)	$f = \dfrac{8PnD^3}{Gd^4} = \dfrac{\pi nD^2\tau}{GdK_1}$	$\tau = \dfrac{8K_1 DP}{\pi d^3} = \dfrac{Gdf K_1}{\pi nD^2}$	$\dfrac{P}{f} = \dfrac{Gd^4}{8nD^3}$
n = number of turns; K_1 = stress factor (see Fig. 3.13)			

Fig. 3.11. Design of basic linear deflection springs.

BASIC ANGULAR DEFLEXION SPRINGS

Shape of spring and mode of loading	Angular deflexion (rad)	Maximum stress (N m⁻²)	Angular spring rate (Nm rad⁻¹)
Torsion bar (Round) 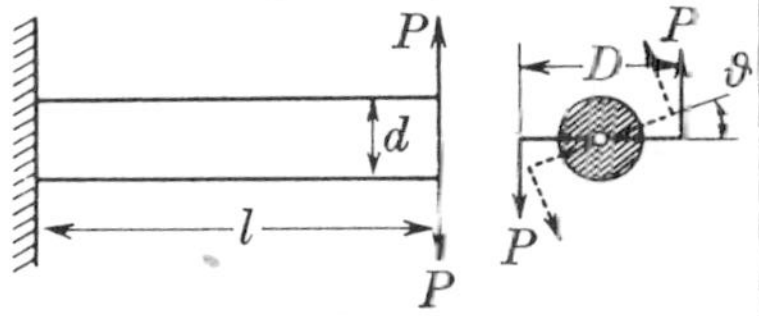	$\vartheta = \dfrac{32PDl}{\pi d^4 G} = \dfrac{2\tau l}{dG}$	$\tau = \dfrac{16PD}{\pi d^3} = \dfrac{dG\vartheta}{2l}$	$\dfrac{PD}{\vartheta} = \dfrac{\pi d^4 G}{32l}$
Torsion bar (Flat) 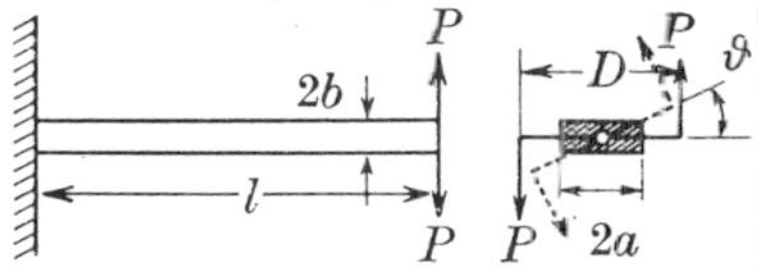	$\vartheta = \dfrac{PDl/Gab^3}{\frac{16}{3}-3\cdot36(b/a)\{1-(b^4/12a^4)\}}$	$\tau = \dfrac{PD(3a+1\cdot8b)}{8a^2b^2}$	$\dfrac{PD}{\vartheta} = \dfrac{Gab^3}{l} \times$ $\times[\frac{16}{3}-3\cdot36(b/a)\{1-(b^4/12a^4)\}]$
Spiral 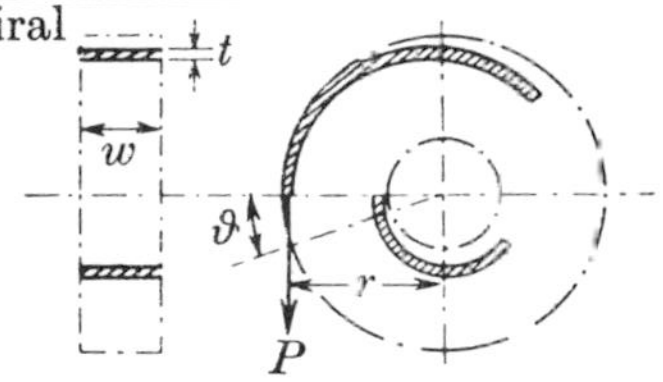	$\vartheta = \dfrac{12\pi Pr^2 n}{Ewt^3} = \dfrac{2\pi rn\sigma}{Pt}$	$\sigma = \dfrac{6Pr}{wt^2} = \dfrac{tE\vartheta}{2\pi rn}$	$\dfrac{Pr}{\vartheta} = \dfrac{wt^3 E}{12\pi rn}$

n = number of turns

Fig. 3.12. Design of basic angular deflexion springs.

seat of the restoring force, and their spring characteristics have to conform with the characteristics discussed in this section. Even in the design of force-balance-type transducers (Chapter 5) which theoretically do not require mechanical springs, they are nevertheless very much in the mind of the designer, since guide elements for the moving parts are invariably required and these often constitute an undesired mechanical spring. This also applies to electrical connections between the moving and stationary parts of this type of transducer.

The relative movements in instrument tranducers are usually small, and engineering approximations for spring deflections are therefore applicable. Fig. 3.11 summarizes well-known design formulae for basic spring configurations with linear deflections, and Fig. 3.12 those for angular deflections. All linear dimensions are given in metres; and forces $P(N)$, Young's moduli E $(N\,m^{-2})$, and moduli of rigidity G $(N\,m^{-2})$ are also measured in SI units. Fig. 3.13 shows the stress factor for helical springs as a function of the diameter ratio D/d. Values of E and G for a selection of spring materials are listed in Table 3.2.

Fig. 3.14 shows examples of spring guides commonly used in instrument tranducers, which are composed either of leaf springs [30] or of spring diaphragms [31]. As to the latter it is obvious that the multiple-start spiral cut produces a more flexible guide than the single spiral, and that the unwanted rotation is smaller. The diaphragms with segment and cantilever cuts are free from rotation, but the former has a very restricted deflection range and soon becomes non-linear. The diaphragm with the cantilever cut is similar to the second of the two parallel spring guides in Fig. 3.14. It has an excellent performance if the diaphragm is perfectly

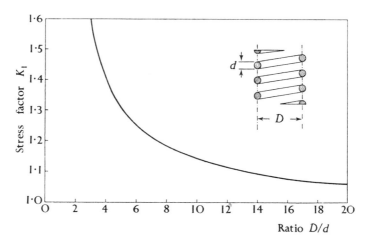

FIG. 3.13. Stress factor K_1 of helical springs as a function of the ratio of mean diameter D to wire diameter d.

TABLE 3.2

Elastic and thermal properties of some spring materials

Material	Composition	Modulus of elasticity, E (N m⁻²)	Modulus of rigidity, G (N m⁻²)	Thermal coefficient of linear expansion α (m per m per °C) (0–100°C)	Thermal coefficient of modulus of elasticity, c (ΔE/E per °C) (−50°C to +50°C)	Thermal coefficient of modulus of rigidity, m (ΔG/G per °C) (−50°C to +50°C)	Remarks
Multiply by		10^9	10^9	10^{-6}	10^{-5}	10^{-5}	
Flat spring steel	Fe, C, Mn	210	—	11.8	−24	−24	General purpose
Piano wire	Fe, C, Mn	210	80·3	11.7	−27	−36	Small helical springs
German silver	Cu, Zn, Ni	110	38	16·2	−35	−37	Corrosion resistant
Phosphor–bronze	Cu, Sn	105	45	17·8	−36	−40	Corrosion resistant, good electrical conductor
Beryllium–copper	Cu, Be	110–30	40–50	16·6	−35	−33	High fatigue life, low hysteresis, good electrical conductor
Invar	Ni, Fe	147	56	1·08	+4·8	—	Low thermal expansion
Modulvar	Ni, Fe, C, Si, Mn, Cr, Si	145	55	0	+4·8	—	Low thermal expansion
Elinvar	Ni, Cr, Si, W, Mn, Fe, C	143	57	0·6	−0·66	−0·72	Constant moduli
Iso-elastic	Ni, Cr, Mn, Fe	180	64	7·2	−3·6-+2·7	—	Fairly constant moduli
Ni-span C	Ni, Cr, Ti, Al, C, Mn	165–85	70	8·1	−1-+1	—	Constant moduli, high strength

PARALLEL SPRING GUIDES

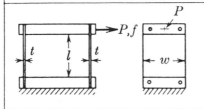

For complete guide
(two springs)

$$\frac{P}{f} = \frac{2Ewt^3}{l^3} \ (\text{N m}^{-1})$$

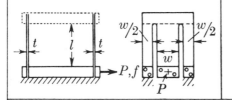

$$\frac{P}{f} = \frac{Ewt^3}{l^3} \ (\text{N m}^{-1})$$

SPRING DIAPHRAGMS

Spiral

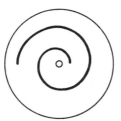

Multiple start spiral

Segments

Cantilevers

FIG. 3.14. Design of parallel spring guides.

flat and free from buckles. All leaf springs and diaphragms must be free from burrs and deformations in order to obtain optimum flexibility and linearity. Spring diaphragms are commonly produced by etching.

Fig. 3.15 shows spring pivots, either consisting of a single-leaf spring or of two pairs of crossed-leaf springs. The crossed-spring pivot is widely used in instrument work and its properties have been investigated both theoretically [32]–[33] and experimentally [34]. At the small angular deflections commonly used in instrument transducers their angular spring rate can be assumed constant.

(b) *Spring materials.* An ideal material for instrument springs should have the following properties (see also section 3.1):

(1) low values of E and G;

(2) high values of stress at limit of proportionality, and of yield strength;

(3) low values of hysteresis, creep, and elastic after-effect;

(4) high values of endurance and fatigue strength;

(5) low values of thermal expansion;

(6) low values of variation with temperature of E and G;

(7) resistance to corrosion;

(8) ease of machining and joining techniques;

(9) simple heat-treatment or none required after manufacture;

(10) low and stable values of electrical restivity;

(11) ease of electrical connections to spring by way of clamping or soft-soldering.

In all applications there is, of course, the demand for low cost and easy availability in convenient form of spring materials. For example, beryllium–copper, one of the best spring materials with respect to (1) to (7) above, falls short of the ideal with respect to (8), (9), and (11), besides being rather expensive. The choice of the most suitable material for a particular application will be a compromise between conflicting requirements.

The main reason for instabilities and inaccuracies in the performance of instrument springs is the variation with temperature of the spring dimensions (coefficient of thermal expansion) and of the moduli of elasticity E and rigidity G. In order to help the designer, Table 3.2 has been compiled, though in specific cases more detailed information may be

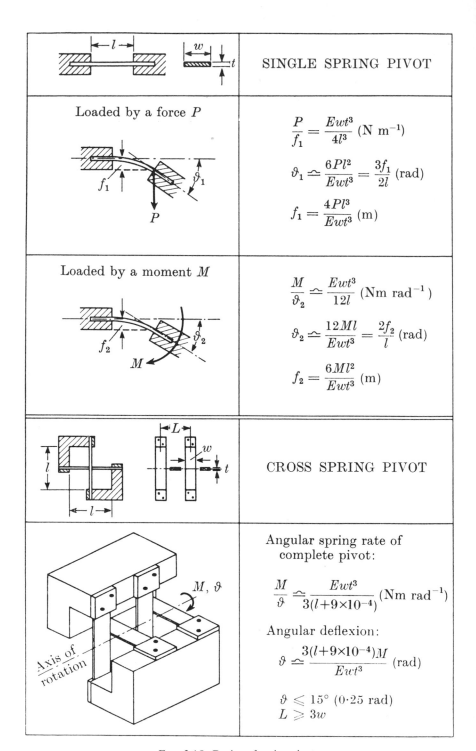

	SINGLE SPRING PIVOT
Loaded by a force P	$\dfrac{P}{f_1} = \dfrac{Ewt^3}{4l^3}$ (N m^{-1}) $\vartheta_1 \simeq \dfrac{6Pl^2}{Ewt^3} = \dfrac{3f_1}{2l}$ (rad) $f_1 = \dfrac{4Pl^3}{Ewt^3}$ (m)
Loaded by a moment M	$\dfrac{M}{\vartheta_2} \simeq \dfrac{Ewt^3}{12l}$ (Nm rad^{-1}) $\vartheta_2 \simeq \dfrac{12Ml}{Ewt^3} = \dfrac{2f_2}{l}$ (rad) $f_2 = \dfrac{6Ml^2}{Ewt^3}$ (m)
	CROSS SPRING PIVOT
	Angular spring rate of complete pivot: $\dfrac{M}{\vartheta} \simeq \dfrac{Ewt^3}{3(l+9\times10^{-4})}$ (Nm rad^{-1}) Angular deflexion: $\vartheta \simeq \dfrac{3(l+9\times10^{-4})M}{Ewt^3}$ (rad) $\vartheta \leqslant 15°$ (0·25 rad) $L \geqslant 3w$

FIG. 3.15. Design of spring pivots.

extracted from specialized sources [35]–[38]. If the basic expressions for the spring rates of different spring configurations are written in a generalized form we have:

$$\text{cantilever:} \quad P/f \propto wt^3E/l^3, \tag{3.42a}$$

$$\text{helical:} \quad P/f \propto d^4G/D^3, \tag{3.42b}$$

$$\text{torsional:} \quad M/\vartheta \propto D^4G/l, \tag{3.42c}$$

and the corresponding percentage errors in spring rate can be shown to be [36]

$$\text{cantilever:} \quad \left\{1 - \frac{1}{(1+\alpha\,\Delta T)(1+c\,\Delta T)}\right\}100 \approx 100\,\Delta T(\alpha+c), \tag{3.43a}$$

$$\text{helical:} \quad \left\{1 - \frac{1}{(1+\alpha\,\Delta T)(1+m\,\Delta T)}\right\}100 \approx 100\,\Delta T(\alpha+m), \tag{3.43b}$$

$$\text{torsional:} \quad \left\{1 - \frac{1}{(1+\alpha\,\Delta T)^3(1+m\,\Delta T)}\right\}100 \approx 100\,\Delta T(3\alpha+m), \tag{3.43c}$$

where
ΔT = temperature variation (°C),

α = thermal coefficient of linear expansion,

c = thermal coefficient of modulus of elasticity E,

m = thermal coefficient of modulus of rigidity G.

The percentage error per °C of spring rate for a number of spring materials, when employed in cantilever, helical, and torsional-type springs, have been computed from eqns (3.43a)–(3.43c) and are listed in Table 3.3. In addition to the spring rate errors of eqn (3.43) and Table 3.3, there is a zero-shift error with temperature which is proportional to $(\alpha\,\Delta T)$. Its magnitude is a function of the length of the spring, its geometry, and the manner in which it is being used.

TABLE 3.3

Percentage errors of spring rate per °C of cantilever, helical, and torsional springs for various spring materials
(−50 °C to +50 °C)

	Cantilever	Helical	Torsional
Piano wire	0·026	0·025	0·022
Phosphor–bronze	0·034	0·038	0·035
Beryllium-copper	0·032	0·031	0·028
Ni-span C	0·0008	0·0008	0·002

3.3.3. *Damping*

The significance of (viscous) damping in the dynamic performance of transducer systems has been discussed in section 3.2. There are a number of apparently different damping notations in common use for both free and sustained oscillations which may cause confusion. A summary of their definitions and co-relations is given below.

(i) *Sustained excitation.* The differential equation of a linear second-order vibratory system with sinusoidal excitation is (eqn 3.26)

$$\ddot{z} + \frac{c}{m}\dot{z} + \frac{k}{m}z = \frac{k}{m}s(t), \tag{3.44a}$$

which can also be written

$$\ddot{z} + 2\,\delta\dot{z} + \omega_0^2 z = \frac{k}{m}s(t) \tag{3.44b}$$

or

$$\ddot{z} + 2h\omega_0\dot{z} + \omega_0^2 z = \frac{k}{m}s(t), \tag{3.44c}$$

and

$$\frac{m}{k}\ddot{z} + \frac{c}{k}\dot{z} + z = s(t) \tag{3.44d}$$

or

$$T_2^2\ddot{z} + T_1\dot{z} + z = s(t), \tag{3.44e}$$

where

$c = $ damping coefficient (N s m^{-1}), $\qquad$ (3.45)

$\delta = c/2m = $ damping factor (s^{-1}), $\qquad$ (3.46)

$h = c/c_c = \delta/\delta_c = $ damping ratio, $\qquad$ (3.47)

$c_c = 2/(km)^{\frac{1}{2}} = 2m\omega_0 = $ critical damping coefficient (N m s^{-1}), $\quad$ (3.48a)

$\delta_c = (k/m)^{\frac{1}{2}} = \omega_0 = $ critical damping factor (s^{-1}), $\qquad$ (3.48b)

$h_c = 1 = $ critical damping ratio $\qquad$ (3.48c)

$\omega_0 = (k/m)^{\frac{1}{2}} = $ undamped natural circular frequency (s^{-1}), $\qquad$ (3.49)

$\omega_r = \omega_0(1-2h^2)^{\frac{1}{2}} = $ resonant circular frequency (s^{-1}), $\qquad$ (3.50)

$Q = 1/2h = $ 'quality factor' of resonant system, $\qquad$ (3.51)

and

$$T_1 = 2h/\omega_0, \qquad T_2 = 1/\omega_0 = \text{time-constants (s)}. \tag{3.52}$$

(ii) *Free oscillations.* After a transient disturbance a linear second-order vibratory system with (small) damping oscillates at the 'damped' natural circular frequency

$$\omega_d = \omega_0(1-h^2)^{\frac{1}{2}} \qquad \text{(for } h < 1\text{)}, \tag{3.53}$$

which is not identical to the resonant frequency ω_r above. The damping of free oscillations can also be obtained from the decay rate of two successive (unidirectional) amplitudes a_1/a_2, in terms of the logarithmic decrement

$$\Theta = \ln(a_1/a_2) = 2\pi h/(1-h^2)^{\frac{1}{2}} \qquad \text{(for } h < 1\text{)}. \qquad (3.54)$$

The ratio ω_d/ω_0 of the damped natural frequency to the undamped natural frequency, and the logarithmic decrement Θ, have been plotted in

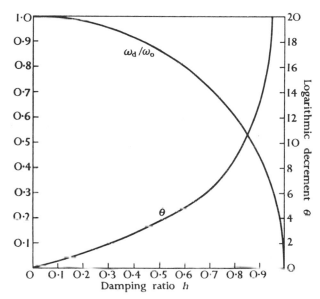

FIG. 3.16. Logarithmic decrement Θ and frequency ratio ω_d/ω_0 as a function of damping ratio h.

Fig. 3.16 against the damping ratio h, the most commonly used damping notation.

In a practical layout of damping devices the designer usually starts by computing the damping force $c\dot{z}$ for a given system, and with the additional information of the undamped circular natural frequency $\omega_0 = (k/m)^{\frac{1}{2}}$, the system performance can conveniently be analysed with the parameters ω_0 and h given (see section 3.2). Eqns (3.44)–(3.54) and Fig. 3.16 will be helpful if alternative parameters have been quoted.

(a) *Air damping.* The conventional air-damping devices, such as vanes and dashpots, are seldom used in instrument transducers since they tend to be large and bulky. Design formulae and typical arrangements of vane- and piston-type air dampers can be found in textbooks on electrical measuring instruments [39]. However, appreciably larger damping forces

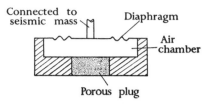

FIG. 3.17. Air damper with porous ceramic plug.

can be obtained in a small space, if the air is forced through a narrow capillary tube or, better still, through a porous ceramic plug. Fig. 3.17 shows schematically the latter arrangement in use with acceleration transducers of the unbonded strain-gauge type [40]. The movement of the seismic mass pumps air from the chamber through the porous plug, and if the volume of the air chamber is small, a high air pressure, and thus an effective flow of air in the plug, can be obtained.

The temperature stability of this type of damper is better than that of an equivalent oil damper, and the resultant frequency-response curve of the acceleration tranducer is virtually constant over a wide temperature range, since air viscosity is little affected by temperature variations. There is also no secondary contribution to the effective mass of the accelerometer because the specific mass of air is negligible (see section 3.1). The viscosity variation with temperature of air (and other gases) follows a fairly complex law though, as a first approximation, air viscosity increases with the square-root of the absolute temperature. This tendency has been plotted in Fig. 3.18 in comparison with viscosity–temperature curves of other damping media, and with eddy-current damping. It is seen that air damping has the smallest variation with temperature, and in contrast to

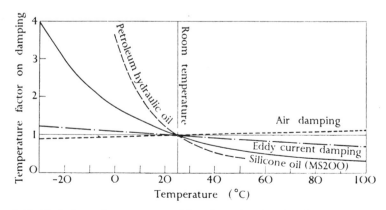

FIG. 3.18. Variation with temperature of damping, related to damping at room temperature (25 °C).

oil and eddy-current damping, air damping increases with increasing temperature.

(b) *Oil damping.* Today the only liquid damping medium in common use is silicone oil [41], mainly by virtue of its fairly stable viscosity–temperature characteristics. Silicone oil for damping purposes is available in 18 nominal (room temperature) viscosities, ranging from 0·65 to 100 000 centistokes $(0·65 \times 10^{-6} - 0·1 \text{ m}^2 \text{ s}^{-1})$, and above 50 centistokes they may be mixed freely to give intermediate viscosities. Below 50 centistokes volatility and vapour pressure increase rapidly with decreasing viscosity and these grades should not be mixed. In the majority of

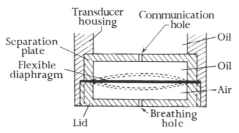

FIG. 3.19. Expansion chamber for oil-filled transducers.

transducer applications the grades with 100–1000 centistokes $(10^{-4}-10^{-3} \text{ m}^2 \text{ s}^{-1})$ are usually sufficient, and Fig. 3.18 shows the average variation of their viscosity with temperature. The improvement of stability is obvious if the curves for silicone oil and for the conventional petroleum hydraulic oil are compared.

The major advantage of liquid damping over eddy-current damping is the almost unlimited damping force obtainable in a very small space. Furthermore, the moving transducer parts can be completely immersed in the liquid, so that no extra space is required for the damping device. However, because of volume expansion of the liquid, an expansion chamber must be provided in order to avoid oil leakage at elevated temperatures and low ambient pressures (high altitudes). For example, silicone oil type MS200 of viscosity 100 centistokes expands by

$$V_T = V_0(1 + \alpha T + \beta T^2), \tag{3.55}$$

where

V_0 = oil volume at 0 °C,

V_T = oil volume at T °C,

$\left. \begin{aligned} \alpha &= 8·77 \times 10^{-4} \\ \beta &= 8·66 \times 10^{-7} \end{aligned} \right\}$ between 0 °C and 100 °C.

The expansion coefficients of other grades are of the same order, except for very low viscosities. Fig. 3.19 shows a typical expansion chamber incorporated in an oil-filled acceleration transducer. The small communication hole in the separation plate provides a passage for the oil from the transducer housing to the expansion chamber, if the ambient temperature rises, but it restricts the unwanted flow of oil due to dynamic acceleration forces. In very small transducers, a piece of foam rubber, a tiny air-filled 'ballon', or even a trapped air bubble may serve a similar purpose.

Oil-filled acceleration transducers offer some further problems connected with the density of the oil. The transducer sensitivity before and after oil-filling are different, since in the latter case the effective seismic mass is lower than in the former, owing to the buoyancy effect of the oil. Pumping of oil in the transducer, especially inside narrow fluctuating 'air' gaps in inductance-type transducers, changes the effective mass, and the dynamic performance of the filled and unfilled transducer cannot be compared unless the effect, which also depends on the viscosity of the oil, has been allowed for (see also section 3.3.1 on seismic mass).

Low values of variation with temperature of viscosity and low values of density and of thermal expansion are thus the major features of a good damping liquid. Other desirable properties include: chemical stability and inertness to metals and other transducer materials, good electrical insulation, and low values of surface tension and of vapour pressure.

(c) *Eddy-current damping.*   Eddy-current damping is probably the most predictable type of damping. The damping forces can be computed with sufficient accuracy from the design parameters of the eddy-current damper at any given ambient temperature. If high-conductivity copper is used for the eddy-current carrier the damping forces decrease by about 0·4 per cent per °C temperature rise, neglecting the temperature coefficient of the magnetic circuit, which is about one order lower. In Fig. 3.18 this variation is indicated for comparison with air and oil damping.

The computation of eddy-current damping forces for common conductor configurations can be found in textbooks on electrical measuring instruments [39]. As an application to instrument transducers the most common and probably the most efficient arrangement of an eddy-current type damper is shown in Fig. 3.20. The permanent magnet produces a

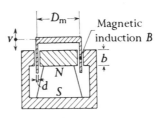

FIG. 3.20. Eddy-current damper with loudspeaker-type magnet and solid copper cup.

magnetic induction B (T = tesla) in the annual air gap of depth b(m). In the solid copper cup of mean diameter D_m(m), wall thickness d(m), and moving with velocity v(m s^{-1}) perpendicular to the magnetic field lines, a voltage

$$e = \pi D_m B v \quad \text{(V)} \tag{3.56}$$

will be generated. The resistance of the cup within the magnetic field ($\rho(\Omega\,\text{m})$ being the resistivity of copper) is

$$R = \pi D_m \rho / bd \quad (\Omega), \tag{3.57}$$

and the current in the cup thus

$$i = e/R = Bbdv/\rho \quad \text{(A)}. \tag{3.58}$$

The damping force $F = c\dot{z}$ (see eqn (3.44a)) therefore becomes

$$F = \pi D_m B i = \pi D_m B^2 bdv/\rho \quad \text{(N)}, \tag{3.59}$$

and with $v = \dot{z}$ the damping coefficient

$$c = F/\dot{z} = \pi D_m B^2 bd/\rho \quad (\text{N s m}^{-1}). \tag{3.60}$$

Evaluating eqn (3.60) for a practical design let

$$B - 1\,\text{T} \qquad\qquad d = 1\,\text{mm},$$
$$b = 5\,\text{mm}, \qquad\qquad D_m = 20\,\text{mm},$$
$$\rho = 1{\cdot}74 \times 10^{-8}\,\Omega\,\text{m},$$

then we have

$$c = 18\,\text{N s m}^{-1}.$$

In a given design the most convenient parameter for adjustment is the thickness d of the copper cup. If fine silver is substituted for high-conductivity copper the damping force is increased by only 8 per cent. If the damping cup does not penetrate the whole gap depth b, the value of b in eqn (3.60) must be reduced to the depth of penetration. Stray fields and non-uniformities of the magnetic field have been neglected here; the measured damping forces are usually slightly higher than those computed from eqn (3.60).

3.4. Pressure-sensitive components

Next to springs, pressure-responding components are the most common elements of instrument transducers [42]. Fig. 3.21(a)–(l) shows a selection of such components in schematic form. (a)–(i) will be discussed in greater detail in this section, with the exception of (c), the so-called catenary type which is said to be the diaphragm type least affected by temperature [43].

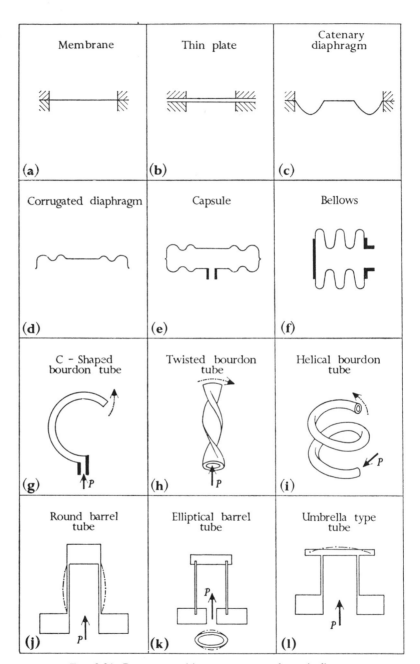

FIG. 3.21. Pressure-sensitive components, schematic diagram.

The elements shown in Fig. 3.21(j)–(l) are normally employed with variable-resistance (strain-gauge) pick-offs; they will be discussed in section 4.2.

3.4.1. Membranes

A membrane is a thin diaphragm under radial tension $S(N\,m^{-1})$. Its stiffness to bending forces is neglected. If pressure is applied to one side the membrane deflects (like a soap bubble) in the shape of a sphere. At low pressure differences the centre deflection $z_{max}(m)$ and the stress $\sigma_{max}(N\,m^{-2})$ are quasi-linear functions of pressure and the appropriate design formulae [44] are listed in Fig. 3.22, together with the formula for the lowest natural frequency of the membrane in vacuum or, nearly enough, in free air. The effect of a thin layer of air (narrow air gap) on the damping of membranes has been investigated by Crandall [45].

3.4.2. Thin plates

In transducer applications we are mainly interested in the static and dynamic behaviour of thin circular plates which are clamped or welded around their circumference. The basic design formulae have again been listed in Fig. 3.22. The reader requiring more detailed information on the static and dynamic theory of membranes, thin plates, and shells of different geometries and boundary conditions should consult the textbooks on elasticity [44]–[46]. A good summary of research on flat diaphragms and circular plates for instrument applications has been given by Wahl [47], together with a comprehensive list of references. If the pressure medium is not air or gas, the formula for the lowest natural frequency in Fig. 3.22 requires modification [48].

The three flat diaphragm types most commonly employed in instrument transducers are shown in Fig. 3.23(a)–(c). In (a) the plate is clamped between two solid rings. Although this arrangement is attractive because of its simplicity and cheapness, it will normally not be free from hysteresis due to friction between the deflecting diaphragm and the clamping rings. Diaphragms of high-class pressure transducers are now stitch-welded or electron-beam-welded to their supports, as shown in (b), thus minimizing hysteresis and avoiding heat stresses connected with soldering of flame-welding. (c) shows a diaphragm machined from a solid blank. Its hysteresis is negligible for small deflections, but it is much more difficult to produce. The main danger here is an initial buckle in the diaphragm which is often caused by the pressure of the cutting tool in the final cut (cold-working). A diameter-to-thickness ratio of 100 is probably the practical limit, unless spark-erosion or electrochemical techniques are used in the machining which exert no tool forces. Even a very tough high-temperature material, such as Nimonic, can be used in this

FLAT CIRCULAR DIAPHRAGMS	CENTRE DEFLECTION (m)	MAXIMUM STRESS (N m^{-2})	LOWEST NATURAL FREQUENCY (Hz)
Membrane 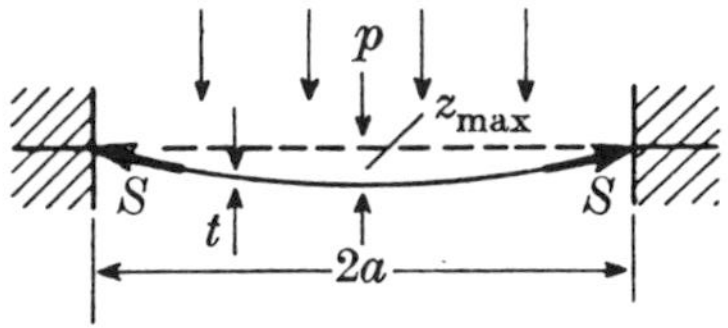	$z_{max} = \dfrac{a^2 p}{4S}$ Linear if $z_{max} \leqslant 0.005a$	$\sigma_{max} \simeq \dfrac{S}{t}$ Uniform over membrane	$f_0 = \dfrac{1{\cdot}20}{\pi a}\sqrt{\left(\dfrac{S}{\mu t}\right)}$
Thin plate 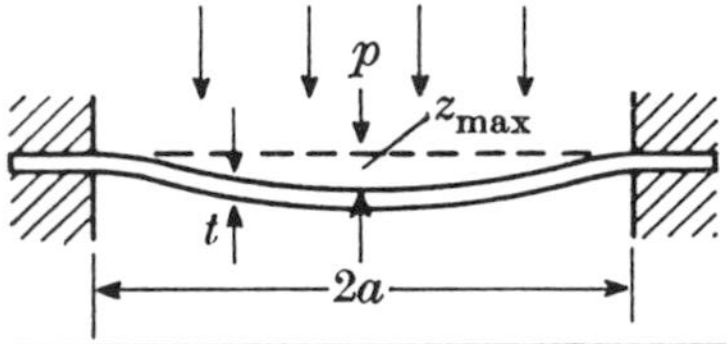	$z_{max} = \dfrac{3(1-\nu^2)a^4 p}{16Et^3}$ Linear if $z_{max} \leqslant 0.5t$	$\sigma_{max} \simeq \dfrac{3a^2 p}{4t^2}$ At circumference	$f_0 = \dfrac{2{\cdot}56t}{\pi a^2}\sqrt{\left(\dfrac{E}{3\mu(1-\nu^2)}\right)}$

p = pressure (N m^{-2})
S = membrane tension (N m^{-1})
E = Young's modulus (N m^{-2})

ν = Poisson's ratio (0·30 for steel)
μ = spec. mass (kg m^{-3})

Fig. 3.22. Design of flat circular diaphragms.

(a)

(b)

(c)

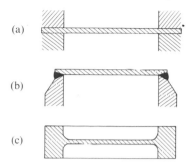

FIG. 3.23. Flat diaphragms: (a) clamped, (b) welded, and (c) machined from solid.

type of design. At the other extreme, diaphragm units of only 3·25 mm diameter and 0·1 mm thickness have been successfully made from an aluminium alloy (see section 4.4.4).

3.4.3. Corrugated diaphragms and capsules

Corrugated diaphragms and capsules are used when the small deflections of flat diaphragms are not sufficient. A typical deflection of a corrugated diaphragm is 2 per cent of its diameter. In order to obtain still larger deflections, capsules composed of two identical corrugated diaphragms and stacks of capsules are used, but in electromechanical instrument transducers, in contrast to direct-reading or recording scientific and industrial instruments, excessively large deflections are seldom required, except perhaps in potentiometer type transducers. Furthermore, the greater resilience of these elements, as compared with flat diaphragms, means greater susceptibility to vibration and acceleration and would normally not be acceptable for other than stationary applications.

Corrugated diaphragms and capsules are used for the measurement of vacuum pressures and up to about 2·5 MN m⁻². In general the flexibility of a corrugated diaphragm increases with the number of corrugations and decreases with their depth. A number of attempts have been made to compute the deflection and stress performance of corrugated diaphragms [49], [50] of which the method devised by Haringx [51] ('smearing out' the effect of the corrugations) is probably the most practical. It permits the design of suitable corrugated diaphragms with specific deflections (with a given non-linearity) and stresses by means of a chart. Materials most commonly used in the manufacture of corrugated diaphragms and capsules are brass, phosphor–bronze, nickel–silver, beryllium–copper, Monel, and stainless steel. In the fabrication of capsules electron-beam-welding is now quite common.

3.4.4. *Bourdon tubes*

Even more than with corrugated diaphragms, the main feature of Bourdon tubes is their large deflection. Conventional C-shaped tubes (Fig. 3.21(g)) of 50 mm bending diameter have a useful travel of typically 3–3·5 mm, and they are therefore more popular with direct-indicating instruments than with electromechanical transducers, except, again, in potentiometer-type pressure transducers. For even larger (angular) travels there are spiral and helical types (Fig. 3.21(i)) which are all susceptible to vibration and acceleration. The twisted type of Fig. 3.21(h), which 'untwists' under internal pressure, provides smaller deflections, usually sensed by a variable-inductance-type pick-off [73], and although it is more rigid than the other types, it needs some additional support at the free end of the tube for optimum performance in an acceleration environment. C-type Bourdon tubes cover ranges from about $35 \, \mathrm{kN \, m^{-2}}$ to $100 \, \mathrm{MN \, m^{-2}}$; the twisted tube has been used for pressure ranges up to $20 \, \mathrm{MN \, m^{-2}}$.

C- and helical-type Bourdon tubes have been investigated theoretically [52], [53]. In general, a large aspect ratio of the tube cross-section gives high sensitivity but weak tubes. Stresses are reduced with decrease of the major axis. The design of C-type tubes has also been facilitated by graphical means [54].

In transducer work small size is almost invariably a major requirement and large tubes can rarely be employed, though they give generally better accuracies. Tubes with bending diameters of about 50 mm are subject to hysteresis errors of 1–2 per cent of full-scale deflection. Special care is required in the anchorage region of the tube. Soft-soldering may cause appreciable hysteresis due to high local stresses. The materials commonly used for Bourdon tube manufacture are brass, phosphor–bronze, beryllium–copper, and steel.† Material selection [55] and engineering aspects of Bourdon tube manufacture [56] cannot be discussed here.

3.4.5. *Bellows*

In instrument transducers metallic bellows are normally used as a flexible pressure seal, rather than as a pressure-sensitive element, the restoring force determining the transducer calibration. This is because the zero position and the stiffness of bellows are not as stable as those of first-class instrument springs; it is therefore advisable to put a reliable instrument spring of appreciably higher stiffness in parallel with the bellows. The stiffness of seamless metallic bellows [57] is proportional to the Young's modulus of the material, and inversely proportional to the

---

† Helical Bourdon tubes made of fused quartz have recently been released by Texas Instruments Ltd., Bedford. They have lower hysteresis, creep, and fatigue values than metal tubes (see also reference [74]).

outside diameter and to the number of convolutions of the bellows. Stiffness also increases with roughly the third power of the wall thickness which, for instrument work, normally lies between 0·1 mm and 0·15 mm. The effective area A of bellows is

$$A = \frac{\pi}{16}(D+d)^2, \qquad (3.61)$$

where D and d are the outside and inside diameters, respectively, of the convolutions. Pressure should be admitted to the outside so that the bellows is under compression [58]. Maximum permissible external pressures for small-diameter brass bellows are round about 1·5 MN m^{-2}. The

FIG. 3.24. Pressure chamber with inlet pipe.

principal material used in bellows manufacture is 80/20 brass, but stainless-steel bellows for higher pressures are also known.

3.5. Effect of pressure inlet pipes

The dynamic performance of mechanical vibratory systems has been dealt with in section 3.2. With respect to pressure transducers the characteristics of these systems apply to the natural frequency and damping of the mechanical transducer components, i.e. stiffness of instrument springs and/or pressure-sensitive elements, seismic mass of moving parts, and the appropriate mechanical damping. There is, however, the much more important problem of the dynamic characteristics of the acoustic system, consisting of the pressure medium in the (inlet) pipes and in the transducer cavity. If, as will often be the case, the natural frequency of this acoustic system is lower than the natural frequency of the mechanical transducer parts, the pressure inlet pipe is the limiting factor in the application to fluctuating (cyclic or transient) pressure measurements. Theoretical investigations on the attenuation of oscillating pressures in pipes have been published [59]–[61], and the results have been applied to practical pressure transducer systems [62]–[64]. However, if only a rough guide to the effect of pipe connections is required, an estimate of natural or resonant frequency, damping ratio, and attenuation can be obtained by considering specific conditions of geometry and operation.

If, in Fig. 3.24, the volume V of the transducer cavity is much smaller than the pipe volume, then the inlet pipe—open at one end and closed at the other by the transducer diaphragm—is most easily excited in 'organ-pipe' fashion at frequencies

$$f_0 = c(1+2n)/4l \quad \text{(Hz)} \qquad (n = 0, 1, 2,...), \tag{3.62}$$

where c (m s^{-1}) is the sound velocity in the gas, and l (m) is the pipe length. Eqn (3.62) is valid for transient and sustained excitations, as long as the pressure fluctuations are small compared with the base pressure. For pressure steps similar to, or larger than, the base pressure, conditions are much more complex; in this case the transducer inlet pipe may even act like a subsidiary shock tube.

More commonly, the pressure inlet pipe of length l (m) and internal radius r (m) will be connected to a sizeable transducer cavity of volume V (m^3), as shown in Fig. 3.24. For small, sustained gas-pressure fluctuations of low frequency $f < c/4l$ (Hz) the distributed pipe parameters can be approximated by lumped parameters and the ratio of the pressure p (N m^{-2}) inside the transducer cavity to the pressure p_0 at the open end of the pipe then becomes

$$|p/p_0| = \{(1-\omega^2 LC)^2 + \omega^2 C^2 R^2\}^{-\frac{1}{2}}$$
$$= \{[1-(\omega/\omega_0)^2]^2 + 4h^2(\omega/\omega_0)^2\}^{-\frac{1}{2}}, \tag{3.63}$$

with the notations [65]:

$$\text{acoustic capacitance of cavity:} \quad C = V/\rho c^2 \quad \text{(m}^5\,\text{N}^{-1}\text{),} \quad \text{(3.64a)}$$

$$\text{acoustic inductance of pipe:} \quad L = 4l\rho/3\pi r^2 \quad \text{(N s}^2\,\text{m}^{-5}\text{),} \quad \text{(3.64b)}$$

$$\text{acoustic resistance of pipe:} \quad R = 8\eta l/\pi r^4 \quad \text{(N s m}^{-5}\text{),} \quad \text{(3.64c)}$$

where

$$\rho = \text{density (kg m}^{-3}\text{),} \tag{3.65a}$$

$$\eta = \text{dynamic viscosity (kg m}^{-1}\,\text{s}^{-1}\text{)} \tag{3.65b}$$

of the gas in the pressure system. Hence, the undamped natural circular frequency is obtained from

$$\omega_0^2 = 1/LC = 3\pi r^2 c^2/4lV \quad \text{(s}^{-2}\text{),} \tag{3.66}$$

and the resonant circular frequency of the damped system (see section 3.3.3) from

$$\omega_r^2 = \omega_0^2(1-2h) \quad \text{(s}^{-2}\text{).} \tag{3.67}$$

The damping ratio is given by

$$h = \tfrac{1}{2}RC\omega_0 = 2\{3\eta(lV)^{\frac{1}{2}}/\pi c\rho r^3\}^{\frac{1}{2}}, \tag{3.68}$$

and the phase angle between p and p_0 by

$$\phi = \tan^{-1}\{-\omega RC/(1-\omega^2 LC)\}$$
$$= \tan^{-1}\{-2h(\omega/\omega_0)/[1-(\omega/\omega_0)^2]\}. \tag{3.69}$$

In case of a very thin capillary tube these results can be simplified further, since the impedance of the tube becomes purely resistive, and we have

$$|p/p_0|_{L\to0} = (1+\omega^2R^2C^2)^{-\frac{1}{2}} = \{1+(\omega T)^2\}^{-\frac{1}{2}},\qquad(3.70)$$

where the time-constant

$$T = CR = 8l\eta V/\pi r^4\rho c^2 \quad (s).\qquad\qquad(3.71)$$

Evaluating eqns (3.66) and (3.68) for dry air at room temperature and atmospheric pressure we substitute

$$c_a = 340 \text{ m s}^{-1},$$

$$\rho_a = 1\cdot2 \text{ kg m}^{-3},$$

$$\eta_a = 1\cdot86\times10^{-5} \text{ kg m}^{-1}\text{s}^{-1},$$

and obtain for the undamped natural frequency

$$f_{0a} = \omega_{0a}/2\pi = 83r/(lV)^{\frac{1}{2}} \quad (\text{Hz}),$$

and for the damping ratio

$$h_a = 89\times10^{-9}(lV)^{\frac{1}{2}}/r^3.$$

Liquid-filled pressure measuring systems, however, can not be treated like gas-filled systems since the compressibility of liquids is very low; in order to arrive at realistic estimates of natural frequency and damping the compliance of the transducer diaphragm—and perhaps even of the tube walls—must be considered. Yet, these compliances are difficult to compute from the design parameters, and therefore rarely known. A useful method which derives the dynamic characteristics of liquid-filled systems from static pressure tests will be described below [66]–[68].

If a static pressure difference Δp is applied to a liquid-filled transducer system, a volume change ΔV can be observed, e.g. by means of a graduated capillary tube. Since

$$\Delta p = k\,\Delta x/A,\qquad\qquad(3.72)$$

where A (m^2) is the area of the (clamped) diaphragm and Δx (m) its centre displacement, and

$$\Delta V = \tfrac{1}{3}A\,\Delta x,\qquad\qquad(3.73)$$

the spring constant k (neglecting for the time being any other compliances) is given by

$$k = \tfrac{1}{3}A^2\,\Delta p/\Delta V \quad (\text{N m}^{-1}).\qquad\qquad(3.74)$$

For laminar flow conditions at low frequencies the equivalent mass of the liquid in the tube can be written [66]

$$m_1 = \tfrac{4}{3}\rho al(A/a)^2 \quad (\text{kg}),\qquad\qquad(3.75)$$

where $a = \pi r^2$ (m^2) is the cross-sectional area of the liquid column in the tube. Adding the mass m_2 (kg) of the moving parts of the transducer, the undamped natural circular frequency is obtained from

$$\omega_0^2 = \frac{k}{m_1+m_2} = \frac{\frac{1}{3}A^2(\Delta p/\Delta V)}{\frac{4}{3}\rho a l(A/a)^2+m_2}. \tag{3.76}$$

Except for very short inlet tubes, $m_2 \ll m_1$, and the undamped natural frequency can be simplified to

$$f_0 = \frac{\omega_0}{2\pi} = \frac{r}{4}\left\{\frac{(\Delta p/\Delta V)}{\pi \rho l}\right\}^{\frac{1}{2}} \quad \text{(Hz)}. \tag{3.77}$$

The damping ratio has the form [66]

$$h = 4\pi l\eta \frac{(A/a)^2}{\{k(m_1+m_2)\}^{\frac{1}{2}}}, \tag{3.78}$$

where η (kg m^{-1} s^{-1}) is now the dynamic viscosity of the liquid in the system. Neglecting again the moving transducer mass m_2, we obtain after some rearrangements

$$h = \frac{6\eta}{r^3}\left\{\frac{l}{\pi\rho(\Delta p/\Delta V)}\right\}^{\frac{1}{2}}. \tag{3.79}$$

With the design parameters of the transducer known and $(\Delta p/\Delta V)$ being determined by a static pressure test as described above, the undamped natural frequency f_0 and the damping ratio h of a liquid-filled transducer system with stiff tube walls can be computed from eqns (3.77) and (3.78), respectively. If, however, the elasticity of the tube walls cannot be neglected, the term $(\Delta p/\Delta V)$ must be modified according to

$$(\Delta p/\Delta V)' = 1/\{(\Delta V/\Delta p)_D + \tfrac{1}{3}(\Delta V/\Delta p)_T\}, \tag{3.80}$$

where the transducer (diaphragm) compliance $(\Delta V/\Delta p)_D$ and the tube compliance $(\Delta V/\Delta p)_T$ are measured separately. Theory and methods of dynamic calibration of pressure transducers are discussed in detail in reference [69].

References

1. LORD RAYLEIGH, Notes on electricity and magnetism III: On the behaviour of iron and steel under the operation of feeble magnetic forces. *Phil. Mag.* Ser. 5, **23**, 225–45 (1887).
2. A. L. KIMBALL and D. E. LOVELL. Internal friction in solids, *Phil. Rev.* 30, 948 (1927).
3. E. BECKER and O. FÖPPL. Dauerversuche zur Bestimmung der Festigkeitseigenschaften; Beziehungen zwischen Baustoffdämpfung und Verformungsgeschwindigkeit. V.D.I. Forschungsheft No. 304 (1928).
4. H. K. P. NEUBERT. Über Fundamentschwingungen mit geschwindigkeitsunabhängiger Dämpfung. *Akust. Z.* **2**, 34–7 (1937).
5. B. J. LAZAN and L. E. GOODMAN. Material and interface damping. In *Shock and*

vibration handbook (ed. C. M. Harris and C. E. Crede), Vol. II, Chap. 36. McGraw-Hill, New York (1961).

6. S. NEUMARK. *Concept of complex stiffness applied to problems of oscillations with viscous and hysteretic damping.* Aeronautical Research Council R&M No. 3269. H.M.S.O., London (1957).
7. H. K. P. NEUBERT, A simple model representing internal damping in solid materials. *Aeronaut. Q.* **14,** 187–210 (1963).
8. V. A. STANTON. The best spring material for high temperatures. *Product Engng* **31,** 44–7 (1960).
9. R. HOUWINK and H. K. DE DECKER. *Elasticity, plasticity and structure of matter.* Cambridge University Press (1971).
10. F. GAROFALO. *Fundamentals of creep and creep rupture in metals.* Macmillan, London (1965).
11. F. K. G. ODQVIST. *Mathematical theory of creep and creep rupture* (2nd edn). Clarendon Press, Oxford (1974).
12. T. H. LIN. *Theory of inelastic structures.* Wiley, New York (1968).
13. A. I. SMITH and A. M. NICOLSON. *Advances in creep design.* Applied Science Publishers, London (1971).
14. A. I. SMITH and A. M. NICHOLSON. *Advances in creep design,* pp. 82 and 103. Applied Science Publishers, London (1971).
15. B. J. STARKEY. *Laplace transforms for electrical engineers.* Iliffe, London (1954).
16. G. DOETSCH. *Anleitung zum praktischen Gebrauch der Laplace Transformation.* Oldenbourg, Munich (1961).
17. J. C. JAEGER. *An introduction to the Laplace transformation with engineering applications.* Methuen, London (1961).
18. F. E. NIXON. *Handbook of Laplace transformation: fundamentals, tables, examples.* Prentice-Hall, Englewood Cliffs, New Jersey (1965).
19. G. E. ROBERTS and H. KAUFMAN. *Tables of Laplace transforms.* W. B. Saunders, Philadelphia, Pennsylvania (1966).
20. W. D. DAY. *Tables of Laplace transforms.* Iliffe, London (1966).
21. M. F. GARDNER and J. L. BARNES. *Transients in linear systems.* Wiley, New York (1942).
22. E. V. BOHN. *The transform analysis of linear systems.* Addison-Wesley, Reading, Massachusetts (1963).
23. J. G. HOLBROOK. *Laplace transform for electrical engineers.* Pergamon Press, Oxford (1966).
24. E. E. ZEPPLER and K. G. NICHOLS. *Transients in electronic engineering.* Chapman & Hall, London (1971).
25. S. LEVY and W. D. KROLL. Response of accelerometers to transient acceleration. *J. Res. natn Bur. Stand.* **45,** paper No. 2138 (1950).
26. R. S. AYRE. Transient response to step and pulse functions. In *Shock and vibration handbook* (ed. C. M. Harris and C. E. Crede), Vol. I, Chap. 8. McGraw-Hill, New York (1961).
27. L. STATHAM. *The mass effect of liquid damping.* Instrument Notes No. 13. Statham Instrument, Oxnard, California (1950).
28. E. J. NESTORIDES. *A handbook on torsional vibrations,* p. 16 Cambridge University Press (1958).
29. G. L. SMITH. Accelerometer errors arising from a bar-type design of seismic mass. *Rev. scient. Instrum.* **23,** 97 (1952).
30. NATIONAL PHYSICAL LABORATORY. *Notes on applied science No. 15.* H.M.S.O., London (1956).
31. D. F. GIBBS. Spring diaphragms. *J. scient. Instrum.* **34,** 34 (1957).
32. J. A. HARINGX, The cross-spring pivot as a constructional element. *Appl. Scient. Res.* **A1,** 313 (1949).
33. W. WUEST. Blattfedergelenke für Messgeräte. *Feinwk-Tech.* **54,** 157–70 (1950).
34. N. W. NICHOLS and H. L. WUNSCH. The design characteristics of cross-spring pivots. *Engineering* **172,** 473 (1951).

35. W. G. FANGEMANN. Instrument springs. *Instrums automn.* **27,** 780–2 (1954).
36. R. GILTIN. How temperature affects instrument accuracy. *Control Enging*, 70–8 (1955).
37. ANON. Spring materials. *J. Instrum. Soc. Am.* **2,** 257–60 (1955).
38. J. T. MILEK. Materials for springs. *Mater. Des. Engng* **50,** 115–26 (1959).
39. G. F. TAGG. *Electrical indicating instruments.* Butterworth, London (1974).
40. C. K. STEDMAN. *Some characteristics of gas-damped accelerometers.* Instrument Notes No. 33. Statham Instruments, Oxnard, California (1958).
41. ANON. *Engineering guide to MS silicone fluids.* Midland Silicones, Ltd., London.
42. W. WUEST. Druckmessung in Flüssigkeiten und Gasen. *Arch. tech. Messen,* V **1330-F6,** 81–6 (1972).
43. Y. T. LI. High-frequency pressure indicators for aerodynamic problems. *Tech. Notes natn advis. Comm., Aeronaut., Wash.* No. 3042 (1953).
44. S. TIMOSHENKO. *Strength of material,* Vol. II. McGraw-Hill, New York (1941).
45. I. B. CRANDALL. Calibration of condensor transmitters; an air-damped vibrating system. *Phys. Rev.* **21,** 449–60 (1918).
46. R. ROARK. *Formulas for stress and strain.* McGraw-Hill, New York (1965).
47. A. M. WAHL. Recent research on flat diaphragms and circular plates with particular reference to instrument application. *Trans. Am. Soc. mech. Engrs* **79,** 83–7 (1957).
48. H. LAMB. On the vibration of an elastic plate in contact with water. *Proc. R. Soc.* **A98,** 205 (1921).
49. W. A. WILDHACK and V. H. GOERKE. Corrugated metal diaphragms for aircraft pressure measuring instruments. *Tech. Notes natn advis. Comm. Aeronaut., Wash.* **738** (1939).
50. W. A. WILDHACK, R. F. DRESSLER, and E. C. LLOYD. Investigation of the properties of corrugated diaphragms. *Trans. Am. Soc. mech. Engrs* **79,** 65–82 (1957).
51. J. A. HARINGX. Design of corrugated diaphragms. *Trans. Am. Soc. mech. Engrs* **79,** 55–64 (1957).
52. A. E. S. MILLS. The design and use of Bourdon tubes and gauges. *Metron* **1,** 321–3 (1969).
53. W. GUEST. Berechnung von Bourdonfedern. V.D.I. Forschungsheft No. 489, Vol. B28 (1962).
54. K. GOITEIN. A dimensional analysis approach to Bourdon tube design. *Instrum. Pract.* **6,** 748 (1952).
55. J. B. GIACOBBE and A. M. BOUNDS. Material selection factor significant in Bourdon tubes, *J. Metals, N.Y.,* **4,** 1147 (1952).
56. F. H. GRAVEL. Fundamental engineering of Bourdon tube pressure indicating gauges. *Instrum. Pract.* **4,** 602 (1950).
57. Metal bellows: (a) Drayton-Hydroflex, Ltd., West Drayton, Middlesex; (b) Accles & Pollok, Ltd., Oldbury, Worcs.; (c) Teddington Bellows, Ltd., Pontardulais, Glamorgan.
58. H. NOTHDURFT. Eigenschaften von Metallbälgen. *Regelungstechnik* **5,** 334–8 (1957).
59. A. S. IBERALL. Attenuation of oscillatory pressures in instrument lines. *Trans. Am. Soc. mech. Engrs* **72,** 689 (1950).
60. G. B. THURSTON. Periodic fluid flow through circular tubes. *J. acoust. Soc. Am.* **24,** 653 (1952).
61. C. K. STEDMAN. *Alternating flow of fluids in tubes.* Instrument Notes No. 30. Statham Instruments, Oxnard, California (1956).
62. J. D. HUMPHREYS. Pressure sensing calculations for aircraft and guided missiles. *Tele-Tech,* April (1953).
63. H. BERGH and H. TIJDEMAN. *Theoretical and experimental results for the dynamic response of pressure measuring systems,* NLR-TR-F 238. National Aero and Astronautical Research Institute, Amsterdam (1965).
64. S. GOLDSCHMIDT and J. F. URY. The influence of transmission line geometry on the measurement of oscillatory pressure. In *The measurement of pulsating flow.* Institute of Measurement and Control, London, 86–9 (1970).
65. H. F. OLSON. *Acoustic engineering.* Van Nostrand, New York (1957).
66. G. E. WHITE. *Liquid-filled pressure gauge systems.* Instrument Notes No. 7. Statham Instruments, Oxnard, California (1949).

67. J. R. BARTON. *A note on the evaluation of designs of transducers for the measurement of dynamic pressures in liquid systems.* Instrument Notes No. 27. Statham Instruments, California (1954).
68. R. C. ANDERSON and D. R. ENGLUND. Liquid-filled transient pressure measuring system; a method of determining frequency response, *NASA tech. Note* No. TN D-6603 (1971).
69. J. L. SCHWEPPE, L. C. EICHBERGER, D. F. MUSTER, E. L. MICHAELS, and G. F. PASKUSZ. Method of dynamic calibration of pressure transducers. *Natn. Bur. Stand. Monogr.* **67,** Washington, D.C. (1963).
70. J. P. DEN HARTOG. *Trans. Am. Soc. mech. Engrs* **53,** APM 53–9, 107–15 (1931).
71. T. A. PERTS and E. S. SHERRARD. *J. Res. natn. Bur. Stand.* **57,** 45–65 (1956).
72. Manufactured by the General Electric Company.
73. Made by Wiancko Engineering Co., Passadena 8, California.
74. *Instrum. control syst.* **41,** 97–8 (1968).

4. Electrical output characteristics

THE electrodynamic generator-type transducer is an 'active' type (see Table 2.2), based on Faraday's law of electromechanical energy conversion. It therefore falls into the category of 'bilateral' transducers (see section 2.3.2) which are capable of converting mechanical energy into electrical energy (receivers, or sensors), as well as electrical energy into mechanical energy (senders, or actuators). In the present context our interest will be concentrated on their characteristics as receivers. Here mechanical energy is extracted from the structure or medium under test, and fed in form of electrical energy into the indicating or recording equipment. However, in the majority of applications the transducer is employed as a voltage generator feeding into a high-impedance load. Although normally the absorbed electrical energy, and thus the mechanical reaction on the structure, appears to be negligible, consideration of the damping energy dissipated in the transducer may warrant attention in special cases (see sections 4.1.2.2. and 4.1.2.5). Likewise, secondary resonances which might severely limit the useful frequency range of an electrodynamic sensor are frequently ignored. Section 4.1.2.6 will deal with this aspect.

The basic principle of electrodynamic generators has been employed over a wide field of electrical measuring instruments and transducers, the majority of which cannot be discussed in this book. Engine-speed indicators, for instance, which are basically a.c. or d.c. generators, are well covered in the literature [1]–[3], and so are eddy-current-type tachometers [4]. Another important and interesting class of generator-type transducers omitted here are electrodynamic microphones, but since they are mainly employed in communication and sound reproduction they, in the first instance, are not strictly instrument transducers, except when used for the measurement of sound levels, sound spectra, etc. They too are adequately covered in the literature [5], [6]. In this section it is our intention first to study Faraday's law of electrodynamics in connection with transducer performance and design. We shall then discuss the dynamic performance of electrodynamic transducers in the light of the unified theory of bilateral transducers as developed in section 2.3.2, and proceed to the design of the magnetic and electric circuits of generator-type transducers, treating seismic vibration sensors in greater detail as a typical application.

4.1.1. The electrodynamic principle

Faraday's law of induction, when written in a general form [7], is

$$\oint \bar{\mathbf{E}}_s \cdot d\bar{\mathbf{s}} = -\frac{d}{dt} \int\int \bar{\mathbf{B}}_n \cdot d\bar{\mathbf{a}}. \qquad (4.1.1)$$

This states that the total e.m.f. $\oint \bar{\mathbf{E}}_s \cdot d\bar{\mathbf{s}}$, round any closed curve, e.g. a loop of wire, is at any moment equal to the rate of diminution of the flux density $\bar{\mathbf{B}}$ through a surface $\bar{\mathbf{a}}$ bounded by this curve. Applying Stoke's theorem we can also obtain a differential relationship between the vectors $\bar{\mathbf{E}}$ and $\bar{\mathbf{B}}$. In a medium at rest the flux only changes so far as the vector $\bar{\mathbf{B}}$ changes and we have

$$\text{curl } \bar{\mathbf{E}} = -\frac{\partial\bar{\mathbf{B}}}{\partial t}. \qquad (4.1.2)$$

If, however, the wire loop is moving with the velocity $\bar{\mathbf{v}}$, the flux variation occurs within an area $\bar{\mathbf{a}}$ which moves with the loop. With the element of area $d\bar{\mathbf{a}}$ moving with velocity $\bar{\mathbf{v}}$, and with div $\bar{\mathbf{B}} = 0$, we can write for a moving medium

$$\text{curl } \bar{\mathbf{E}} = -\left(\frac{\partial\bar{\mathbf{B}}}{\partial t} - \text{curl}[\bar{\mathbf{v}}\bar{\mathbf{B}}]\right). \qquad (4.1.3)$$

With respect to the two terms on the right-hand side, eqn (4.1.3) clearly shows that the total e.m.f. in the wire loop can have two possible causes:

(a) the rate of change of $\bar{\mathbf{B}}$ with time only (such as in a.c. transformers);

(b) the motion of the wire in a constant field $\bar{\mathbf{B}}$ (such as in alternators).

The structure of the second term indicates, by way of being a vector product of $\bar{\mathbf{v}}$ and $\bar{\mathbf{B}}$, the corkscrew rule as applied to the mutually perpendicular vectors of velocity, magnetic field, and e.m.f. (Lenz's law).

Eqn (4.1.3), which is quite general in terms of the geometry of the conductor and magnetic field relationship, is limited in its validity in two aspects:

(a) 'displacement' currents in the insulation of the conductor are neglected;

(b) velocity $\mathbf{v}$ is assumed small compared with the velocity of light.

Both these conditions are easily satisfied in practical applications. Also, in transducer work the geometrical relationship between field and conductor is normally simple, i.e. perpendicular relative movement of conductor and field, or rotation of a rectangular-shaped conductor loop in a homogeneous field. Then the voltage e(V) induced in a wire of length l(m) moving at a velocity v (m s^{-1}) in a constant and homogeneous magnetic field of strength B (T or Wb per m^2) is (Fig. 4.1.1(a)),

$$e = -Blv \quad (V), \tag{4.1.4a}$$

and the force f(N) exerted by a current i(A) flowing in a wire of length l immersed in a similar magnetic field is (Fig. 4.1.1(b)),

$$f = Bli \quad (N). \tag{4.1.4b}$$

Incidentally, combining eqns (4.1.4a) and (4.1.4b) yields

$$\frac{e}{v} = -\frac{f}{i}. \tag{4.1.5}$$

This relationship permits a static calibation of electrodynamic generator-type transducers in terms of V s m^{-1} by way of measuring the wire, or coil,

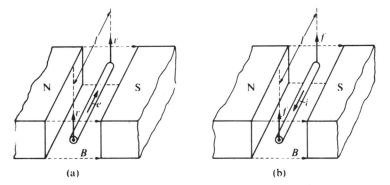

(a)   (b)

FIG. 4.1.1.(a) Electrodynamic voltage generator; (b) electrodynamic force generator, schematic diagram.

force f (N) at a given current i (A). A complete calibration curve can be plotted when f/i has been determined at a number of coil displacement settings.

The electrodynamic generator principle, of course, is not restricted to single insulated wire conductors. Fig. 4.1.2(a) shows a solid conducting strip of metal moving in its own plane and normal to the magnetic field. Two stationary brushes are arranged as shown. An e.m.f. proportional to the strip velocity appears across the brushes, but since eddy currents are induced in the plate the pattern of currents is rather complex. If the strip is replaced by a rotating metal disc the device is known as a 'homopolar' or 'unipolar' generator [8], which is capable of delivering direct current without commutation. Faraday's original disc dynamo was of this design.

The induction flowmeter is a more recent application of the electrodynamic principle although the linear motion of sea water through a magnetic field was used to measure the speed of a vessel as far back as

1917. The basic arrangement is shown schematically in Fig. 4.1.2(b). The theory of flowmeters with static and alternating magnetic fields is well known and its hydrodynamic and electrical aspects have been studied in great detail [9]–[11]. It can be used with liquids of conductivities as low as that of commercially distilled water.

Complications in electrodynamic generators occur if the conductors are loaded, i.e. if current is allowed to flow through the coil. The current, then, generates a magnetic field of its own which in accordance with

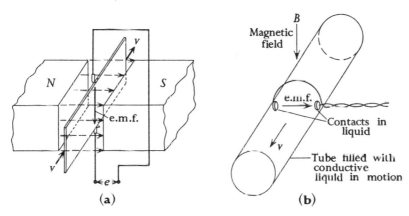

(a)   (b)

FIG. 4.1.2. Electrodynamic generator principle: (a) solid conductor plate; (b) liquid conductor.

Lenz's law opposes the original field. At higher frequencies of conductor movements, and at sizeable currents, the superimposed a.c. field can penetrate the solid pole pieces of the magnet assembly and cause eddy currents to flow. Under these conditions it is advisable to laminate the pole-piece sections which are adjacent to the air gap.

4.1.2. Dynamic performance

The electrodynamic sensor is responsive to dynamic input quantities only; static or quasi-static input signals do not produce an output. The configuration chosen here for detail investigation is the most common moving-coil generator employed as a 'receiver', or sensor (Fig. 4.1.3).

4.1.2.1. *The transfer matrix*

With the basic transducer eqns (4.1.4a) and (4.1.4b), and the conventions of Fig. 2.2, force equilibrium yields [12], [13]

$$f = Z_m v + Bli \qquad (4.1.6a)$$

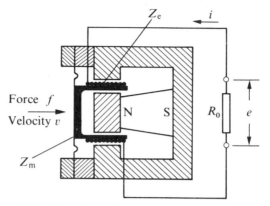

FIG. 4.1.3. Electrodynamic transducer, schematic diagram.

and mesh-voltage zero condition

$$e = -Blv + Z_e i, \qquad (4.1.6b)$$

where Z_m and Z_e are the mechanical and the electrical transducer impedances, respectively (see section 2.3.2). Comparison of eqns (4.1.6) above with eqns (2.26) and (2.28) shows that a voltage-source representation is appropriate, with

$$M = Bl \, (\mathrm{N \ A^{-1}}) \quad \text{and} \quad -M^* = -Bl \, (\mathrm{V \ s \ m^{-1}}) \qquad (4.1.7)$$

The transfer matrix equation (2.29) of the electrodynamic transducer then takes the form

$$\begin{bmatrix} f \\ v \end{bmatrix} = \begin{bmatrix} Bl + Z_m Z_e/Bl & -Z_m/Bl \\ Z_e/Bl & -1/Bl \end{bmatrix} \begin{bmatrix} i \\ e \end{bmatrix}. \qquad (4.1.8)$$

The electrical impedance can be written

$$Z_e = R + \mathrm{j}\omega L, \qquad (4.1.9)$$

where $R(\Omega)$ is the moving coil resistance and $L(\mathrm{H})$ the coil inductance. The electrodynamic transducer can then be represented by Fig. 2.3(a), where it is convenient to include the reactive component of Z_e, i.e. the coil inductance L, in the 'characteristic' two-port. Also if we extract the mechanical impedance Z_m by writing in eqn (4.1.8)

$$f_t = f - Z_m v, \qquad (4.1.10)$$

the characteristic transfer matrix equation of the electrodynamic transducer assumes the form

$$\begin{bmatrix} f_t \\ v \end{bmatrix} = \begin{bmatrix} Bl & 0 \\ \mathrm{j}\omega L/Bl & -1/Bl \end{bmatrix} \begin{bmatrix} i \\ e \end{bmatrix} \qquad (4.1.11)$$

The characteristic transfer matrix equation (4.1.11) of the electrodynamic transducer—together with those of other (bilateral) electromechanical transducers to be discussed later—have been inserted in Fig. 2.3 for convenient comparison.

4.1.2.2. Mechanical input impedance of electrodynamic force sensor

The mechanical input impedance Z_{mi} is given by eqn (2.35) and with $M^2 = (Bl)^2$ we have

$$Z_{mi} = Z_m + Z_{m0} + (Bl)^2/(Z_e + Z_{e0}) \quad (\text{N s m}^{-1}). \qquad (4.1.12)$$

Evaluating eqn (4.1.12), let the mechanical transducer impedance

$$Z_m = c + j(\omega m - k/\omega), \qquad (4.1.13)$$

where

$c = $ mechanical damping factor (N s m^{-1}),

$m = $ mass (kg),

$k = $ mechanical stiffness (N m^{-1}),

$\omega = $ circular excitation frequency (s^{-1}),

and Z_e is the electrical impedance given by eqn (4.1.9). Since, in a force receiver, the mechanical source impedance Z_{m0} can usually be neglected, and also since the electrical load impedance Z_{e0} is normally represented by an indicator resistance R_0, the mechanical input impedance of an electrodynamic force sensor becomes

$$Z_{mi} = c + j(\omega m - k/\omega) + (Bl)^2/(R_0 + R + j\omega L). \qquad (4.1.14)$$

In an electrodynamic transducer the coil inductance is small, so that at sizeable values of R_0, and in the main frequency regime, $\omega L \ll (R_0 + R)$. Eqn (4.1.14) then simplifies to

$$Z_{mi} = c + \frac{(Bl)^2}{R'} + j\left[\frac{k}{\omega}\left\{\left(\frac{\omega}{\omega_0}\right)^2 - 1\right\} - \omega L\left(\frac{Bl}{R'}\right)^2\right], \qquad (4.1.15)$$

where $R' = R_0 + R$, and $\omega_0^2 = k/m$ is the mechanical circular natural frequency. The second term on the right-hand side of eqn (4.1.15) represents the electrical damping factor, which often predominates the mechanical damping c. It is inversely proportional to the electrical resistance R'. The effect of $\omega L(Bl/R')^2$ on the reactive component is usually much smaller than that of $(Bl)^2/R'$ on the real component because of $\omega L/R' \ll 1$.

4.1.2.3. Transfer function of electrodynamic force, or pressure, sensor.

If the input is a (complex) force f_0, then the transfer function, in terms of output current i, of the electrodynamic force receiver is given by eqn (2.41), with $M^* = Bl$ and $M^2 = (Bl)^2$,

$$\begin{aligned} G_r &= (-i/f_0) \\ &= -Bl/\{(Bl)^2 + (Z_m + Z_{m0})(Z_e + Z_{e0})\} \quad (\text{A N}^{-1}), \end{aligned} \qquad (4.1.16)$$

or, with the notations of section 4.1.2.2,

$$\frac{-i}{f_0} = \frac{-Bl}{(Bl)^2 + \{c + j(\omega m - k/\omega)\}(R_0 + R + j\omega L)},$$ (4.1.17)

where ω is the excitation frequency related to f_0.

The most common form of a force receiver is perhaps the pressure transducer (Fig. 4.1.4). Its voltage sensitivity at a given pressure input $p_0 = f_0/a$ (N m^{-2}) and at high indicator resistance R_0 becomes

$$\frac{e}{p_0} = \frac{-aBl}{(Bl)^2/R' + \{c + j(\omega m - k/\omega)\}}   (V\ m^2\ N^{-1}),$$ (4.1.18)

where $a(m^2)$ is the pressure-sensitive area of the diaphragm, and $R' = (R_0 + R) \gg j\omega L$.

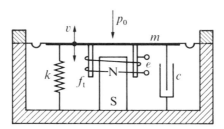

FIG. 4.1.4. Electrodynamic pressure transducer.

Considering eqn (4.1.18), the response of the electrodynamic pressure sensor can be devided into four frequency ranges.

(a) At low frequencies ω the stiffness term (k/ω) in the denominator prevails; the voltage output rises proportionally with frequency.

(b) Near resonance $(\omega m \simeq k/\omega)$, the response is determined by the damping term $\{c + (Bl)^2/R'\}$ in the denominator. This is the most useful frequency range of the electrodynamic pressure transducer; it can be made fairly flat by a suitable choice of damping, i.e. mainly by R_0.

(c) At high frequencies the inertia term predominates; the response declines with increasing frequency.

(d) A further drop occurs at a still higher frequency, where $\omega L \simeq R'$, which is caused by the electrical reactance of the transducer coil (eqn (4.1.17)).

4.1.2.4. *Transfer function of electrodynamic velocity sensor*

The common electrodynamic 'vibration transducer' (Fig. 4.1.5) is a velocity receiver, i.e. an input velocity v_0 is applied to the transducer housing and in the operational frequency range the output voltage is expected to be proportional to velocity. The mechanical input conditions, therefore, differ from those of the force sensor.

Eqn (2.33a) for $e_0 = 0$, and first disregarding the mechanical input impedance, yields

$$f_t = Mi, \tag{4.1.19a}$$

$$v_t = \frac{Z_e + Z_{e0}}{M^*}\, i. \tag{4.1.19b}$$

From eqn (4.1.19a) the ratio of the output current $(-i)$ to the input

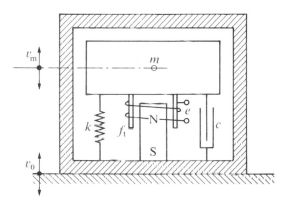

FIG. 4.1.5. Electrodynamic vibration transducer.

velocity v_t of the transducer without the mechanical impedance is

$$(-i/v_t) = -M^*/(Z_e + Z_{e0}). \tag{4.1.20}$$

We now need a relationship between v_t and the input velocity, v_0, applied to the transducer body. From Fig. 4.1.5, v_t is seen to be the difference between the input velocity v_0 and the velocity v_m of the seismic mass,

$$v_t = v_0 - v_m. \tag{4.1.21}$$

Force equilibrium then requires

$$v_t(c - jk/\omega) + f_t = j\omega m v_m, \tag{4.1.22}$$

where f_t is the reaction force exerted by the electrical side of the transducer. Rearranging eqn (4.1.22) and introducing the undamped circular natural frequency $\omega_0^2 = k/m$, we have

$$v_0 = \left\{1 - (\omega_0/\omega)^2 + \frac{c + f_t/v_t}{j\omega m}\right\} v_t, \tag{4.1.23}$$

where, from eqns (4.1.19a) and (4.1.19b),

$$f_t/v_t = M^2/(Z_e + Z_{e0}). \tag{4.1.24}$$

Substituting eqns (4.1.23) and (4.1.24) into eqn (4.1.20), and letting $M^* = Bl$ and $M^2 = (Bl)^2$, the transfer function of the electrodynamic velocity sensor becomes

$$\frac{-i}{v_0} = \frac{-Bl}{\dfrac{(Bl)^2}{j\omega m} + \left\{1 - \left(\dfrac{\omega_0}{\omega}\right)^2 + \dfrac{c}{j\omega m}\right\}(Z_e + Z_{e0})} \quad (\text{A s m}^{-1}). \tag{4.1.25}$$

Assuming again that for $(Z_e + Z_{e0}) = (R' + j\omega L)$ the indicator resistance $R' \gg j\omega L$, eqn (4.1.25), rewritten for the voltage output, then simplifies to

$$\frac{e}{v_0} = \frac{-Bl}{1 - \left(\dfrac{\omega_0}{\omega}\right)^2 + \dfrac{c + (Bl^2)/R'}{j\omega m}} \quad (\text{V s m}^{-1}). \tag{4.1.26}$$

The frequency response of the electrodynamic vibration transducer, therefore, follows a pattern different from that of the pressure transducer, namely:

(a) at very low excitation frequencies ω the square term $(\omega_0/\omega)^2$ in the denominator of eqn (4.1.26) prevails—in this range the output grows with ω^2;

(b) at low frequencies above resonance the term $\{cR' + (Bl)^2/j\omega mR'\}$ comes into play—the response increases proportionally with ω;

(c) above the frequency $\{cR' + (Bl)^2\}/mR'$ the voltage–velocity ratio is constant—the electrodynamic vibration transducer behaves like an ideal velocity receiver;

(d) a final decline occurs again at $\omega \simeq R'/L$, due to the effect of the coil resistance.

In a practical performance test the transition produced by the damping term (item (b) above) may easily be masked by the mechanical resonance. Anyway, its occurrence is not suspected from elementary theory which disregards mechanical damping and electromechanical interaction forces.

4.1.2.5. The effect of eddy-current damping

Eddy-current damping in moving-coil vibration transducers is usually obtained by winding the transducer coil on a high-conductivity metal former, such as copper or silver (see section 4.1.3.3). An estimate of its effect on transducer performance is given in this section; a more detailed analysis can be found in reference [14].

The equivalent circuit of Fig. 4.1.6(a) shows the generator voltage e and the internal impedance, represented by R_1 and L_1. Inductively coupled to the coil is the damping former of self-inductance L_2 and resistance R_2. The coupling coefficient between coil and former is k. Since the load impedance is assumed to be high in comparison with the internal impedance, a change in output can only be caused by change in the generated voltage e. Let

$$e = knv, \tag{4.1.27}$$

where

k = constant, depending on geometry and on magnetic field strength,

v = velocity of coil relative to field,

n = number of turns of coil.

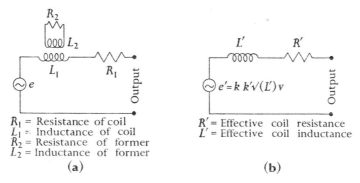

R_1 = Resistance of coil
L_1 = Inductance of coil
R_2 = Resistance of former
L_2 = Inductance of former

(a)

R' = Effective coil resistance
L' = Effective coil inductance

(b)

FIG. 4.1.6. (a) and (b) Equivalent circuits of generator-type transducer with conductive coil former for eddy-current damping.

Fig. 4.1.6(b) shows an equivalent circuit based on the efféctive resistance R' and the effective inductance L'. Since

$$n = k\sqrt{L'},$$

k' being a constant, we have

$$e' = kk'\sqrt{L'}\, v. \tag{4.1.28}$$

The output voltage thus depends upon the effective inductance L'. From Fig. 4.1.6(b) we obtain

$$R' = R_1\left(1 + \frac{R_2}{R_1}\cdot\frac{k^2\omega^2 L_1 L_2}{R_2^2+\omega^2 L_2^2}\right) = R_1\left(1 + \frac{k^2 Q_1 Q_2}{1+Q_2^2}\right), \tag{4.1.29a}$$

$$L' = L_1\left(1 - \frac{\omega^2 L_2^2 k^2}{R_2^2+\omega^2 L_2^2}\right) = L_1\left(1 - \frac{k^2 Q_2^2}{1+Q_2^2}\right), \tag{4.1.29b}$$

where Q_1 and Q_2 are the storage factors of coil and coil former, respectively. A variation with frequency of R' may be ignored since R' is small compared with the load resistance. The variation with frequency of the effective inductance L' can be shown to be

$$\frac{dL'}{d\omega} = \frac{-2L_1 k^2 Q_2^2}{\omega(1+Q_2^2)^2}.$$ (4.1.30)

Since the factor $Q_2^2/(1+Q_2^2)^2$ is of sizeable value only in the neighbourhood of $Q_2 = 1$, $dL'/d\omega$ is inversely proportional to Q_2^2 for values of

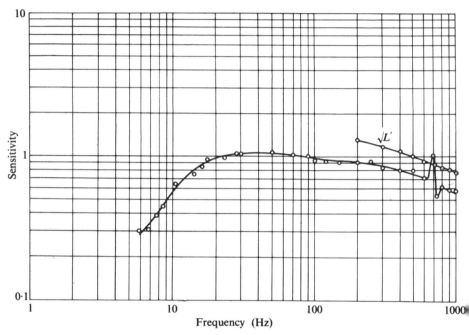

FIG. 4.1.7. Frequency response curve of electrodynamic vibration transducer; copper coil former; fundamental undamped natural frequency at ~17 Hz.

$Q_2 \gg 1$. Some improvement is thus expected, when using high-conductivity materials, such as electro-formed silver coil formers, or if the adjacent parts of the magnet assembly are laminated in order to reduce the iron losses. Otherwise, little can be done, unless any coupling between coil and former is avoided ($k = 0$), i.e. if damping former and coil are operating in separate air gaps.

According to Fig. 4.1.6(b) and eqn (4.1.28) the output voltage should vary proportionally to $\sqrt{L'}$. In order to prove the point, the $\sqrt{L'}$ values of a number of transducers have been measured at various frequencies; Fig. 4.1.7 shows an example. It is seen that the high-frequency end of the response curve follows quite closely the shape of the $\sqrt{L'}$ curve in this log–log presentation. (Incidentally, the coil inductance must be measured with the armature locked in order to prevent motional impedance influencing the measurements.)

4.1.2.6. Secondary resonances

The frequency response curve of Fig. 4.1.7 shows a secondary resonance at about 700 Hz. Although its shape is typical for the majority of

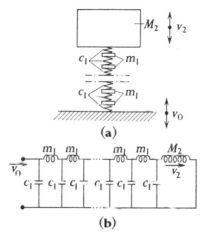

FIG. 4.1.8. (a) and (b) Mechanical and electrical equivalents of a vibratory system with a spring having distributed mass and resilience.

secondary resonances, other shapes also occur. Since secondary resonances limit the useful frequency range of vibration transducers, it is important to understand their mechanism in order to eliminate them. In the following, three different types of secondary resonance will be investigated theoretically and experimentally.

(a) *Spring with self-mass*. Considering the general case first, the spring of a practical seismic system has compliance and mass distributed over its length, as shown in Fig. 4.1.8(a). The analogue electrical circuit is given in Fig. 4.1.8(b). Let

$$m_1 = \text{self-mass per unit length of spring,}$$
$$c_1 = \text{compliance per unit length of spring,}$$
$$M_2 = \text{main mass,}$$
$$l = \text{length of spring,}$$
$$v_0 = \text{forcing velocity at housing end of spring,}$$
$$v_2 = \text{velocity of main mass.}$$

By analogy with transmission-line theory [15]

$$v_0 = v_2 \cosh \gamma l + \frac{\mathrm{j}\omega M_2 v_0}{Z_0} \sinh \gamma l, \qquad (4.1.31)$$

where

$$\gamma = \mathrm{j}\omega(m_1 c_1)^{\frac{1}{2}},$$
$$Z_0 = (m_1 c_1)^{\frac{1}{2}}, \qquad (4.1.32)$$
$$\omega = 2\pi \times \text{frequency},$$

and response

$$\frac{v_2 - v_0}{v_0} = \frac{1}{\cosh \gamma l + (\mathrm{j}\omega M_2/Z_0)\sinh \gamma l} - 1$$

$$= \frac{1}{\cos bl - (\omega M_2/Z_0)\sin bl} - 1, \qquad (4.1.33)$$

where

$$b = \omega(m_1 c_1)^{\frac{1}{2}}. \qquad (4.1.34)$$

Resonances occur at values of ω satisfying the relation

$$\cos bl - \frac{\omega M_2}{Z_0}\sin bl = 0, \qquad (4.1.35)$$

and the response falls to zero at values satisfying the relation

$$\cos bl - \frac{\omega M_2}{Z_0}\sin bl = 1. \qquad (4.1.36)$$

The numerical evaluation of the frequency response of such a system is relatively complex and thus of little use for the present purpose. If we, however, assume that the self-mass of the spring is lumped in the centre of its length we have the simpler system shown in Fig. 4.1.9(a). The 'direct' analogy (see section 2.3.1) of such a system is shown in Fig. 4.1.9(b), while Fig. 4.1.9(c) indicates that for the 'current' ratio $(v_2 - v_0)/v_0$, the equivalent circuit of Fig. 4.1.9(b) behaves as a four-terminal network. Let

$M_1 = $ lumped spring mass, located at centre of spring,

$C_1 = C_2 = $ compliance of half length of spring,

$M_2 = $ main mass,

$v_1 = $ velocity of M_1,

$v_2 = $ velocity of M_2,

$v_0 = $ forcing velocity.

From the electrical network of Fig. 4.1.9(b) we obtain

$$v_2 = v_1/(1 - \omega^2 M_2 C_2) \qquad (4.1.37)$$

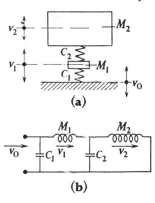

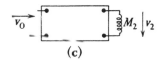

FIG. 4.1.9. (a), (b) and (c) Mechanical and electrical equivalents of a vibratory system with self mass of spring lumped in the centre.

and

$$v_0 = v_1\left\{1+\frac{C_1}{C_2}-\omega^2 M_1 C_1+\frac{C_1}{C_2(\omega^2 M_2 C_2-1)}\right\},\qquad(4.1.38)$$

from which the frequency response

$$\frac{v_2-v_0}{v_0}=\frac{1}{(1-\omega^2 M_1 C_1)(1-\omega^2 M_2 C_2)-\omega^2 M_2 C_1}-1.\qquad(4.1.39)$$

Resonances occur at values of ω satisfying the relation

$$(1-\omega^2 M_1 C_1)(1-\omega^2 M_2 C_2)-\omega^2 M_2 C_1 = 0,\qquad(4.1.40)$$

and the response falls to zero at values of ω satisfying the relation

$$(1-\omega^2 M_1 C_1)(1-\omega^2 M_2 C_2)-\omega^2 M_2 C_1 = 1.\qquad(4.1.41)$$

The characteristic shape of the frequency-response curve can be obtained by evaluation of eqn (4.1.39) using the following numerical relationships:

$$M_1 = M_2/10 = 10^{-3}\,\text{kg},$$
$$M_2 = 10^{-2}\,\text{kg},\qquad(4.1.42)$$
$$C_1 = C_2 = 10^{-2}\,\text{m N}^{-1}.$$

Fig. 4.1.10 indicates that the frequency response has its first (fundamental) resonance at $11\cdot1$ Hz and its second resonance at $72\cdot1$ Hz. The response is zero at $73\cdot0$ Hz. The main characteristic of this type of

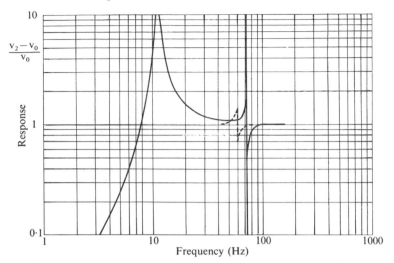

FIG. 4.1.10. Frequency response of vibratory system of Fig. 4.1.9.

secondary resonance is that it follows the fundamental resonance, separated by a part of the curve which is asymptotic to unity response. This
behaviour agrees with the theory of a four-terminal network which
permits resonances to follow each other. Zero occurs immediately after
the secondary resonance.

To verify experimentally this curve in the case of a seismic vibration
transducer of the electrodynamic generator-type a mass of $1 \cdot 2 \times 10^{-3}$ kg
was attached to the centre of the spring. The main mass was $0 \cdot 026$ kg.
The dotted line in Fig. 4.1.10 shows the experimental frequency response
for these conditions. It follows the features of the theoretical curve but
has finite values at resonance and at 'zero', owing to damping. The
frequencies of the theoretical and the experimental curve are not related.

The secondary resonance shown in the experimental curve of Fig. 4.1.7
is of the same character, i.e. that the secondary resonance exhibited in
that particular transducer was due to the self-mass of the spring.

(b) *Secondary mass, linked to main mass.* A different type of secondary resonance occurs if a small mass M_2 is attached to main mass M_1 by
an elastic medium C_2 (Fig. 4.1.11). Let

M_1 = main mass,

C_1 = compliance of main spring,

M_2 = small mass attached to M_1,

C_2 = compliance of link between M_1 and M_2,

v_1 = velocity of M_1,

v_2 = velocity of M_2,

v_0 = forcing velocity.

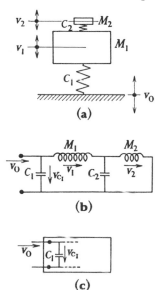

FIG. 4.1.11. (a), (b), and (c) Mechanical and electrical equivalents of a vibratory system with secondary mass linked to the main mass.

We have

$$\frac{v_{C_1}}{v_0} = \frac{v_1 - v_0}{v_0} = \frac{1 - \omega^2 M_2 C_2}{(1 - \omega^2 M_1 C_1)(1 - \omega^2 M_2 C_2) - \omega^2 M_2 C_1} - 1, \quad (4.1.43)$$

with the resonance condition for

$$(1 - \omega^2 M_1 C_1)(1 - \omega^2 M_2 C_2) - \omega^2 M_2 C_1 = 0, \quad (4.1.44)$$

and the zero condition for

$$\omega^2 M_1 C_1 (1 - \omega^2 M_2 C_2) + \omega^2 M_2 C_1 = 0. \quad (4.1.45)$$

The following values have been substituted in eqn (4.1.43) to obtain the frequency-response curve shown in Fig. 4.1.12:

$$M_1 = 10^{-2} \text{ kg}, \qquad\qquad M_2 = M_1/10 = 10^{-3} \text{ kg},$$
$$C_1 = 2 \times 10^{-2} \text{ m N}^{-1}, \qquad C_2 = 5 \times 10^{-3} \text{ m N}^{-1}. \qquad (4.1.46)$$

Fig. 4.1.12 shows that after the fundamental resonance at 10·7 Hz the response curve reaches zero at 74·7 Hz, closely followed by a secondary resonance at 74·8 Hz. The main feature of this type of secondary resonance is the zero occurring between the fundamental and secondary resonance, i.e. resonances and zeros alternate, an essential characteristic of two-terminal networks. The seismic vibration transducer already mentioned in (a) above was modified and a small mass of about $0·5 \times 10^{-3}$ kg was suspended from the main mass by means of a short cantilever. The

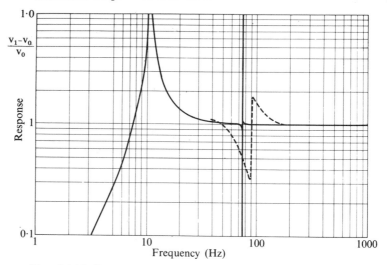

FIG. 4.1.12. Frequency response of vibratory system of Fig. 4.1.11.

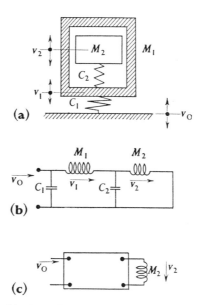

FIG. 4.1.13. (a), (b), and (c) Mechanical and electrical equivalents of a vibratory system with flexible mounting.

dotted experimental secondary resonance curve is shown in Fig. 4.1.12. The frequencies of the two curves are not related, but the general shapes are similar.

(c) *Contact resonance of transducer on structure.* In a third case the frequency-response curve may comprise a secondary resonance if the transducer housing is not rigidly fixed to the vibrating structure under test. Fig. 4.1.13(a)–(c) shows the mechanical system and the electrical equivalent circuit, emphasizing its four-terminal character. Let

$$M_1 = \text{mass of transducer housing,}$$

$$C_1 = \text{compliance of medium between transducer housing and vibrating structure,}$$

$$M_2 = \text{main mass,}$$

$$C_2 = \text{compliance of main spring,}$$

$$v_1 = \text{velocity of } M_1,$$

$$v_2 = \text{velocity of } M_2,$$

$$v_0 = \text{forcing velocity,}$$

then we have

$$\frac{v_2 - v_1}{v_0} = \frac{-\omega^2 M_2 C_2}{(1 - \omega^2 M_1 C_1)(1 - \omega^2 M_2 C_2) - \omega^2 M_2 C_1}. \tag{4.1.47}$$

The resonance condition is

$$(1 - \omega^2 M_1 C_1)(1 - \omega^2 M_2 C_2) - \omega^2 M_2 C_1 = 0, \tag{4.1.48}$$

and the zero condition is

$$\frac{\omega^2 M_2 C_2}{(1 - \omega^2 M_1 C_1)(1 - \omega^2 M_2 C_2) - \omega^2 M_2 C_1} = 0. \tag{4.1.49}$$

The frequency response plotted in Fig. 4.1.14 is obtained when substituting into eqn (4.1.47) the following values:

$$M_2 = 10^{-2} \text{ kg,} \qquad M_1 = 10 M_2 = 10^{-1} \text{ kg,}$$

$$C_2 = 2 \times 10^{-2} \text{ m N}^{-1}, \qquad C_1 = 5 \times 10^{-5} \text{ m N}^{-1}.$$

The fundamental resonance occurs at 11·2 Hz, and this is followed (as is to be expected for a four-terminal network) by a secondary resonance at 71·3 Hz. Above this frequency the response curve drops sharply towards zero without further recovery, since at frequencies higher than the secondary resonant frequency the whole transducer 'floats' and there is no relative movement between the seismic mass and the housing. The experimental curve, shown dotted in Fig. 4.1.14, was obtained by mounting the transducer on an elastic pad. The total mass of the transducer was

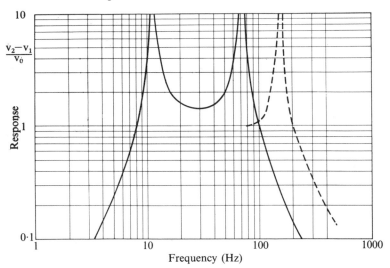

FIG. 4.1.14. Frequency response of vibratory system of Fig. 4.1.13.

0·37 kg and the seismic mass 0·026 kg. Again, the experimental curve is similar in character to the theoretical curve, but the frequencies are not related.

4.1.3. Design of electrodynamic transducers

4.1.3.1. Design of the magnetic path

The magnetization curve of a typical permanent magnetic material is shown in Fig. 4.1.15. From a demagnetized state at the origin, the

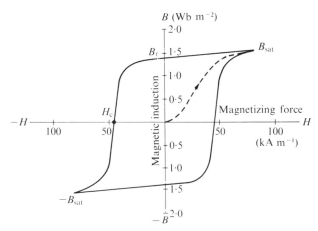

FIG. 4.1.15. Total magnetization curve of a permanent magnet (Ticonal 'G').

magnetic induction attains a saturation value B_{sat} at corresponding values of the magnetic field strength H (dotted curve). At gradually decreasing values of H, a 'residual' induction or remanence B_r occurs at $H = 0$, and in a reversed magnetic field the induction is further reduced and becomes zero at the coercive force H_c. The rest of the hysteresis loop can be easily explained by way of Fig. 4.1.15. For the designer of permanent-magnet devices the 'demagnetization curve' between $+B_r$ and $-H_c$ is of major importance. Fig. 4.1.16 shows a demagnetization curve of a typical permanent magnet material. On the right the product BH has been

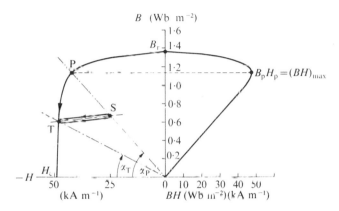

FIG. 4.1.16. Demagnetization curve and curve of stored magnetic energy of permanent magnet (Ticonal 'G').

plotted for all points on the demagnetization curve. The 'peak energy product' occurs at values of B_p and H_p and represents the maximum external energy $(BH)_{max}$ that can be produced per unit volume of a certain material. It thus serves us a measure for the potential strength of a magnetic material and it is obvious that in a magnetic circuit with a constant air gap the size of the magnet will be a minimum if it is designed to work at the peak energy point, i.e. if the so-called permeance coefficient (slope B/H) intersects the demagnetization curve at this point.

Under practical conditions, however, magnet assemblies are subjected to demagnetizing influences so that the working point does not remain stationary. Small variations may result from the influence of external fields, but if the air gap is appreciably increased by the removal of a keeper or armature, the field strength H may even approach zero. On return to the original conditions the magnetic inductance B no longer corresponds to that indicated by the demagnetization curve, but operates on a minor hysteresis loop. In Fig. 4.1.16, P is the point of the peak energy product. The slope of the permeance line is α_P. If the air gap is increased, yielding the new slope α_T, the working point attains the position T on the demagnetization curve. Reduction of the air gap to its original magnitude, corresponding to slope α_P, does not restore the working point P, but moves along a minor loop to S at the intersection with the original permeance line. (The slope of any minor hysteresis loop is approximately the same as the slope of a tangent to the magnetization curve at B_r.) The magnetic energy available at S is clearly smaller than the maximum value at P, but it is easily seen that stability is improved at the expense of energy, because the change in magnetic induction over a cycle T–S is smaller than that corresponding to P–T. High energy values and simultaneously good stability can be achieved if the magnetic circuit has been designed for α-values slightly greater than α_P (which means relatively long permanent magnets, as will be shown presently), in order to operate on minor loops in the neighbourhood of P.

To compute the major dimensions of a permanent magnet for a given magnetic path we require a magneto-motive force $F(A)$ at a magnetic induction B_g (Wb m^{-2}) in an air gap of length l_g(m) ($\mu_0 = 4\pi \times 10^{-7}$ H m^{-1} being the permeability of empty space),

$$F = l_g B_g / \mu_0. \qquad (4.1.51)$$

This magneto-motive force must be produced by a permanent magnet of length l_m(m) and working at a field strength H_m (A m^{-1}):

$$l_m = F/H_m. \qquad (4.1.52)$$

Combining eqns (4.1.51) and (4.1.52) and introducing a correction factor k_1 for the losses in pole pieces, joints, etc., we have the required length of the permanent magnet

$$l_m = k_1 l_g B_g / \mu_0 H_m \quad \text{(m)}. \qquad (4.1.53)$$

The second relationship essential for the design of permanent magnets can be derived from the required air-gap flux $\Phi_g = B_g A_g$. If B_m (Wb m^{-2}) is the working magnetic induction corresponding to H_m (A m^{-1}) on the demagnetization curve the required cross-section area of the magnet is

$$A_m = k_a \Phi_g / B_m = k_a A_g B_g / B_m \quad \text{(m}^2\text{)}. \qquad (4.1.54)$$

k_a is a correction factor for the leakages in the magnetic path. With l_m (eqn (4.1.53)) and A_m (eqn (4.1.54)) the design of a permanent magnet is complete. In most cases B_m and H_m will be equal, or nearly equal, to the peak energy values B_p and H_p respectively. With devices having a variable air gap, it is convenient to compute the slope of the permeance line which, with reference to Fig. 4.1.16, is

$$\tan \alpha = \frac{B_m}{H_m} = \frac{k_a l_m A_g \mu_0}{k_1 l_g A_m}. \qquad (4.1.55)$$

Since l_g and B_g have two sets of extreme values in a variable air-gap device, two values of α can be computed from eqn (4.1.55) thus enabling the designer to plot the two permeance lines and the minor hysteresis loop, as shown in Fig. 4.1.16.

As to the correction factors k_1 and k_a, an estimated loss of magnetomotive force in pole pieces and other parts of the magnetic circuit of 10–30 per cent is appropriate, or as an average, $k_1 \simeq 1 \cdot 2$. The amount of leakage, factor k_a, is more difficult to assess. Computation of the leakage fluxes of a particular design can be executed with good approximation by substituting equivalent leakage paths of simple geometrical shapes which can be

TABLE 4.1.1
Leakage factor k_a [19]

Application	k_a
(a) Moving-coil meters with rectangular magnets	3
Moving-coil meters with semicircular magnets	2
Moving-coil meters with internal core magnets	1·5
(b) Loudspeakers or microphones with small air gaps, and	
central magnet, 20 mm coil, up to 0·7 Wb m^{-2}	2
central magnet, 25 mm coil, up to 0·8 Wb m^{-2}	2
ring magnet, 25 mm coil, up to 1·2 Wb m^{-2}	3
ring magnet, 25 mm coil, up to 1 6 Wb m^{-2}	6
ring magnet, 40 mm coil, up to 1·6 Wb m^{-2}	5
(c) Generator-type vibration transducers with loud- ⎫ speaker-type magnet assembly, but larger air gaps ⎭	⎧Twice the values of (b)⎭
(d) Motors and generators, two-pole type	2
Motors and generators, four-pole type.	4

evaluated correctly [16]–[18]. For an estimate of the leakage factor k_a, Table 4.1.1 gives, for various applications, typical values based on practical experience. They apply to conventional shapes of permanent-magnet assemblies; shapes related to electrodynamic transducer configurations are easily recognized. In doubtful cases the designer is well advised to check the performance of his magnet assembly in a prototype model prior to finalizing his design.

In the layout of a magnetic assembly the permanent magnet should be situated as closely as possible to the air gap it is serving, in order to avoid excessive leakages. Fig. 4.1.17(a)–(d) shows four possible arrangements of a basic moving-coil circuit with progressively increasing magnetic efficiency (reduced leakage). In the design of cylindrical loudspeaker-type magnets, which are widely used in electrodynamic transducers for the measurement of linear vibration, there are two alternatives for the positioning of the permanent magnet. The arrangement of Fig. 4.1.18(a) is more economical than that of Fig. 4.1.18(b) because the external leakage (3) is almost non-existent. Also, since the magnet is not accessible

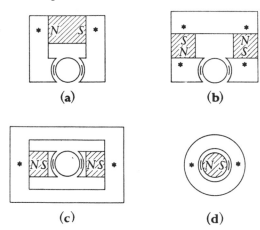

FIG. 4.1.17. Magnet assemblies for moving-coil-type instruments: (a), (b), and (c) with external magnet(s); (d) with internal magnet.

to accidental demagnetization by screwdrivers, etc., it tends to be more stable. On the other hand, it is easier to obtain a sufficiently large cross-section of the magnet in Fig. 4.1.18(b) for a given outside diameter, but the larger leakage of this design usually makes the apparent advantage very doubtful. Another source of avoidable leakage frequently occurs in those parts of the yokes and pole pieces which are close to the permanent magnet. Danger spots of high magnetic flux densities are marked by a star in Figs 4.1.17 and 4.1.18. These sections have to carry the useful and the leakage flux and it is therefore advisable to choose materials of the highest possible saturation values such as pure iron (Swedish, electrolytic, or Armco iron) or a 20–50 per cent cobalt–iron. In ordinary cold-rolled mild steel, when used for yokes and pole pieces the permeability μ drops sharply with increasing flux density. Mild-steel pole pieces and yokes therefore must have generous dimensions if saturation, and thus leakage, is to be avoided. Bad joints between sections of the magnetic circuit may also represent appreciably reluctances. A useful guide to the dimensioning of a variety of practical permanent-magnet assemblies has been given by Tyrrell [20].

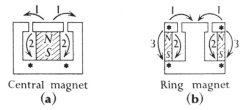

Central magnet
(a)

Ring magnet
(b)

FIG. 4.1.18. Flux distribution in loudspeaker-type magnet assemblies: (a) internal magnet, and (b) external magnet.

4.1.3.2. Permanent magnet materials

Modern permanent magnets for use in instrument transducers are commercially available in a great variety of shapes and sizes; they are conveniently listed in manufacturers' catalogues. Table 4.1.2 provides an introduction to the magnetic properties of some typical materials, i.e. two cast metallic Ticonals and two sintered ceramic Magnadurs, both types made by Mullard Ltd., London [21], and a new experimental material, a sintered intermetallic compound of composition $SmCo_5$ [22]. Demagnetization curves of these three classes of modern magnetic material are shown in Fig. 4.1.19.

TABLE 4.1.2

Properties of some permanent-magnet materials

Material	$(BH)_{max}$ $(Wb\ m^{-2})(kA\ m^{-1})$	D_r $Wb\ m^{-2}$	H_e $kA\ m^{-1}$	B_p $Wb\ m^{-2}$	H_p $kA\ m^{-1}$	Remarks
Ticonal 'G'	45·5	1·35	46·6	1·1	41·4	Cast
Ticonal 'XX'	71·8	1·06	111	0·82	87·6	metallic magnets
Magnadur 3	22·9	0·34	239	0·18	127	Sintered
Magnadur 6	25·7	0·37	239	0·19	135	ceramic magnets
$SmCo_5$	170	0·90	620	0·50	300	Sintered intermetallic magnets

Ticonal 'G' represents modern metallic material for general-purpose permanent magnets [23]. Its nominal composition is 7 per cent aluminium, 14·5 per cent nickel, 15·5 per cent cobalt, and 1·5 per cent copper. These magnets are cast in the presence of a strong magnetic field which determines its magnetic axis. They can be shaped only by grinding. AlNiCo magnets are made with varying properties in large numbers by several manufacturers in this country and abroad.

Magnadur 3 stands for a family of ceramic permanent magnetic material [24] (nominal compositions either $BaFe_{12}O_{19}$ or $SrFe_{12}O_{19}$) which is produced by sintering, some anisotropic materials in a strong field. Its peak energy product (BH_{max}) is only about half that of Ticonal 'G', but its high coercive force results in magnets whose performance is virtually indestructible, even if the magnetic circuit is completely broken. This is due to the fact that any conceivable minor loop is almost coincident with the demagnetization curve. Being a ceramic, Magnadur magnets should not be exposed to sudden temperature changes which might lead to cracks. Ceramic permanent magnets are also offered by a number of manufacturers.

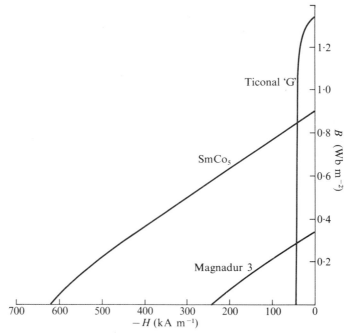

FIG. 4.1.19. Demagnetization curves of three typical permanent magnet materials.

The SmCo$_5$ material represents compounds of the general form (RE)Co$_5$, where RE stands for a variety of intermetallic elements which have been investigated in recent years. SmCo$_5$ magnets, with their extremely high energy product $BH_{max} = 170(\text{Wb m}^{-2})(\text{kA m}^{-1})$, are now available under a variety of commercial names, such as SAMCOMAG (UK), RECOMA (Switzerland), and RAECO (USA). They will be of great interest for use in miniature transducers, where space is at a premium.

The cyclic temperature coefficient between $-40\,^\circ$C and $+100\,^\circ$C of AlNiCo-magnets is about $-0\cdot02$ per cent per $^\circ$C, and of the ceramic magnets about 10 times higher. The value for SmCo$_5$-material is about $-0\cdot04$ per cent per $^\circ$C. Compensation can be achieved either by means of a temperature-sensitive magnetic shunt across the air gap, or by a temperature-sensitive resistor in the electric circuit. The magnetic shunt is either made of temperature-sensitive magnetic material, such as JAE-metal (Henry Wiggins & Co. Ltd.), or by a bimetallic strip operating a soft-iron armature.

In order to avoid flux losses caused by external magnetic fields, permanent magnets should be stabilized by applying a demagnetizing field prior to use. Any interfering field lower than the demagnetizing field, then, will result in no changes of the magnetic performance. Stabilized magnets are also immune to severe mechanical vibrations.

The technique of magnetizing permanent magnets cannot be discussed here in detail. The supplier will usually undertake the job if the magnet assembly is reasonably convenient in size and shape, or he will furnish the relevant information. It normally amounts to finding a convenient and economical way of applying a magnetic flux density of up to $1 \cdot 7$ Wb m^{-2} in the magnet along its magnetic axis. This requires quite high magnetomotive forces since the permeability of modern permanent magnet materials is not very much higher than that of air. The energization of the magnetizer involves the supply and control of very heavy currents over short periods. With respect to the materials listed in Table 4.1.2, their successful magnetizations requires the following minimum magnetic field strengths, measured in ampere-turns per metre length of the magnet:

Ticonal 'G':	24×10^4 A m^{-1},
Ticonal 'XX':	48×10^4 A m^{-1},
Magnadur 3 and 6:	132×10^4 A m^{-1}.
SmCo$_5$:	630×10^4 A m^{-1}

For highest efficiency of the process the magnetic circuit should be closed during magnetization by an iron return path capable of carrying the saturation flux.

Small permanent magnets (up to 1 kg mass) can be partially demagnetized (stabilized) by use of an open coil energized from the 50 Hz mains supply. Complete demagnetization (for handling and assembly) of ceramic permanent magnets is best done by raising their temperature above their Curie point (about 450 °C). AlNiCo-type permanent magnets have their Curie point at about 850 °C, but since their magnetic performance is permanently ruined at only 600 °C, they cannot be demagnetized in this fashion. Their demagnetization by alternating fields is somewhat difficult and the user should seek the advice, or service, of the manufacturer, if required. Anyway, suppliers normally ship permanent magnets in the demagnetized state.

4.1.3.3. Design of the coil unit

The e.m.f. generated in a coil moving in a stationary magnetic field is proportional to the flux density of the field, the conductor velocity, and the length and number of conductors in the field. Consider a typical 'loudspeaker'-type electrodynamic generator of Fig. 4.1.20. In the case of a transducer for the measurement of linear vibration the coil unit (copper wire wound on a copper or silver damping cylinder) represents the seismic mass. It is suspended by a suitable spring and guided by longitudinal bearings or spider diaphragms to move freely in the direction of its axis in the air gap of the magnet assembly. Assume, for example, a coil former of

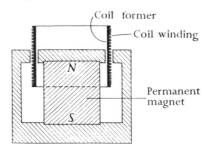

FIG. 4.1.20. Magnet and coil arrangement of electrodynamic generator-type transducer for the measurement of linear vibration.

copper with walls 0·4 mm thick, an external insulation layer of 0·1 mm and a coil winding depth of 0·5 mm. If the air gap is 2 mm wide there is thus a space of 0·5 mm on either side of the coil assembly which is normally quite adequate. The depth of the gap should be about 4 times its radial width, i.e. 8 mm. Deeper gaps cause problems of coil alignment and shallower gaps produce non-uniform magnetic fields along the gap depth. Even at a depth–width ratio of 4:1 the magnetic flux density is by no means uniform, as can be seen from Fig. 4.1.21 (as a loudspeaker assembly, its gap of only 1 mm is sufficient, but not so for a vibration

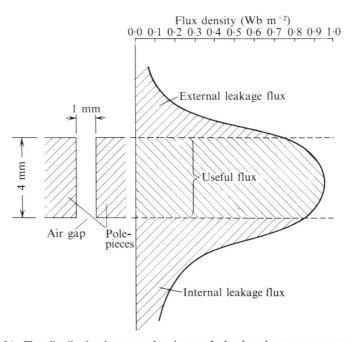

FIG. 4.1.21. Flux distribution in an annular air gap of a loudspeaker-type magnet assembly.

transducer). The coil former diameter should be about 10–15 times the radial air-gap width. Much larger diameters lead to difficult coil alignment and small coil diameters give low output voltages or may even forbid the use of a permanent magnet of sufficient cross-section. With the gap dimensions decided upon the size and shape of the permanent magnet and the remaining parts of the magnetic circuit can be computed as described in section 4.1.3.1 above, assuming a reasonable flux density in the gap of $0 \cdot 5 - 1 \cdot 0$ Wb m^{-2}. The length of the coil depends upon the required travel of the seismic mass. In order to avoid excessive non-linearity due to non-uniform leakage flux distribution (Fig. 4.1.21) its basic length should be increased by about 30 per cent,

$$l_{tot} = 1 \cdot 3(l_{gap} + l_{travel}).\qquad(4.1.56)$$

On the other hand, with coils longer than air-gap depth there is a danger of distortion of the flux distribution in those parts of the perma-nent magnet that are close to the gap. This happens if the coil field, due to coil current, penetrates the magnet. At the normally low coil currents of generator-type vibration transducers the effect is negligible, but in force-balance-type transducers (see Chapter 5), where the moving-coil system represents the servo-motor, the effect can be quite sizeable. A symmetrical twin-magnet arrangement, or a compensation coil wound on the magnet and connected in series with the moving coil, are effective remedies.

For the purpose of a preliminary estimate of the output voltage consider only that part of the coil winding which is inside the gap,

$$n_{gap} = n_{tot} \cdot \frac{l_{gap}}{l_{tot}},\qquad(4.1.57)$$

and with an average flux density B_{uv} (Wb m^{-2}), in the gap we have the instantaneous e.m.f. e (V) at an instantaneous coil velocity v (m s^{-1}),

$$e = n_{gap} l_{coll} B_{av} v,\qquad(4.1.58)$$

where l_{coil} is the circumferential length of one coil turn. If, for the purpose of a high voltage output, the coil has more than one layer it must be wound with constant pitch in uniform layers and the wire must be well secured to the coil former.

The coil former can be used as a damping device for the seismic system. However, this is not recommended when the transducer is intended to cover a wide frequency range, because the damping cylinder introduces attenuation of the output voltage at the high-frequency end of the spectrum (see section 4.1.2.5). In this case damping can be obtained by use of a suitable external damping resistance, or by an independent damping cylinder in a second magnetic gap.

The damping force exerted by a solid copper cylinder in an annular air gap can be computed as follows. The induced voltage in the cylinder is

$$e = \pi d_c B_{av} v \quad \text{(V)}, \tag{4.1.59}$$

where d_c (m) is the mean diameter of the cylinder, B_{av} (Wb m^{-2}) the average flux density in the air gap, and v (m s^{-1}) the axial linear velocity of the cylinder. The resistance of the cylinder is

$$r = \frac{\pi d_c \rho}{l_{gap} t}, \tag{4.1.60}$$

where l_{gap} is the depth (m) of the air gap, t (m) is the wall thickness of the cylinder, and ρ is the resistivity of the cylinder material, usually high-conductivity copper or aluminium (Ωm) (because of the cylinder parts outside the gap, the actual resistance is somewhat lower). From eqn (4.1.4b) the damping force P_d (N) is thus

$$P_d = \pi d_c e B_{av}/r = \pi d_c l_{gap} t B_{av}^2 v/\rho \quad \text{(N)}. \tag{4.1.61}$$

Damping devices consisting of solid plates or discs moving in magnetic fields have been dealt with elsewhere [25]. In section 3.3.3 the characteristics of various damping methods (air, oil, eddy-current damping) have been discussed in general terms, including the effect of temperature changes.

4.1.4. Construction

As with almost all transducers, size and weight are at a premium when designing electrodynamic generator-type transducers. In the case of a seismic vibration sensor weight is possibly even more important than size because a heavy transducer, when attached to a vibrating structure, might easily falsify the dynamic test conditions. Fortunately, modern permanent magnets with their high-energy products lead to much smaller sizes and lower weights than earlier types (the density of AlNiCo and ceramic magnets are 7.3×10^3 kg m^{-3} and 4.8×10^3 kg m^{-3}, respectively). If a fixed datum (e.g. the floor of the laboratory) is available as a reference, a 'non-seismic' transducer of much smaller moving mass can be used, the magnet being stationary and attached to the floor. Fig. 4.1.22(a) and (b) illustrate the two schemes.

The transducer housing is normally made of light alloy. Temperature stability of dimensions is less important than in transducer types operating on the measurement of small air-gap changes, such as variable capacitance and inductance transducers.

The construction of permanent-magnet sub-assemblies offers no appreciable difficulties. The (demagnetized) magnet can be fixed in position between soft-iron pole pieces by means of non-magnetic screws. Where screws are undesirable the magnet can be secured by soft-soldering, or 'Aralditing', it to the soft-iron parts. Modern permanent magnets are produced in their final shape either by casting or sintering processes. Later adjustments are normally possible only by way of grinding, though ultrasonic machining can be employed with the ceramic types. The alloy types have also successfully been machined by the spark-erosion technique, particularly for the purpose of drilling fixing holes, etc. Soft magnetic materials, normally low-carbon irons, are standard engineering materials for the pole pieces. It should be, however, remembered that they rust easily and thus should be protected by, for example, cadmium-plating.

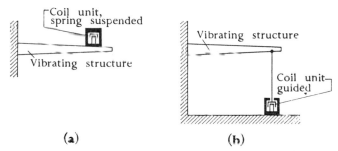

FiG. 4.1.22. (a) Seismic and (b) non-seismic arrangement of an electrodynamic vibration transducer.

The coil unit consists of the coil former, which is frequently also the damping cylinder, and the coil winding. In some cases it may be advantageous to use laminated pole pieces in the neighbourhood of the air gap, in order to avoid eddy currents. High-conductivity copper is used for the coil winding. Depending on the ambient temperatures at which the transducer is expected to work the wire insulation can be selected from a variety of materials, starting with oil-base enamel for relatively low temperatures above room temperature, synthetic resin-base enamels for medium–high temperatures, and silicone-base enamel or PTFE insulation for temperatures up to about 200 °C. Still higher temperatures can be met by ceramic and glass insulations. The coil must be secured to the coil former by cements of high strength and high insulation properties at the ambient temperatures, unless self-bonding wire is used (see section 4.3.7.2). When eddy-current damping is not intended, the coil former is made of insulation material of suitable strength and form stability. If, however, the former is to provide eddy-current damping a material of the highest possible conductivity should be used, in order to keep its mass small. Since fine silver has a conductivity only about 8 per cent higher than that of high-conductivity copper, the latter is usually preferred, chiefly because of its better machineability when alloyed with a small amount of tellurium. Damping formers made by the electro-forming process are often produced in fine silver.

Any electrodynamic transducer must comprise mechanical components to guide the moving parts in a linear or angular fashion, as the case may be. Angular movement is readily controlled by a suitable type of bearing. If the transducer under consideration is of the seismic type, damping forces are expected to be proportional to velocity, whilst friction forces are not. Appreciable friction damping would result in a frequency response which does not agree with the familiar theoretical curves based on viscous damping, and it may even be, or become with time, erratic. Bearing noise is generated by rough bearings, and skidding of ball- and roller-type bearings in their races. Since the output voltage of the electrodynamic transducer is proportional to velocity, high-frequency ripples, i.e. noise, are emphasized by differentiation. Bearing-life is another important aspect of the choice of material and of design.

The design of guiding elements for large linear movements is a difficult task if friction and noise have to be kept low. The two basic types of linear guides are:

(a) springs, such as spider diaphragms;

(b) guide rods, possibly with longitudinal ball or roller bearings.

Spring guides are commonly used if the travel of the mass is not in excess of, say 3 mm.. With very light seismic masses the stiffness of even a rather flimsy spider diaphragm may be too high for a given natural frequency of only a few hertz. The transverse stiffness of such a diaphragm is rather low, and there is the danger of collapse under appreciably transverse acceleration. A number of spring strip guides and flexure devices for linear and angular movement have been discussed in section 3.3.2. The fatigue strength of the guide springs is of great importance, since deflections are relatively large and frequencies may be quite high. Beryllium–copper is a favourite material.

With rigid guides, such as rods or rails, friction between the guides and guide shoes at different temperatures often determines the threshold of the transducer range. In a particular case synthetic sapphire rods and gold-plated bushes have been used with a very small mass, thus limiting the useful range of the instrument to vibrations above 1 g. Another design used lognitudinal ball bearings to support a somewhat heavier mass with amplitudes up to ±16 mm. The main advantage of rigid guides is, of course, that transverse forces cannot harm the delicate coil unit, but on the other hand friction may easily increase beyond acceptable magnitudes if the transverse forces are of steady-state character and occur during the period of measurement.

When measuring angular vibration, slip rings are most likely required and may be integrated with the transducer. The aspect of slip-ring design will be discussed briefly in Section 4.3 in connection with resistance strain gauges.

References

1. T. T. BAKER. Electrical tachometers. *Electl. Rev., Lond.* **128**, 179 (1940).
2. W. H. COULTHARD. *Aircraft instrument design*, pp. 227–33. Pitman, London (1952).
3. B. RICHTER. Drehzahlmessgeneratoren. *Arch. tech. Messen J.* 162–167 (1951).
4. B. RICHTER. Wirbelstrom-Drehzahlmesser. *Arch. tech. Messen J.* 162–7 (1951).
5. H. F. OLSON. *Acoustic engineering*, Chap. 8. Van Nostrand, New York (1957).
6. M. L. GAYFORD. *Electroacoustics; microphones, earphones and loudspeakers.* Butterworth, London (1970).
7. M. ABRAHAM and R. BECKER. *The classical theory of electricity and magnetism*, p. 145. Blackie, London (1950).
8. C. C. HAWKINS. *The dynamo*, pp. 24 and 127. Pitman, London (1922).
9. E. J. WILLIAMS. The introduction of electro-motive forces in moving liquids. *Proc. phys. Soc.* **42**, 466 (1930).
10. V. CUSHING. Induction flowmeter. *Rev. scient. Instrum.* **29**, 692 7 (1958).
11. J. A. SHERCLIFF. *The theory of electro-magnetic flow-measurements.* Cambridge University Press (1962).
12. L. CREMER and M. HECKL. *Körperschall*, p. 34–43 Springer, Berlin (1967).
13. H. K. P. NEUBERT. Bilateral electro-mechanical transducers; a unified theory. R.A.E. Tech. Rep. No. TR68248 (1968).
14. A. T. DENNISON. The design of electromagnetic geophones. *Geophys. Prospect.* **1**, 3–28 (1953).
15. W. L. EVERITT. *Communication engineering*, p. 157 McGraw-Hill, New York (1937)
16. H. C. ROTERS. *Electromagnetic devices.* Wiley, New York (1941).
17. F. G. SPEADBURY. *Permanent magnets.* Pitman, London (1949)
18. C. A. MAYNARD. Analysis and design of permanent magnet assemblies, *Mach. Des.* **29**, 122–34.
19. ANON. *Handbook of permanent magnets*, p. 19. Mullard Ltd., London (1953).
20. A. J. TYRRELL. The design and application of modern permanent magnets, *J. Br. Instn. Radio Engrs.* **6**, 178 (1946)
21. ANON. *Mullard technical handbook No. 3*, Part 2. Mullard Ltd., London (1971).
22. K. BACHMANN. Permanent magnets. *Lingny. Mater. Des.* **17**, 19–22 (1973).
23. M. G. SAY. Permanent ferrites. In *Magnetic alloys and ferrites*, p. 148. Newnes, London (1954).
24. K. J. STANDLEY. *Oxide magnetic materials* (2nd edn.). Clarendon Press, Oxford (1972).

4.2. VARIABLE-RESISTANCE TRANSDUCERS

IN the previous section a transducer type was discussed which converts a physical quantity (velocity) into an electrical signal (voltage). This was an example of direct energy conversion requiring no auxiliary source of energy. Variable-resistance transducers—the subject of this section—are typical 'passive' transducers; they require an energizing source. There are many versions of this transducer type, and although they all work on the same basic principle, they differ considerably in their design and characteristics. Fig. 4.2.1 shows examples of typical variable-resistance

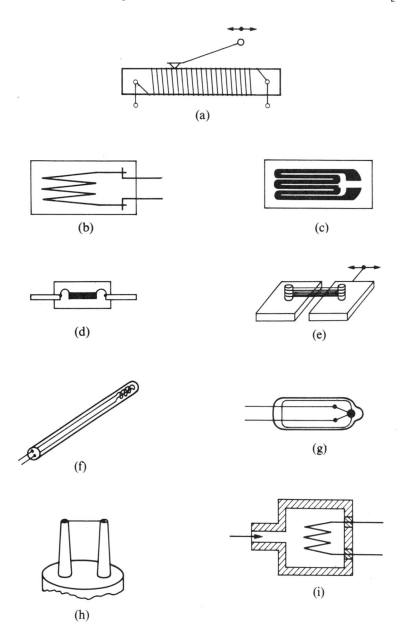

FIG. 4.2.1. Some examples of variable-resistance sensors: (a) Potentiometer, (b) wire strain gauges, (c) foil strain gauge, (d) semiconductor strain gauge, (e) unbonded strain gauge, (f) wire resistance thermometer, (g) thermistor, (h) hot-wire anemometer, and (i) wire pressure sensor.

arrangements. They may roughly be divided into two classes:

(a) transducers operating on large changes of resistance, employing mainly potentiometer circuits;

(b) transducers operating on small changes of resistance, employing mainly bridge circuits.

The two classes are concerned with different characteristics. For example, resolution and noise are important with transducers in the first class, while in the second class high sensitivities are wanted, though non-linearity and temperature sensitivity may be a problem, e.g. in semiconductor strain gauges, and may require elaborate compensating artifices.

4.2.1. The equivalent circuit

Although the majority of variable-resistance transducers operate on d.c., a.c. energization is sometimes used for two main reasons:

(a) elimination of contact potentials and thermoelectric voltages between dissimilar metals in the transducer or its leads;

(b) possibility of employing a.c. amplification and signal conditioning which, in general, provide improved stability at less effort.

We shall, therefore, at the beginning of this chapter, also discuss the general a.c. characteristics of resistors.

The d.c. resistance $R(\Omega)$ of a metal wire is

$$R = \rho l/A, \tag{4.2.1}$$

where l(m) is the length of the wire, A(m^2) its cross-section, and ρ(Ωm) its resistivity. The d.c. resistance of all pure metal conductors and the majority of resistance-wire alloys depends appreciably on temperature, except for a few alloys especially developed for small or negligible temperature coefficient. However, it should be borne in mind that these desirable low values of temperature coefficient are only obtainable within limited temperature ranges. Let R_1 be the resistance at a temperature t_1 and R_2 the resistance at a temperature t_2, then we have

$$R_2 = R_1\{1+\alpha(t_2-t_1)\}. \tag{4.2.2}$$

This simple relationship only holds as long as α can be considered a constant. A more detailed discussion of variation of resistance with temperature will be found in section 4.2.3.1.

The a.c. resistance of a metal wire depends on frequency, and is always greater than its d.c. resistance. This fact is due to eddy currents which reduce the effective cross-section of the wire (skin effect). From theory [1]

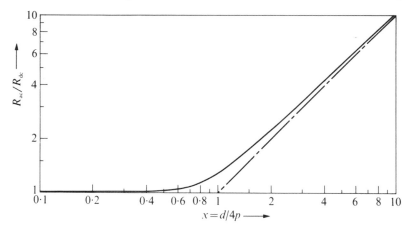

FIG. 4.2.2. Ratio of a.c. to d.c. resistance of copper wire as a function of eddy-current
factor x.

the eddy-current factor

$$x = \frac{d}{4}\left(\frac{\pi f \mu \mu_0}{\rho}\right)^{\frac{1}{2}} = \frac{d}{4p}, \qquad (4.2.3)$$

where

d = diameter of metal wire (m),

f = frequency (Hz),

μ = relative permeability of wire material ($\mu = 1$ for air
and most alloys, except ferromagnetic materials),

$\mu_0 = 4\pi \times 10^{-7}$ = permeability of empty space (H m^{-1}),

ρ = resistivity of wire material (Ωm),

p = 'depth of penetration' of eddy currents (m).

Then, the a.c.–d.c. resistance ratio becomes approximately

$$\left.\begin{aligned}
\frac{R_{ac}}{R_{dc}} &= x + \frac{1}{4} + \frac{3}{64x} \quad \text{for} \quad x > 1, \\
\frac{R_{ac}}{R_{dc}} &= 1 + \frac{x^4}{3} \qquad\quad \text{for} \quad x < 1.
\end{aligned}\right\} \qquad (4.2.4)$$

Eqn (4.2.4) has been plotted in Fig. 4.2.2 as a function of the (dimensionless) eddy-current factor $x = d/4p$. For copper wire the depth of penetration can be computed from

$$p_{Cu} = 2 \cdot 11/\sqrt{f} \quad \text{(mm)}, \qquad (4.2.5)$$

if the frequency f is measured in kHz.

The eddy-current distribution in a solid wire is also affected by adjacent conductors (proximity effect) and may account for an additional increase in effective resistance of up to 30 per cent, in closely wound coils at high frequencies [2]. Likewise, magnetic material in the neighbourhood of the resistance wire also affects its a.c. resistance. This might be significant in a.c.-operated strain gauges attached to ferro-magnetic substrates.

A resistor, when wound in a coil, has also reactive components which affect the real component (the 'resistance') of its impedance. Consider the equivalent circuit of an a.c. resistor, as shown in Table 4.2.1(A). Let R be the effective a.c. resistance of the wire and C its lumped self-capacitance, representing the distributed self-capacitance of the resistor and the terminal capacitance. L is the self-inductance of the resistor and its leads, if any. This circuit can be represented either by a series combination with the resistive component R_s and the reactive component X_s (Table

TABLE 4.2.1
Equivalent circuits

Equivalent circuit	Frequency: $\omega = 2\pi f$ Natural frequency: $\omega_0 = 1/\sqrt{(LC)}$ Frequency ratio: $\omega/\omega_0 = \omega\sqrt{(LC)}$ Dissipation factor (at resonance): $D_0 = R/\sqrt{(L/C)}$ Time constant: $T = (\tan\phi)/\omega$ Phase angle: ϕ	
A — series R, L with parallel C		
	RESISTANCE	**REACTANCE**
Series circuit **B** R_s, X_s	$R_s = \dfrac{R}{(1-\omega^2 LC)^2+(R\omega C)^2}$ $\dfrac{R_s}{R} = \dfrac{1}{\left\{1-\left(\dfrac{\omega}{\omega_0}\right)^2\right\}^2+\left(\dfrac{\omega}{\omega_0}\right)^2 D_0^2}$	$X_s = \dfrac{\omega\{L(1-\omega^2 LC)-R^2 C\}}{(1-\omega^2 LC)^2+(R\omega C)^2}$ $\dfrac{X_s}{R} = \dfrac{\dfrac{\omega}{\omega_0}\left[\dfrac{1}{D_0}\left\{1-\left(\dfrac{\omega}{\omega_0}\right)^2\right\}-D_0\right]}{\left\{1-\left(\dfrac{\omega}{\omega_0}\right)^2\right\}^2+\left(\dfrac{\omega}{\omega_0}\right)^2 D_0^2}$
Parallel circuit **C** R_p, X_p	$R_p = R\left\{1+\omega^2\left(\dfrac{L}{R}\right)^2\right\}$ $\dfrac{R_p}{R} = 1+\left(\dfrac{\omega}{\omega_0}\right)\dfrac{1}{D_0^2}$	$X_p = \dfrac{1}{\omega C - \dfrac{1}{\omega L\left\{1+\dfrac{1}{\omega^2}\left(\dfrac{R}{L}\right)^2\right\}}}$ $\dfrac{X_p}{R} = \dfrac{\dfrac{\omega}{\omega_0}\cdot\dfrac{1}{D_0}+\dfrac{\omega_0}{\omega}D_0}{\left(\dfrac{\omega}{\omega_0}\right)^2+D_0^2-1}$
Time constant and phase angle **D**	$\omega T = \tan\phi = \omega\left\{\dfrac{L}{R}(1-\omega^2 LC)-RC\right\}$ $\quad = \dfrac{\omega}{\omega_0}\left[\dfrac{1}{D_0}\left\{1-\left(\dfrac{\omega}{\omega_0}\right)^2\right\}-D_0\right]$	

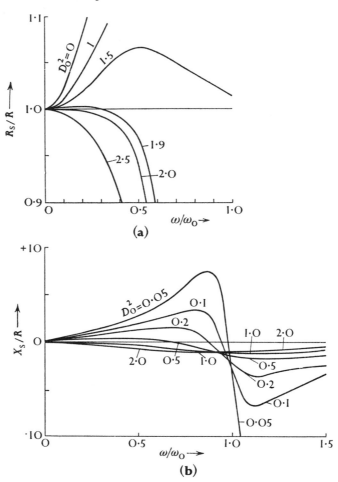

FIG. 4.2.3. (a) Resistance and (b) reactance ratio of a series equivalent circuit of resistor, plotted against frequency ratio.

4.2.1(B)), or by a parallel combination with resistive component R_p and reactive component X_p (Table 4.2.1(C)). The choice depends mainly on convenience in computation or representation. Table 4.2.1 shows the components as computed for the network, including their dimensionless ratios with the resistance R. It also includes the expressions for the time-constant T (Table 4.2.1(D)) and the phase angle ϕ between the reactive and the resistive components. It is seen from the table that the real component of a resistor depends greatly upon its inductance and capacitance characteristics, especially at higher frequencies. Fig. 4.2.3 illustrates the frequency characteristics of R_s/R and X_s/R, respectively. At low frequencies where terms with ω^2 as a factor may be neglected we have

$$R_s = R; \qquad X_s = \omega(L - R^2 C), \left.\vphantom{\frac{1}{\omega}}\right\}$$
$$R_p = R; \qquad X_p = \frac{1}{\omega(C - L/R^2)}. \right\} \qquad (4.2.6)$$

Here the network behaves like a resistance R in series with an inductance of value $(L - R^2 C)$, or in parallel with a capacitance of value $(C - L/R^2)$. For high values of resistance the capacitance dominates, while for low values of resistance the inductance is in control.

In the design of a variable-resistance transducer which is expected to work over the widest possible frequency range there is an optimum combination of the residual parameters for any given resistance value. For resistors with small L and C values and at low frequencies we have $\omega^2 LC \ll 1$, and the phase angle ϕ is given, approximately, by

$$\phi \simeq \tan \phi = \omega(L/R - RC). \qquad (4.2.7)$$

ϕ vanishes for $R^2 = L/C$. However, if applied to potentiometers, this condition does not necessarily hold for arbitrary wiper settings.

4.2.2. Large changes in resistance; potentiometers

Under this heading we shall deal exclusively with potentiometer type transducers. Other types with large changes in resistance, such as semiconductor strain gauges, or thermistors, have characteristics entirely different from metal-wire potentiometers and will be treated separately. Our main interest will be the behaviour and design of wire-wound precision potentiometers as applied to instrument transducers, which have a wide field of application, in spite of many shortcomings, mainly because of their high output. Continuous-track potentiometers, using conductive plastics, have only recently come into use [98].

4.2.2.1. *Resolution and noise*

The linked problems of resolution and noise are peculiar to potentiometer transducers and are here discussed before general performance characteristics because of the influence they have on sensitivity, linearity, and associated circuit requirements. Consider a linear wire-wound potentiometer [3]. A cross-section of 10 turns is shown in Fig. 4.2.4(a) together with a wiper gliding over the bare wires. If the wiper could be arranged to touch only one wire at a time the voltage resolution would be

$$\Delta V = V/n, \qquad (4.2.8)$$

where V is the total voltage across the potentiometer, say 10 V, and n the number of turns, say 10, the apparent resolution of such a potentiometer would thus be $\Delta V = 1$ V, or 10 per cent, of the total voltage. In

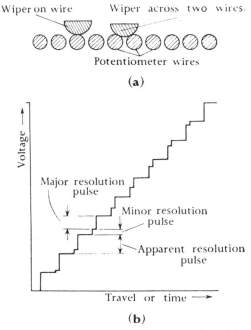

FIG. 4.2.4. (a) Linear potentiometer (schematic diagram) and (b) voltage diagram.

reality, however, the wiper is bound to short-circuit adjacent wires at some time of its travel across the winding. This causes $(n-2)$ minor resolution pulses, since no change in output voltage occurs if the two first and the two last turns of the winding are short-circuited. The total number of resolution pulses is therefore equal to $(2n-2)$. Fig. 4.2.4(b) shows the distribution of major and minor resolution pulses along the wiper travel across the potentiometer of Fig. 4.2.4(a). The amplitude ΔV_{mi}, of a minor pulse as we short turn x to turn $(x+1)$, is

$$V_{mi} = Vx\left(\frac{1}{n-1} - \frac{1}{n}\right), \tag{4.2.9}$$

and of a major pulse ΔV_{ma} is the difference between the apparent resolution ΔV and the minor resolution,

$$\Delta V_{ma} = \Delta V - \Delta V_{mi} = \frac{V}{n} - Vx\left(\frac{1}{n-1} - \frac{1}{n}\right). \tag{4.2.10}$$

The ratio of the duration times of the major and the minor resolution pulses depends upon the geometry of wire winding and wiper profiles, i.e. in the first instance upon the ratio of the wiper radius to the wire radius. These considerations will be disturbed in practice by flats developing on wire and wiper when the potentiometer has been in use for some time. A small wiper diameter, especially when used with relatively soft alloys, favours the early development of flats, and may even tear the wire track. On the other hand, an excessively large radius only needs a small amount of wear to short-circuit three, or even more, turns of the potentiometer. This results in reduced precision of the potentiometer, an effect which is aggravated if the track is 'bumpy'. In fine-wire potentiometers of standard design a wiper–wire radius ratio of 10 will normally give satisfactory results. Details of wire winding, coils and formers, and some wiper patterns, will be discussed in section 4.2.2.3, dealing with the construction of potentiometer-type pick-offs. The normally bad resolution of low-resistance potentiometers can be avoided by shunting a low-value fixed resistor with a high-resistance potentiometer of good resolution which will give satisfactory performance if little current is drawn from the potentiometer–resistor combination.

Electric noise differs from resolution irregularities by its random character and its peculiar frequency spectrum. Any fixed resistor is a noise generator [4]. The electron current has a noise current superimposed upon it, caused by the random motion of free electrons, which are in equilibrium with the thermal motion of the molecules. This type of noise has a uniform frequency spectrum, and its magnitude depends on resistance and temperature only. The wider the frequency bandwidth of the measuring equipment following the resistor, the greater the amount of noise transmitted. This so-called Johnson noise is normally very small, except in extreme cases of high resistance, high temperature, and wide frequency bandwidth. In carbon-composition fixed resistors, and in some continuous-track compound potentiometers, there is an additional type of noise which is generated when a current is passed through the resistor. It is thought to be caused by random changes in the resistance material and its characteristics. Its magnitude depends in a complicated way on frequency, on material and shape of the resistor, and, of course, on the transmitted bandwidth. Typical noise values for fixed cracked-carbon resistors are

$0 \cdot 05 \, \mu V$ per V for 1000 Ω resistor,

$0 \cdot 5 \, \mu V$ per V for 1 MΩ resistor.

Wire-wound resistors are free from this type of noise.

Variable resistors suffer from further types of noise which are generated by the wiper travelling along the potentiometer track. In transducer work they are of great importance. The contact noise, generated in the wiper–wire contact area, is caused by variation of the contact resistance. Contributors are contact area variation and pressure fluctuation, especially in the presence of foreign particles on the track. This noise is probably the most important single factor in potentiometer noise. It tends to increase with the lifetime of a variable resistance owing to wear and to contamination and oxidization of track and wiper. Major criticisms of potentiometer transducers are based on this progressive increase of noise generation. Another noise source is the generation of small voltages by the rubbing action between dissimilar metals of wiper and track. However, by suitable choice of materials this may be reduced to 100–300 μV. Thermoelectric effects which may come under this heading should also be

TABLE 4.2.2

Suitable combinations of materials for wiper and wire of potentiometers

Wiper material	Winding-wire material
Rhodium, rhodium plating 40% copper–palladium Gold–silver Osmium–iridium 10% ruthenium–platinum Gold	80% nickel, 20% chromium (Nichrome)
2–5% graphite in silver 10% graphite in copper 40% nickel, 60% silver	Nichrome Constantan Manganin
Gold Gold–silver	55% copper, 45% nickel (Constantan)
Platinum–iridium	Silver–palladium Platinum–iridium
10% gold, 13% copper, 30% silver, 47% palladium	Silver–palladium

watched especially in d.c. applications at high environmental temperatures. Combinations of wiper and wire materials which have been found satisfactory are given in Table 4.2.2 [5]. Finally, there is vibrational noise, or high-velocity noise, which is caused by a jumping or bouncing movement of the wiper. Its magnitude, when it occurs, is comparable with the full-scale voltage variation, since it is caused by the temporary disengagement of wiper and track. It must be avoided at all costs. For a given wiper–track combination there is a maximum wiper speed which must not be exceeded. The remedy is often an improved wiper design which should include a revision of the contact pressure and oscillatory characteristics of the wiper structure.

4.2.2.2. Sensitivity and linearity

The sensitivity of an unloaded potentiometer is given in volts for the full-scale mechanical travel of the wiper. This voltage is ideally equal to the input voltage across the total potentiometer winding, though there may be slight irregularities at the extreme ends of the track. The input

voltage is set by the dissipated wattage which causes the temperature of
the winding to rise to, but not beyond, a specified level. Its magnitude
thus depends on the cooling conditions under which the potentiometer
has to work which, in turn, depend on the thermal characteristics of the
materials used in the pick-off, and on the transducer design, including its
shape and size. In general, the current density in the winding wire should
not exceed $10 \, \text{A} \, \text{mm}^{-2}$. This rough guide is applicable to small fine-wire
pick-offs wound on anodized aluminium formers. Under worse conditions
of heat conduction, e.g. for coils on formers of solid insulating materials,
the current density should probably not be greater than $5 \, \text{A} \, \text{mm}^{-2}$.

In order to meet a specified linearity the minimum resolution is fixed. If
the apparent resolution, as discussed above, is n per cent of full scale, the

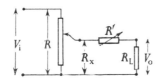

FIG. 4.2.5. Potentiometer circuit.

linearity error cannot be smaller than $\pm\frac{1}{2}n$ per cent of full scale. Since
there are additional errors affecting linearity—such as mechanical inaccuracies in the wiper movement, irregularities in winding pitch, variations
in wire and former dimensions, etc. –in a practical design the computed
number of turns required to achieve a specified linearity should be
multiplied by 2 for coil formers of circular cross-section, and by about 3
for other than circular cross-sections, such as for phenolic card formers.
For potentiometers of large travel, especially for multi-turn helical types,
a safe factor on the required turn number is probably 4–5. In general it
can also be said that the finer the wire the greater the factor to achieve a
specified linearity.

Apart from the above considerations which limit the 'inherent' linearity
of a potentiometer transducer in the presence of finite resolution and
noise, with loaded potentiometers there is what we might call its 'circuit
non-linearity'. This, in the first instance, is determined by the ratio of
total potentiometer resistance R to the load resistance R_L. In Fig. 4.2.5,
V_i is the input voltage and V_o the output voltage. R_x is the portion of the
total potentiometer resistance R which occurs between wiper and 'bottom' end of R at the potentiometer setting $x = R_x/R$. In uniformly
wound potentiometers x is identical with the relative travel t, which is
zero for $R_x = 0$ and unity for $R_x = R$. R' is a variable resistance in series
with the load resistance R_L, which will concern us later on. For a simple

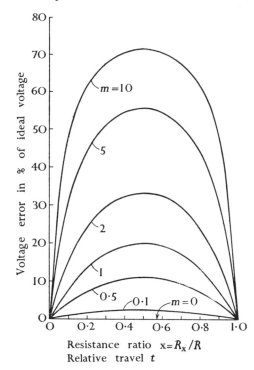

FIG. 4.2.6. Percentage voltage error of a loaded linear potentiometer at various resistance ratios $m = R/R_L$, plotted against resistance ratio $x = R_x/R$, or relative travel t.

circuit, where $R' = 0$, it can be shown that the voltage ratio

$$\frac{V_0}{V_i} = \frac{x}{1+mx(1-x)}, \qquad (4.2.11)$$

where m is the ratio of total potentiometer resistance R to the load resistance R_L, i.e. $m = R/R_L$. In general, if $V_{xl}/V = x$ the voltage ratio of the perfectly linear potentiometer, then the percentage error ε of V_x/V with reference to V_{xl}/V, is

$$\varepsilon = \frac{100(V_{xl}-V_x)}{V_{xl}} = \left\{ 1 - \frac{1}{1+mx(1-x)} \right\}100. \qquad (4.2.12)$$

A family of error curves for some resistance ratios m have been plotted in Fig. 4.2.6 against the resistance ratio x, which is identical with the relative wiper travel t if the potentiometer is uniformly wound. It is obvious from these curves that the load resistance R_L must be at least, say, 10–20 times higher than the total potentiometer resistance R in order to keep the non-linearity within 1–2 per cent of full scale. If, however, this condition cannot be met, linearity can be improved in two

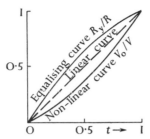

FIG. 4.2.7. Non-linear potentiometer characteristic with equalization, schematic diagram.

different ways:

(a) by a non-linear potentiometer characteristic of suitable shape;

(b) by the introduction of a non-linear variable resistance R' of suitable characteristic in series with the load R_L (Fig. 4.2.5).

Fig. 4.2.7 shows schematically the first method applied to a potentiometer circuit of given non-linearity. Linearity is obtained [6] by means of a potentiometer winding which follows a law given by the inverse function of eqn (4.2.11), i.e.

$$\frac{R_y}{R} = \frac{mt-1}{2mt} + \left\{ \left(\frac{mt-1}{2mt} \right)^2 + \frac{1}{m} \right\}^{\frac{1}{2}}. \qquad (4.2.13)$$

The two curves are in mirror symmetry to the ideal straight line at 45°. The non-linear resistance function of eqn (4.2.13) can be realized by a card-wound potentiometer of a variable winding height h, as shown in Fig. 4.2.8. Now it can be shown that the function $h = h_0(t)$ must follow the differential curve of R_y/R for full equalization of the original non-linearity. Thus

$$h = \frac{d(R_y/R)}{dt} = \frac{1}{2mt^2} \left[1 + \frac{mt-1}{\{(mt-1)^2 + 4mt^2\}^{\frac{1}{2}}} \right]. \qquad (4.2.14)$$

The second method employs a variable series resistance (Fig. 4.2.5) of maximum value R'. The wiper is coupled with the wiper of the potentiometer, and it is convenient to make its travel of the same length as the

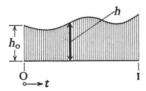

FIG. 4.2.8. Scheme of winding card for the purpose of equalizing potentiometer non-linearities.

potentiometer travel. A closer study of the conditions for equalization by this method shows [6] that the following two equations are valid:

$$R' = R/4, \qquad (4.2.15a)$$

$$R'_t/R' = (2t-1)^2. \qquad (4.2.15b)$$

The resistance characteristic of eqn (4.2.15b) is a parabola with its axis of symmetry at $t = 0.5$. The conditions of eqns (4.2.15a) and (4.2.15b) are affected by the potentiometer resistance R only, and are independent of the load resistance R_L. However, the maximum voltage across R_L for

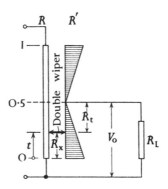

FIG. 4.2.9. Schematic arrangement of potentiometer R with equalizing resistance R' in series with load resistance R_L.

$t = 1$ is $V_{0max} = 4V/(4+m)$, and thus depends on R_L by way of m. The profile of the winding card, again, can be computed from the differential curve to eqn (4.2.15b), i.e.

$$h' = \frac{d(R'_t/R')}{dt} = 4(2t-1). \qquad (4.2.16)$$

This represents an oblique straight line which intersects the t-axis at $t = 0.5$. A practical arrangement of a potentiometer with a variable series resistance for the purpose of equalization is shown in Fig. 4.2.9.

In either method steep slopes of the profile of winding cards may in some extreme cases defy all efforts to wind at a uniform pitch. Very long potentiometers, and potentiometers which do not need a high degree of resolution, have been wound on tiny horizontal steps cut into the continuous profile.

Potentiometers working in a push–pull fashion, as shown in Fig. 4.2.10, have smaller non-linearities due to loading than single-ended potentiometers. At small m-ratios the improvement may approach a factor 2. The reader is invited to derive the error equations for the circuit of Fig. 4.2.10, or may refer to charts published by Parker [7].

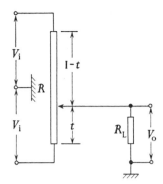

FIG. 4.2.10. Push–pull potentiometer circuit.

4.2.2.3. Construction

In transducer work we are concerned almost exclusively with small straight or curved potentiometers of round or rectangular cross-section. The wider aspects of potentiometer design and construction have been dealt with in Dummer's book [5] on variable resistors, which also contains a comprehensive bibliography. The potentiometer itself is only one of several components of a complete instrument transducer. It may be actuated by the moving mass of an accelerometer, by a pressure-sensitive Bourdon tube, by a rotating shaft, or in many other ways. In instrument transducers it must be of a 'precision' type, and its design and construction must be viewed in this light. In previous sub-sections of section 4.2 we have discussed the conditions under which noise and non-linearity are reduced to a minimum and resolution is an optimum. These, with the addition of stability in adverse environments, are the primary requirements in transducer performance.

It is convenient to discuss the construction of potentiometers under three headings:

> winding wire and wire winding,
>
> winding former,
>
> wiper.

Briefly, the following are the more important points.

Winding wire for precision potentiometers is precision-drawn and annealed in a reducing atmosphere to avoid surface oxidization. Its resistivity may vary between $0.4~\mu\Omega$ m and $1.3~\mu\Omega$ m and its temperature coefficient of resistance between 0·002 per cent per °C and 0·01 per cent per °C. (See also Table 4.2.3, p. 120, on wire resistance strain gauges.)

The thermoelectric e.m.f. against copper should be as small as possible. Fine wire may have to be protected from surface corrosion by enamelling or oxidization. In small gauges it must be strong and ductile enough to stand winding round small-radius corners. The uniformity of wire diameter and resistance is controlled by BS 115 and 117, calling, for instance, for a tolerance of 5 per cent on resistance of S.W.G.50 (0·025 mm diameter) wire, or finer tolerances down to ±1 per cent, according to grade. The reader will notice that these tolerances are rather wide with respect to resolution and linearity requirements. Higher-precision wire is obtainable at higher cost. The resistance stability with time depends upon annealing and upon the ability of the wire material to withstand corrosion. Resistance wire should be soldered or welded with ease, but in difficult cases the terminal ends may be copper-plated and silver-soldered. The wire material should have a high melting-point to avoid creep at high working temperatures.

The following families of resistance alloys are in common use.

(a) Copper–nickel alloys (Constantan, Ferry, Advance, Eureka) have the lowest thermal coefficient of resistance and a medium–high resistivity. Their mechanical strength is adequate, but their thermoelectrical potential with copper is high.

(b) Nickel–chromium alloys (Nichrome V, Karma) have high resistivity and fairly low thermal coefficient of resistance. Their resistivity depends on the state of annealing. They can be obtained with a thin film of oxide for insulation in closely wound potentiometers. They have a high mechanical strength and high ultimate working temperatures. Karma is a material of somewhat higher resistivity and lower thermal coefficient of resistance than Nichrome V.

(c) Nickel–chromium–iron alloys (several) are cheaper than nickel–chromium alloys but are ferromagnetic at room temperature and have higher thermal coefficients of resistance.

(d) Silver–palladium alloys have high resistance to corrosion and thus lower contact resistance, but are otherwise similar to the alloys under (b) above.

The choice of the wire material will also be affected by the wiper material and Table 4.2.2 should be consulted.

The windings of bare wire must be spaced, but enamelled or oxidized wire can be wound touching. In this case the track for the wiper must be cleaned either mechanically or chemically. The mechanical method is normally preferred since it is almost impossible to dissolve the enamel on the track without softening the remaining insulation between and underneath the windings.

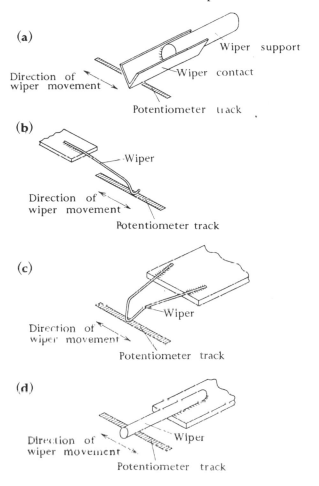

FIG. 4.2.11. Some wiper constructions.

Wire-winding techniques cannot be discussed here in any detail. Potentiometer winding is a specialized field and must be learned like a craft. The principles of some winding machines for linear, toroidal, and helical potentiometers are explained in Dummer's book [5], and further references are given.

The main requirements for winding formers are form stability and high surface-insulation resistance. Ceramic bodies are probably the most suitable, but may have to be treated with silicone varnish if highly porous; they are rather expensive unless wanted in large quantities. Some ceramic bodies can be shaped in their 'green' stage before firing (Steatite, Lava), and winding groves for bare wire can be produced by means of a blind winding of slightly larger-diameter wire.

Some materials can be machined even in their fired state [99]. Very stable winding formers can also be made from aluminium which has been anodized by a special process. This latter type of former also permits higher current densities because of its good heat conductivity. Satisfactory formers have also been made of polymethyl methacrylate (Perspex), phenolic resins (Bakelite), and moulded epoxies (Araldite). These materials are easily machinable and fairly stable. A smooth and accurate finish of all coil formers is of great importance.

The wiper construction is frequently responsible for the noise level of a potentiometer. Suitable combinations of wiper and wire materials are listed in Table 4.2.2. A great variety of diverse shapes have been tried but the results obtained from tests are anything but conclusive. Some common constructions are shown in Fig. 4.2.11. The flexibility of the wiper arm has to provide the contact pressure, and, together with the head mass of the wiper, constitutes a vibratory system which may cause high-velocity noise when unsuitably dimensioned. The contact pressure required depends on the materials employed in the design, the ratio of wire and wiper diameter, and the expected life of the potentiometer. Practical values vary between 5 mN and 50 mN.

4.2.3. Small changes in resistance

This section deals with small resistance changes due to variable strain or variable temperature. One of the major difficulties in application is the mutual interference between the two effects, i.e. strain gauges are sensitive to temperature and certain resistance thermometers are sensitive to strain. Suppression of the one or the other effect is essential when accurate results are expected, but since the magnitudes of both influences are of similar order this is not an easy task.

4.2.3.1. Wire and foil strain gauges

The application of fine wire to the measurement of strain was established in the U.S.A. during the Second World War but the basic principle has been known for a long time [8] and results of experimental investigations had been published at earlier dates [9]. In recent years the strain-gauge literature has grown beyond limits. The subject has also been treated in monographs [10]–[15] in greater detail than is possible or desirable for the purpose of this book. In the present section our aim is to cover the elements of wire and foil strain gauges (and of semiconductor strain gauges in the section to follow) thoroughly enough to enable the reader to understand and assess strain-gauge performance, and to design strain-gauge transducers for particular applications (see section 4.2.3.4). It will, however, be impossible to discuss all the finer points of strain-gauge design, manufacture, application, and analysis.

The modern wire resistance strain gauge usually consists of a fine resistance wire of, say, 0·025 mm diameter which is arranged in the form of a grid in order to obtain higher resistances. The grid is bonded to the structure under test with an insulation layer between the two (Fig. 4.2.12(a)). The bonding layer transmits the surface strain in the structure to the wire, thus producing a change of resistance in the wire, which is proportional to strain. Since the bonded-resistance strain gauge is not an extensometer its length does not matter in analysis and need not be

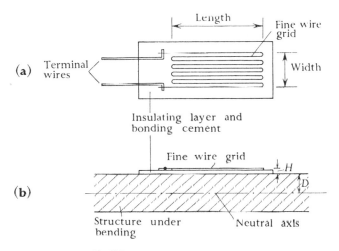

FIG. 4.2.12. Wire resistance strain gauge (flat-grid type).

known accurately. In very thin structures, such as sheet metal, under bending stress (Fig. 4.2.12(b)) a correction factor may have to be applied which is

$$\varepsilon = \varepsilon_a \frac{D}{D+H},$$
(4.2.17)

where

ε = actual surface strain,

ε_a = apparent strain measured,

D = distance between surface and neutral axis of structure,

H = height of gauge wire above structure surface.

Also, since the strain-gauge grid has a finite width, the gauge has a sensitivity to transverse strains [16] which may amount to 0·5–2 per cent of the longitudinal sensitivity [14]. The gauge factor quoted by the manufacturer is normally obtained on a test rod under tension and is thus related to uniaxial stress involving both longitudinal and transverse strains.

4.2.3.1.1. Sensitivity and linearity.   The resistance of a single long wire is

$$R = \rho l / a \quad (\Omega) \tag{4.2.18}$$

where

ρ = resistivity (Ωm),

l = length of wire (m),

a = cross-sectional area of wire (m^2).

If a uniform stress σ ($N\,m^{-2}$) is applied to this wire along its total length the variation in resistance is

$$\frac{dR}{d\sigma} = \frac{d(\rho l / a)}{d\sigma} = \frac{\rho}{a} \cdot \frac{\partial l}{\partial \sigma} - \frac{\rho l}{a^2} \cdot \frac{\partial a}{\partial \sigma} + \frac{l}{a} \cdot \frac{\partial \rho}{\partial \sigma}, \tag{4.2.19a}$$

or, if referred to the initial resistance R,

$$\frac{1}{R} \cdot \frac{dR}{d\sigma} = \frac{1}{l} \cdot \frac{\partial l}{\partial \sigma} - \frac{1}{a} \cdot \frac{\partial a}{\partial \sigma} + \frac{1}{\rho} \cdot \frac{\partial \rho}{\partial \sigma}, \tag{4.2.19b}$$

thus

$$\frac{dR}{R} = \frac{\partial l}{l} - \frac{\partial a}{a} + \frac{\partial \rho}{\rho}. \tag{4.2.19c}$$

Eqns (4.2.19) show that for a small but finite stress variation the total resistance change is due to:

(a) fractional change in length, $\Delta l / l$;   (4.2.20a)

(b) fraction change in cross-section a, or in wire diameter d, $-\Delta a / a \simeq -2\Delta d / d$;   (4.2.20b)

(c) fractional change in resistivity, $\Delta \rho / \rho$.   (4.2.20c)

With Poisson's ratio ν the lateral contraction is

$$\Delta d / d = -\nu \, \Delta l / l, \tag{4.2.21}$$

and we have with eqn (4.2.19c)

$$\frac{\Delta R}{R} = \frac{\Delta l}{l}(1 + 2\nu) + \frac{\Delta \rho}{\rho}. \tag{4.2.22}$$

Eqn (4.2.22) yields the strain sensitivity ('gauge factor')

$$K = \frac{\Delta R / R}{\Delta l / l} = 1 + 2\nu + \frac{\Delta \rho / \rho}{\Delta l / l}, \tag{4.2.23}$$

and with Poisson's ratio $\nu = 0 \cdot 3$ for most metals, eqn (4.2.23) becomes

$$K = 1 \cdot 6 + \frac{\Delta \rho / \rho}{\Delta l / l}. \tag{4.2.24}$$

It is known, from a great number of experiments, that most materials under elastic strain have gauge factors different from 1·6, which implies that a change in resistivity must be involved. In the purely plastic region of deformation of any wire material no volume change is possible ($\Delta V = 0$), as the wire cannot store energy under these conditions. For the same reason no change in resistivity can be expected ($\Delta\rho/\rho = 0$). Therefore, since Poisson's ratio in the plastic constant-volume case is $\nu = 0·5$,

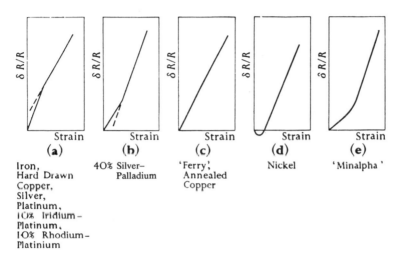

Strain	Strain	Strain	Strain	Strain
(a)	**(b)**	**(c)**	**(d)**	**(e)**
Iron, Hard Drawn Copper, Silver, Platinum, 10% Iridium – Platinum, 10% Rhodium – Platinium	40% Silver– Palladium	'Ferry', Annealed Copper	Nickel	'Minalpha '

FIG. 4.2.13. Typical strain-sensitivity curves.

the gauge factor of eqn (4.2.23) becomes approximately

$$K_{pl} = 1 + 2\times 0·5 = 2. \tag{4.2.25}$$

If some materials exhibit a gauge factor of approximately 2 also in the elastic region of deformation, this means that the volume and the resistivity changes are 'accidentally' cancelled [16]. Such a material will have an almost constant gauge factor over its elastic and plastic strain range, thus providing a wide linear stress–strain relationship. Some annealed nickel–copper alloys are of this type.

Fig. 4.2.13 shows schematically the strain sensitivity for some wire materials in an annealed as well as in a hard drawn state [17]. The 'knees' in these curves indicate the change-over points between essentially elastic and essentially plastic strain. However, not all materials follow the simple patterns of Fig. 4.2.13(a), (b), and (c). Nickel (d) has quite an unexpected change-over from negative to positive values of strain sensitivity, and Minalpha (e) exhibits a very gradual transition period. For practical uses type (c) of Fig. 4.2.13 has the most attractive characteristic, covering a

TABLE 4.2.3

Characteristics of some materials for use in metal resistance strain gauges

Alloy	Nominal composition	Gauge factor	Resistivity ($\mu\Omega$ m)	Thermal coefficient of resistance (per cent per °C)	Ultimate strength (MN m^{-2})
Constantan		2·1	0.48		
Ferry	45% nickel, 55% copper	2·2	0·45	±0·002	460
Advance		2·1	0·45		
Karma	75% nickel, 20% chromium, etc.	2·1	1·25	0·002	1000
Iso-elastic	36% nickel, 8% chromium, 0·5% molybdenum, etc.	3·6	1·05	0·0175	1250
Nichrome V	80% nickel, 20% chromium	2·5	1·0	0·01	800
Alloy No. 479	92% platinum, 8% tungsten	4·7	0·62	0·024	2000
Nickel	—	-12	0·065	0·68	400
Platinum	—	4·8	0·1	0·4	200

wide range of a practically linear stress–strain relationship. The strain sensitivity is nearly 2, as would be expected from eqn (4.2.25). A fair number of wire materials have a higher gauge factor in the elastic range (e.g. type (a)), but their application is normally limited to dynamic measurements because of higher thermal coefficients of resistance. Nickel is employed in surface thermometry because of its high ratio of thermal to strain sensitivity.

Table 4.2.3 gives a survey of the materials most commonly used in wire and foil strain gauges [15]. Ferry and Advance are copper–nickel alloys of the Constantan type. Apart from the wide range of almost uniform gauge factor they have a low thermal coefficient of resistance but should not be used at temperatures above 400 °C because of corrosion. Karma has a higher resistivity and its temperature coefficient is low at temperatures below 150 °C. With respect to corrosion it can be used up to 1000 °C. It is also somewhat stronger than Constantan. Iso-elastic wire has a higher gauge factor, but because of its high temperature coefficient it is normally employed only in dynamic strain measurements, and so is Nichrome V wire, at temperatures up to 1200 °C. The alloy No. 479 is used in unbonded strain-gauge transducers because of its high mechanical strength and strain sensitivity. Nominal values for nickel and platinum have been added for comparison.

4.2.3.1.2. Construction and bonding.   In the construction of resistance strain gauges we have to consider six particular aspects of gauges and gauge application:

(a) gauge wire material;

(b) shape of gauges, and gauge manufacture;

(c) gauge backing;

(d) cements;

(e) connecting leads;

(f) protection of gauges.

(a) Gauge-wire materials have been discussed in the previous section. From the manufacturing point of view it is desirable to employ materials which can be handled with ease as a fine wire. Materials for foil gauges must respond to the appropriate etching processes. In both cases the gauge material must weld to suitable lead material without danger of electrolytic corrosion.

The most common shapes of metal resistance strain gauges are shown in Fig. 4.2.14. The flat grid construction is normally preferred to the wrap-around type because the grid is closer to the surface of the

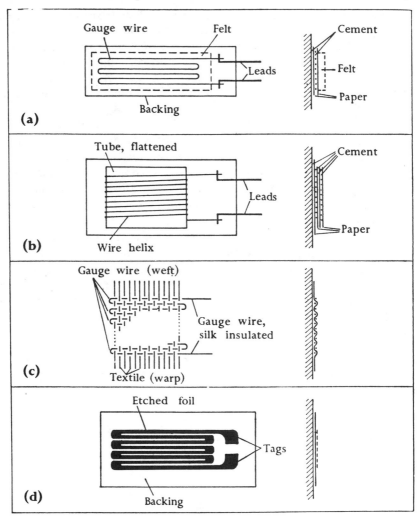

FIG. 4.2.14. Some types of resistance strain gauges. (a) flat grid; (b) wrap-around; (c) woven; (d) etched foil.

structure under strain. This results in better strain transmission and higher stability (smaller hysteresis and creep) and gives more accurate results on thin structures. The woven type has been used mainly for the measurement of high strain values in fabrics and similar application, but is of no great interest to the transducer designer. The etched foil gauge is a later development [18] with some advantages over the wire gauges. The high surface/cross-section ratio of the individual conductors gives it heat transfer properties better than those of round wire gauges. This results in a higher current-carrying capacity unless the structure to which the gauge is attached has very low thermal capacity. The photo-chemical process employed in the manufacture of these gauges permits reduction of the dimensions of a gauge pattern down to gauge lengths less than 3 mm. It also promotes mass-production of uniform gauges. An important variant of the etched foil gauge is the transferable type for use at high ambient temperatures. Here the grid is attached to a temporary backing. The grid

TABLE 4.2.4

Summary of application techniques for bonded strain gauges

Gauge backing	Adhesive	Gauge material	Remarks
Paper	Cellulose–acetone	Copper–nickel	Useful up to 60 °C; up to about 100 °C with increased drift.
Non-paper	Acrylate (pressure-sensitive)	Copper–nickel	Useful up to 80 °C. Limited life
Paper or epoxy type	Polyester or cold-setting epoxy	Copper–nickel	Useful up to 80 °C. Limited life.
Phenolic or epoxy type	Heat-setting phenolic or epoxy, respectively	Copper–nickel	Up to 180 °C; for short periods up to 250 °C Requires bonding pressure.
Glass weave or none (transfer gauge)	Ceramic cements (phosphate, oxide, glass)	Nickel chromium (various alloys)	400 °C and above. Check insulation resistance. Mainly for dynamic strains.

can be transferred to the structure which has been coated with a high-temperature ceramic cement. The backing is peeled off in the process of transfer (see also section 4.2.3.1.4 below on application of wire and foil gauges). The so-called thin-film strain gauge will be discussed in section 4.2.3.3.

(c) and (d) lead us into the centre of the much discussed problem of gauge backing and cements. The backing must provide a strong bond between structure and grid. Its choice depends therefore largely upon the cement which must suit both the material of the structure and the gauge material. Table 4.2.4 gives a brief summary of present day practice in strain-gauge application [14], [15]. Gauge bonding is a specialist's field and the reader is advised to consult a recent survey [19] of the subject which covers most application problems.

(e) Connecting leads can also cause instability of the gauge. The connection point between the thin gauge wire (or foil) and the thick leads is the most likely location of fatigue fracture. This may be brought about by stress concentrations due to change of cross-section or by material changes in the wire due to heating (soldering or welding), or by both. A 'dual-lead' arrangement is said to improve upon the hysteresis, creep, and fatigue characteristics of the gauge. The choice of a suitable lead material rests with the gauge manufacturer and depends upon resistivity, upon ability to produce reliable soldered or welded joints, and upon corrosion resistance.

(f) The main problem in the protection of strain gauges at normal temperatures is moisture. A large number of materials have been investigated and practical methods have been discussed elsewhere [14], [19]. Arranged in rough order of permissible ambient temperatures they are: wax; bitumen; natural, synthetic, and silicone rubber; thiocol polyester;

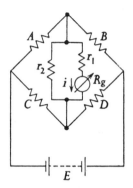

FIG. 4.2.15. General arrangement of strain-gauge bridge.

epoxy and ceramic cements. Moisture-absorbing materials such as silica gel have occasionally been enclosed with the moisture-proving can.

4.2.3.1.3. Associated circuits.   Strain gauges are almost invariably used with bridge circuits. The bridge technique, employing d.c. energization, is so well established [20] that we need discuss here only those aspects which are specifically related to strain gauges and strain-gauge transducers. In the neighbourhood of balance the bridge output current (Fig. 4.2.15) is

$$i = \frac{Er_2(AD-BC)}{r_2(R_g+r_1)(A+B)+\{ABC+ABD+ACD+BCD\}(R_g+r_1+r_2)}, \quad (4.2.26)$$

where

A, B, C, D = resistance of the bridge arms,

R_g = load (galvanometer) resistance,

r_1 = series resistance to load,

r_2 = shunt resistance to load,

E = input voltage.

The significance of the resistances r_1 and r_2 will be appreciated later
when the damping conditions of indicator galvanometers are being discussed. The output current i vanishes for $AD = BC$ (balance condition).
In strain-gauge application there are three practical possibilities for the
arrangement of active gauges in the bridge: one, two, or four active
gauges. For a convenient analysis Fig. 4.2.15 has been redrawn in Fig.
4.2.16(a)–(c). At small fractional changes in resistance ε_1, ε_2, ε_3, ε_4 in A,
B, C, D, the bridge output into any load is proportional to

$$\varepsilon = \varepsilon_1 - \varepsilon_2 - \varepsilon_3 + \varepsilon_4. \tag{4.2.27}$$

R_1 is the ratio arm resistance which, for one or two active gauges, may be

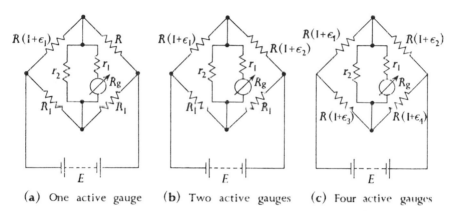

(a) One active gauge   (b) Two active gauges   (c) Four active gauges

FIG. 4.2.16. Practical bridge circuit for strain gauges.

different from the gauge resistance R. Now, eqn (4.2.26) simplifies to

$$i_{1,2} = \frac{E\varepsilon r_2}{2\{R_g(R+R_1+2r_2)+2r_1r_2+(r_1+r_2)(R+R_1)\}} \tag{4.2.28a}$$

and applies in this form to cases with one active gauge, where $\varepsilon = \varepsilon_1$, and
to cases with two active gauges, where $\varepsilon = \varepsilon_1 - \varepsilon_2$. ε_2 in eqn (4.2.27) is
here negative because of the push–pull action of the gauges as, for
instance, obtained on a bending beam. With $\varepsilon_1 = -\varepsilon_2$ we thus have
$\varepsilon = 2\varepsilon_1$. If the gauges A and D are active, we have $\varepsilon = \varepsilon_1 + \varepsilon_4$ and with
$\varepsilon_1 = \varepsilon_4$ the total effect again is $\varepsilon = 2\varepsilon_1$. A further simplification occurs for
four active gauges; this is the most important case in strain-gauge
transducer work. We have

$$i_4 = \frac{E\varepsilon r_2}{4\{R_g(R+r_2)+r_1r_2+R(r_1+r_2)\}}. \tag{4.2.28b}$$

If, as usually will be the case, the four gauges are working in two push–pull pairs, we have from eqn (4.2.27) with $\varepsilon_1 = -\varepsilon_2 = -\varepsilon_3 = \varepsilon_4$ the total

$$\varepsilon = 4\varepsilon_1. \tag{4.2.27a}$$

Torsional strain can also be measured by means of resistance strain gauges in an arrangement similar to Fig. 4.2.17, normally with two gauges in push–pull at 45° to the axis of torsion. This method [21] is based on the fact that shear in a plane normal to the axis of the shaft is accompanied by principal strains (tension and compression) in planes at 45° to the axis. Basically, one gauge would suffice, but the advantage of using two (or four) active gauges is obvious. The strain at 45° to the shaft axis can be

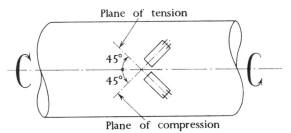

FIG. 4.2.17. Arrangement of two resistance strain gauges for the measurement of torque.

computed from (see section 4.2.3.4.1, p. 140)

$$\varepsilon_{45} = \frac{16M}{\pi d^3 E}(1+\nu), \tag{4.2.27b}$$

where $M(\mathrm{N\,m})$ is the torque moment, $d(\mathrm{m})$ is the shaft diameter, $E(\mathrm{N\,m^{-2}})$ is Young's modulus, and ν is Poisson's ratio of the shaft material. The measurement of torsional strain in rotating shafts requires the gauges to be connected via slip-rings. Here it is advantageous if the gauges form a complete bridge on the shaft, in order to avoid the slip-rings being in series with the individual strain gauges. Variations in slip-ring resistance are thus not competing directly with the resistance variations due to strain. Nevertheless, slip-ring noise may still be trouble-some, except in installations using optimum slip-ring design in combina-tion with a high degree of care and maintenance [22].

The application of strain gauges in various geometrical patterns (roset-tes) for the determination of the principal strains and stresses of two-dimensional strain fields in complex structures can only be mentioned here. To facilitate analysis simple patterns of 45°, 60°, and 90° angles between gauges have been evolved, and formulae for analysis have been worked out ([14], [15], [23], [24]), including appropriate computer pro-grams [25].

In some cases it is convenient to feed the output of a strain-gauge bridge with, say, four active gauges direct into a mirror galvanometer representing one channel of a multi-channel recorder. In order to obtain a flat response of the galvanometer over as wide a frequency range as possible the galvanometer should operate at about 70 per cent of critical damping. Galvanometer damping is mainly determined by the back-current which varies with the bridge resistance as 'seen' by the galvanometer. The optimum damping resistance R_D can be adjusted by choice of suitable series and shunt resistance r_1 and r_2, as shown in Fig. 4.2.15. Also the over-all sensitivity as recorded by the galvanometer depends upon the gauge resistance which is a design feature of the strain-gauge transducer. The problem is further complicated by the fact that in a particular strain-gauge transducer one (or more) of the following quantities may be limiting factors:

(a) the transducer current I;

(b) the power consumption of the transducer $W = I^2 R$;

(c) the available input voltage $E = IR$.

Now with r_1 and r_2 chosen to make the resistance external to the galvanometer equal to the required damping resistance value R_D, we have

$$R_D = r_1 + r_2 R/(R + r_2), \qquad (4.2.29)$$

with the following range of adjustment for r_1 and r_2,

$$\left. \begin{array}{l} r_1 = R_D - R \\[2mm] r_2 = \infty \end{array} \right\} \qquad \text{for} \quad R < R_D, \qquad (4.2.30)$$

or

$$\left. \begin{array}{l} r_1 = 0 \\[2mm] r_2 = RR_D/(R - R_D) \end{array} \right\} \qquad \text{for} \quad R > R_D. \qquad (4.2.31)$$

(a) *The transducer current is limited.* From eqn (4.2.28b) and Fig. 4.2.16(c) with the gauge current limited to I the output current is

$$i = IR\varepsilon/4(R + R_g), \qquad (4.2.32a)$$

if the galvanometer is used directly without additional damping ($r_1 = 0$, $r_2 = \infty$). With adjustment of damping

$$\left. \begin{array}{ll} i = IR\varepsilon/4(R_D + R_g) & \text{for} \quad R < R_D, \\[2mm] i = IR_D\varepsilon/4(R_D + R_g) & \text{for} \quad R > R_D. \end{array} \right\} \qquad (4.2.33a)$$

(b) *The power consumption of the transducer is limited.* Here $W = I^2 R$, hence without damping adjustment,

$$i = W^{\frac{1}{2}} R^{\frac{1}{2}} \varepsilon/4(R + R_g), \qquad (4.2.32b)$$

and with damping adjustment,

$$i = W^{\frac{1}{2}}R^{\frac{1}{2}}\varepsilon/4(R_D + R_g) \qquad \text{for} \quad R < R_D,$$
$$i = W^{\frac{1}{2}}R_D\varepsilon/4R^{\frac{1}{2}}(R_D + R_g) \quad \text{for} \quad R > R_D. \qquad (4.2.33b)$$

(c) *The available voltage is limited.* With $E = IR$, here we have without damping adjustment,

$$i = E\varepsilon/4(R + R_g), \qquad (4.2.32c)$$

and with damping adjustment,

$$i = E\varepsilon/4(R_D + R_g) \qquad \text{for} \quad R < R_D,$$
$$i = E\varepsilon R_D/4R(R_D + R_g) \quad \text{for} \quad R > R_D. \qquad (4.2.33c)$$

If, however, the strain-gauge bridge feeds into a high impedance load, say, an amplifier input circuit, then from Fig. 4.2.16(a) and eqn (4.2.28a) we have with $r_1 = 0$ and $r_2 = \infty$,

$$i = E\varepsilon/2(2R_g + R + R_1), \qquad (4.2.34)$$

or the voltage across resistance R_g,

$$e = iR_g = \frac{E\varepsilon}{2\{2 + (R + R_1)/R_g\}}. \qquad (4.2.35)$$

At very high load resistances $R_g \rightarrow \infty$ and the output voltage is thus

$$e_{R_g \rightarrow \infty} = \tfrac{1}{4}E\varepsilon \qquad (4.2.36)$$

with $\varepsilon = \varepsilon_1 - \varepsilon_2 - \varepsilon_3 + \varepsilon_4$.

Temperature compensation in strain-gauge bridge circuits is in the first instance concerned with 'zero-stability,' i.e. the error produced by variation of ambient temperatures irrespective of actual strain. Apart from this error there may be a temperature error in sensitivity, i.e. in the slope of the calibration curve, which is mainly due to a thermal change of the Young's modulus of the strained structure.

We shall here, however, consider only the task of compensating for zero-instabilities; sensitivity compensation can be obtained by inserting a temperature-sensitive resistor into either the bridge-supply line or into the load circuit.

In order to eliminate the resistance variation at variable ambient temperatures consider a strain gauge of resistance R and with a thermal coefficient of resistance, α_1 in the unstrained state. Its fractional resistance change is

$$\Delta R/R = \alpha_1 \Delta t, \qquad (4.2.37)$$

where Δt is the temperature variation $(t - t_0)$ about the reference temperature t_0. With a thermal coefficient of linear expansion α_2, its change in length over the same temperature interval is

$$\Delta l/l = \alpha_2 \, \Delta t. \tag{4.2.38}$$

If the gauge is bonded to a structure, which almost invariably is much more substantial than the gauge, the structure takes charge, altering the length of the gauge grid with temperature to the same extent as would a strain of the same magnitude

$$\Delta l/l = (\alpha_3 - \alpha_2) \, \Delta t, \tag{4.2.39}$$

α_3 being the thermal coefficient of linear expansion of the structure. The resultant fractional change in gauge resistance due to this 'apparent' strain is thus

$$\Delta R/R = K \, \Delta l/l = K(\alpha_3 - \alpha_2) \, \Delta t, \tag{4.2.40}$$

where K is the gauge factor $(\Delta R/R)/(\Delta l/l)$ and the total fractional change in resistance becomes eventually

$$\Delta R/R = \{\alpha_1 + K(\alpha_3 - \alpha_2)\} \, \Delta t = \alpha \, \Delta t, \tag{4.2.41}$$

or the effective temperature coefficient of the bonded strain gauge is

$$\alpha = \alpha_1 + K(\alpha_3 - \alpha_2). \tag{4.2.42}$$

Approximate values of thermal coefficients of linear expansion for a number of materials common in transducer work are given in Table 4.2.5. Together with Table 4.2.3 for gauge factor and temperature coefficient of resistance the reader will be able to evaluate eqn (4.2.42) by choice of various combinations of gauge and structural materials with a view to minimizing the temperature error. Unfortunately, it will be seen that in most cases this method is ineffective in strain-gauge work, because the useful resistance variations due to the strain to be measured will be swamped to the same degree as the apparent strain due to temperature variation. In certain applications, however, the expression $(\alpha_3 - \alpha_2)$ in eqn (4.2.42) can become negative (e.g. for copper–nickel gauges on steel we have $(11 - 14 \cdot 4) \times 10^{-6} = -3 \cdot 4 \times 10^{-6}$), and the effective temperature coefficient α assumes a small positive value, which in some temperature regions may even turn negative. There are commercially available strain gauges which have been corrected for use on certain structures, such as mild steel, aluminium, titanium, and quartz. They are essentially a series arrangement of copper–nickel and nickel wires or foils on a suitable backing. Since nickel has a negative gauge factor, a positive temperature coefficient α_1 can be compensated according to $\alpha_1 - K(\alpha_3 - \alpha_2) = 0$. Over a certain temperature range their thermal stability is better than that of the uncompensated gauge but not as good as that

TABLE 4.2.5
*Thermal coefficient of linear expansion per °C
of some materials at room temperature*

Material	Thermal coefficient of linear expansion $\times 10^6$	
Advance	14·4	
Aluminium	25·5	
Aluminium–bronze	17·0	
Brass	18·9	
Constantan	17·0	
Copper	16·7	
Duralumin	22·6	
Ferry	12·5	
Invar	0·9	
Iso-elastic	4·0	
Karma	10·0	
Nichrome V	13·2	
Nickel	12·8	
Platinum	8·9	
Steel (carbon)	11 ⎫	
Steel (stainless)	11 ⎬ approximately	
Titanium	9 ⎭	

obtained with compensating or dummy gauges in a well-designed installation. In contrast to wire and foil strain gauges certain types of semiconductor gauges have large negative gauge factors, and this method will become a major artifice in the temperature compensation of semiconductor strain-gauge bridge circuits (see section 4.2.3.2).

Resistance changes with temperature in the leads of strain-gauge installations are compensated if active push–pull gauges or dummies in close proximity are used, as shown in Fig. 4.2.18(a). Uncompensated single gauges or temperature-compensated single gauges, which would need two leads only, may suffer appreciable resistance variations in the non-symmetrical leads (Fig. 4.2.18(b)). A better scheme is shown in Fig. 4.2.18(c), featuring a three-lead arrangement. Two lead resistances are in adjacent bridge arms and thus compensated. Resistance changes in the third, the bridge-supply lead, does not affect the bridge balance. As an alternative an extended galvanometer lead can be substituted for a battery lead (Fig. 4.2.18(d)).

Besides d.c. energized strain-gauge bridges, a.c. operated installations have been winning ground in recent years. In fact, most commercial strain-measuring equipment is now of the a.c. carrier type. The main reasons are freedom from contact potential and thermal voltage errors bedevilling d.c. systems, but, perhaps even more important, a.c. amplifiers

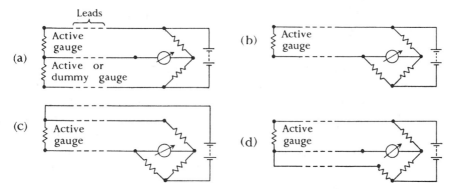

FIG. 4.2.18. Lead arrangements for strain-gauge bridges.

are simpler, cheaper, and generally more reliable than their d.c. counter-
parts of similar performance, and a.c. bridges are not much more complex
than d.c. bridges. The a.c. 'resistor' representing the strain gauge comprises
reactive components (see section 4.2.1), and a.c. bridges strictly require
balancing with respect to both the in-phase and the quadrature component. A.c. bridge techniques will be discussed in greater detail in sections
4.3 and 4.4 on inductance and capacitance transducers, respectively.

4.2.3.1.1. Application of wire and foil strain gauges. Reference [26]
offers a fairly recent survey of resistance strain gauges with particular
emphasis on wire and foil types. It should be read in conjunction with
reference [19] on modern strain-gauge adhesives and gauge protection
materials. Apart from ordinary general-purpose strain gauges, the user
may be interested in gauges for specific and rather exacting applications.

(a) *High-stability strain gauges.* Long-term stability of mounted gauges,
i.e. freedom from drift over long periods (months, or even years), remains
a most difficult problem. Such requirements occur for multiple gauge
installations in aerospace research (aircraft structures, wind-tunnel balances, etc.), but also in industrial applications, such as pressure transducers, load cells, and dynamometers. Copper–nickel wire and foil gauges on
phenolic and epoxy bases—and attached with the greatest care by experienced operators—are in common use.

(b) *High- and low-temperature strain gauges.* At environmental temperatures well above 200 °C nickel–chromium gauges, often in combination
with ceramic cements, are suitable for dynamic strain measurements up to
800 °C, and for static measurements up to about 400 °C. Platinum–
tungsten dual gauges (with a platinum temperature sensor incorporated)
have been used up to 500 °C with acceptable accuracy. These gauges also
come in stainless-steel sheaths filled with a ceramic medium; they are

normally welded to the test structure. Dual gauges have also been employed in flight trials at extremely low environmental temperatures (−250 °C).

(c) *High-elongation strain gauges.* So-called post-yield strain gauges, consisting of well-annealed copper–nickel wire or foil, have been used to measure strains up to 10 per cent, with fair accuracies. Because of work-hardening of the gauge material, post-yield gauges can normally be used only once. Their calibration curves are non-linear with strain.

(d) *Fatigue strain gauges.* The normally unwanted cumulative hysteresis in cyclic strain tests of copper–nickel wire and foil gauges can be exploited for the measurement of fatigue. The sum-total of the cyclic zero-shifts, then, is a (non-linear) measure of fatigue, since it represents the product of energy loss (i.e. area of hysteresis loop) and number of cycles.

These specialized gauges described above are commercially available from a number of suppliers in this country and abroad. Application of precision wire and foil gauges, apart from strain-gauging proper, is mainly in instrument transducers for the measurement of physical quantities other than strain, which will be discussed in section 4.2.3.4.

4.2.3.2. Semiconductor strain gauges

The search for strain gauges with higher gauge factors led, in the mid-1950s, to the development of the semiconductor strain gauge ([27], [28]). Gauge factors of at least 50 times higher than those of wire and foil gauges were obtained with thin strips cut from single-crystal silicon and germanium, though it soon became clear that linearity and temperature stability were distinctly inferior to those of conventional gauges; in fact, they required sophisticated compensation artifices in order to make them acceptable as predictable strain sensors.

The literature on semiconductor strain gauges is vast and may give the unsuspecting reader the impression that they have replaced conventional types in all practical strain-gauging. However, from a balanced assessment of their attractions and shortcomings ([29], [30]) it emerges that, though they offer a welcome solution to many measuring problems which are not accessible to conventional gauges, semiconductor strain gauges play rather a supplementary role in the wide and varied field of strain-gauge applications, including their use as sensing elements in instrument transducers, but they have also opened up new fields of strain sensing by permitting the diffusion of suitable impurities into selected small areas of monolithic pieces of silicon, thus producing strain-sensitive cantilevers and diaphragms ([31], [32]). These strain sensors may even be integrated with diffused electronic components providing signal modification [33].

4.2.3.2.1. Sensitivity ([30], [34]–[36]). The high strain sensitivity of semiconductor strain gauges can, in essence, be reduced to anisotropic variations of the carrier mobility in strips of silicon and germanium, the most common materials in use. Briefly, a valence electron can be excited into the conduction band. In an electric field it participates in the conduction of electric current; the vacancy ('hole') travels like a positive charge. However, the excitation energy of about 1 eV is relatively high and the resistivity of these 'intrinsic' semiconductors therefore is also high.

The situation changes dramatically if pure silicon or germanium, belonging to the fourth group of the periodic system, are 'doped' with minute quantities of atoms of the third or fifth group.

(a) *n-conduction.* Atoms of the fifth group (e.g. phosphorus) produce four perfect bonds with the silicon atom. The fifth is only weakly attached, and an energy of about 0·05 eV suffices to excite it into the conduction band. This process by *negative* electrons is known as n-conduction.

(b) *p-conduction.* With an atom of the third group (e.g. boron) there are three perfect bonds. An energy of only 0·08 eV will fill the hole at the fourth bond. Conduction by *positive* vacancies, or holes, is known as p-conduction.

The resistivity of both the n- and p-conduction processes is lower than that of intrinsic conduction in pure crystals because of the smaller energies required to initiate the processes. Generally, the thermal coefficient of resistivity of doped semiconductor material decreases with increasing degrees of doping.

Similar to eqn (4.2.23) the gauge factor of semiconductor gauges can be written

$$K = \frac{\Delta R/R}{\Delta l/l} = 1 + 2\nu + m. \qquad (4.2.43)$$

Here $m = (\Delta \rho/\rho)/(\Delta l/l)$ is the dominant term and can be computed from

$$m = \pi E, \qquad (4.2.44)$$

where $\pi \, (\text{m}^2 \, \text{N}^{-1})$ is the appropriate piezoresistive coefficient and $E (\text{N m}^{-2})$ is the Young's modulus, both depending on material, crystal orientation, and relative position of the directions of current and stress (e.g. tension, shear, or hydraulic compression). Table 4.2.6 provides a brief survey of the relevant characteristics of lightly doped silicon and germanium at low levels of strain and constant room temperature. Generally the gauge factor of doped semiconductors decreases with increasing degrees of doping, and so does its thermal coefficient.

TABLE 4.2.6

Average strain characteristics of lightly doped silicon and germanium at low strain levels and constant room temperature

	Units	Factor	p-Si	n-Si	p-Ge	n-Ge
Crystal orientation	—	—	(111)	(100)	(111)	(111)
Young's modulus	N m^{-2}	10^9	187	130	155	155
Poisson's ratio	—	—	0·180	0·278	0·156	0·156
Unstrained resistivity	Ω m	10^{-3}	78	117	150	166
π_{11}	m^2 N^{-1}	10^{-12}	+66	−1022	−106	−52
π_{12}	m^2 N^{-1}	10^{-12}	−11	+534	+50	+55
π_{44}	m^2 N^{-1}	10^{-12}	+1381	−136	+986	−1387
m_1	—	—	+175	−133	+102	−157

In practical semiconductor strain gauges both the stress and the current are along the length of the gauge. Then, the longitudinal piezoresistance coefficient can be shown to be

$$\pi_1 = \pi_{11} - 2(\pi_{11} - \pi_{12} - \pi_{44})(r_1^2 s_1^2 + r_1^2 t_1^2 + s_1^2 t_1^2), \qquad (4.2.45)$$

where π_{11}, π_{12}, π_{44} are the fundamental piezoresistive coefficients given in Table 4.2.6, and r_1, s_1, t_1 the direction cosines of the current with respect to the crystallographic axes. Maximum values of π_1 occur for crystal orientations as indicated in Table 4.2.6. The m_1 values listed there are valid for longitudinal stress and have been computed according to eqns (4.2.44) and (4.2.45). For shear stress normal to the current direction the transverse piezoresistance coefficient would take the form [37]

$$\pi_t = \pi_{12} + (\pi_{11} - \pi_{12} - \pi_{44})(r_1^2 r_2^2 + s_1^2 s_2^2 + t_1^2 t_2^2), \qquad (4.2.46)$$

where the subscripts refer to the current and stress directions, respectively. A more detailed discussion of (longitudinal) gauge factor as a function of crystal orientation and resistivity (degree of doping) can be found in reference [38].

From the above it is seen that a semiconductor gauge filament requires description with respect to:

(a) basic material (now almost invariably silicon); (b) conduction process (p- or n-carriers); (c) crystal orientation according to Miller index ((111) or (100)); (d) room-temperature resistivity ρ_0 (indicating degree of doping); (e) shape and size.

For practical application we also need:

(f) gauge factor (positive or negative); (g) gauge resistance; (h) gauge length; (i) temperature compensation (if any); (j) backing or encapsulation (if any); (k) bonding (cementing or welding); (l) lead geometry.

Fig. 4.2.19 shows the most common shapes. Their manufacture cannot be discussed here [34], nor can we go into details of bonding techniques. The reader is advised to consult reference [19] and follow the procedure suggested by the suppliers. An inorganic bonding method for semiconductor strain gauges has been described in reference [39].

4.2.3.2.2. Linearity. In contrast to wire and foil gauges, semiconductor strain gauges are distinctly non-linear. The fractional resistance variation with strain can be written generally

$$\Delta R/R = C_1\varepsilon + C_2\varepsilon^2 + C_3\varepsilon^3 + ... \quad . \qquad (4.2.47)$$

For instance, the lightly doped p-type silicon of Table 4.2.6, at constant

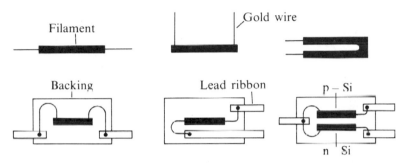

FIG. 4.2.19. Some types of semiconductor strain gauges.

room temperature, follows the law [36]

$$\Delta R/R = +175\varepsilon + 72\,625\varepsilon^2. \qquad (4.2.48)$$

This relationship applies only to moderate levels of tensile stress. At compressional strains greater than, say 0.5×10^{-3} m per m, the non-linear term may be much higher and also temperature dependent [40]. When using pre-stressed gauges in order to eliminate this disability the user should note that R in $\Delta R/R$ now differs from the unstrained resistance. Related errors may also occur in bonded gauges experiencing shrinkage during the curing process.

Non-linearity can be improved appreciably by using heavily doped material of lower resistivity. A p-silicon gauge with a resistivity of about 0.2×10^{-3} Ω m shows a much improved linearity, namely,

$$\Delta R/R = +119.5\varepsilon + 4000\varepsilon^2, \qquad (4.2.49)$$

though the sensitivity is appreciably lower. For comparison, an n-type silicon gauge of resistivity 0.31×10^{-3} Ω m would give

$$\Delta R/R = -110\varepsilon + 10\,000\varepsilon^2. \qquad (4.2.50)$$

Note that there is near-mirror symmetry in the more linear regions of eqns (4.2.49) and (4.2.50).

4.2.3.2.3. Temperature effects. Environmental temperature changes affect both the resistance and the gauge factor of semiconductor strain gauges to an extent which depends in a particular material on the level of doping [41]. Temperature instability and non-linearity are the two major shortcomings of semiconductor strain gauges. Fig. 4.2.20(a) shows

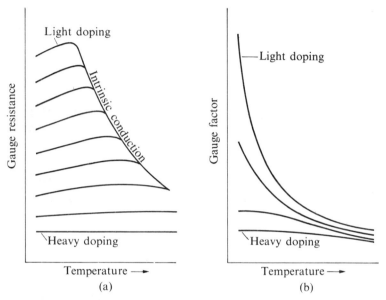

FIG. 4.2.20. Variation with temperature of (a) gauge resistance and (b) gauge factor of p-type silicon at various degrees of doping.

schematically the response of gauge resistance to temperature variations. Lightly doped silicon has high resistivity and a positive thermal coefficient of resistance up to about 100 °C. Above this temperature the material reverts to an intrinsic conduction mode with a high negative temperature coefficient. Heavily doped gauges have both low resistivity and thermal coefficient of resistance over a wide temperature range.

The variation of gauge factor with temperature is shown in Fig. 4.2.20(b). Here again heavy doping improves temperature stability, at the expense of strain sensitivity.

Resistance variation in bonded gauges due to different values of thermal expansion of gauge material and of substructure ('apparent strain') has been discussed in section 4.2.3.1.3 in connection with metal gauges. This effect has been exploited for self-compensation of n-type silicon gauges by making use of their negative gauge factor [42].

Gauges matched to a variety of substrate materials are commercially available. Further compensation techniques will be discussed in the section to follow.

4.2.3.2.4. Associated circuits [43]. Since the operational resistance variation of semiconductor strain gauges is large, the non-linearities of bridge circuits comprising these gauges cannot be ignored. With one active strain gauge of initial resistance R connected in a bridge circuit with three similar resistances the output–input voltage ratio is

$$\frac{V_0}{V_i} = \frac{\Delta R}{4R}\left(1 - \frac{\Delta R}{2R} + \dots - \dots\right). \qquad (4.2.51)$$

On the other hand, it was shown in section 4.2.3.2.2 that the non-linear calibration curve of p-type silicon gauges follows the general pattern

$$\Delta R/R = \varepsilon(C_1 + C_2\varepsilon + \dots). \qquad (4.2.52)$$

Therefore, since the two non-linear terms in eqns (4.2.51) and (4.2.52) have opposite signs, compensation can be achieved by making the two ratio arms either smaller or larger than R, thus matching the bridge non-linearity to that of the strain gauge.

Two (or four) semiconductor strain gauges operating in push–pull fashion (tension and compression, respectively†) result in fairly good, though not perfect, linearity. If compressional strain is not available, or too high, the side-by-side use of p- and n-type semiconductor strain gauges with gauge factors of similar magnitude but opposite sign are recommended. Finally, since the linearity of p-type silicon gauges improves with increasing tension, and that of n-type gauges with increasing compression, linearity is improved if the two matched gauges are mounted with the appropriate strain biasses.

Zero-shift due to temperature variation in single-gauge installations is reduced by employing 'self-compensated' n-type silicon gauges matched to their substrate [42], as explained in section 4.2.3.2.3 above, or by inserting short lengths of a wire with a high thermal coefficient of resistance into the appropriate bridge arms, though the use of push–pull pairs of gauges is generally superior.

The temperature coefficient of the gauge factor of semiconductor strain gauges is negative (Fig. 4.2.20(b)). But since a large part of this loss in sensitivity at higher environmental temperatures is caused by the positive temperature coefficient of resistance (Fig. 4.2.20(a)), resulting in an increased gauge resistance, the gauge factor variation can be reduced by operating the bridge at constant current rather than at constant voltage.

† But avoid compression strains in excess of $0 \cdot 5 \times 10^{-3}$ m per m (see section 4.2.3.2.2 on gauge linearity).

This is now common practice. Either constant-current bridge supplies are used, or (temperature-sensitive) resistors are inserted in the supply line of a conventional constant-voltage supply.

Obviously, the accuracy of a strain-measuring installation using semiconductor strain gauges depends on the often delicate balance of individually non-linear and temperature-sensitive gauge and circuit characteristics. An attempt to 'zero' such a bridge, or to eliminate the output from an irrelevant pre-load, by introducing extra resistance, may easily spoil the compensation. In this case bridge balance can be obtained by using a second bridge in parallel with conventional balancing elements and giving an opposing output voltage which just cancels the zero-signal output of the strain-gauge bridge.

4.2.3.2.5. Application of semiconductor strain gauges.   The major advantages of semiconductor strain gauges, as compared with metal gauges, are their vastly higher gauge factor and their smaller size, the latter particularly in diffused strain sensors for the measurement of force and pressure. Their disadvantages are non-linearity and temperature instability, both requiring sophisticated compensation techniques.

A detailed comparison of the mechanical and electrical properties also shows [44] that the range of elastic strain and fatigue life are high in single-crystal silicon gauges, while in conventional gauges they are limited by plastic flow and hysteresis in the metal. The smallest permissible bending radius of semiconductor gauges depends on their breaking strength; standard gauges cut from solid crystals may be bent to a 70–80 mm radius, filaments made from very thin 'whiskers' permit bending radii of 3–5 mm. The latter figure indicates that their attachment need not be more difficult than that of wire and foil gauges. The current-carrying capacity and the noise generation do not substantially differ from those of conventional gauges, but semiconductor material is photosensitive and gauges must be protected from strong fluctuating light.

Although the inherent non-linearity and temperature instability of semiconductor strain gauges can be reduced—at the expense of strain sensitivity—by choice of heavily doped materials, the extra care that semiconductor gauges need is easier provided in strain sensors for use in transducers than in multiple-gauge installations for stress-analysis work, and it is therefore not surprising that the former kind of application predominates (see section 4.2.3.4 on strain-gauge transducers). As to future trends in strain-gauge development generally, the field of semiconductors is still wide open, while conventional gauges have perhaps reached their level of perfection.

4.2.3.3. Thin-film strain gauges [16]

The thin-film strain gauge was conceived as early as 1951 [45]. Laboratory work on the strain sensitivity of vacuum-deposited thin metal films ([46]–[48]) established the basic characteristics shown in Fig. 4.2.21 which are well supported by theoretical work ([49], [50]). Relatively thick gold films deposited on glass slides have gauge factors somewhat below that of the bulk metal, while gauge factors of thinner (discontinuous) films rise to values of 100; they would thus be comparable with those of semiconductors. In fact, discontinuous films involve conduction through the 'insulating' substrate; they have negative thermal coefficients of resistivity like intrinsic semiconductors.

However, for practical use in strain-gauging the discontinuous films of pure metals in particular exhibit unpredictable variations of resistance

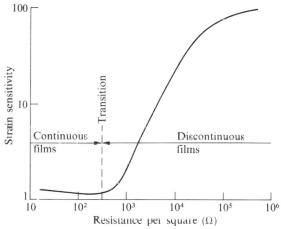

FIG. 4.2.21. Experimental strain sensitivity of gold film on glass.

with temperature and time. Useful work on thin-film strain gauges is restricted, at least at present, to continuous films of pure metals and some alloys with conventional gauge factors; they are deposited mainly by sputtering techniques [51]. A pressure transducer with (continuous) film sensors of an (unspecified) alloy [52] will be discussed in section 4.2.3.4 on strain-gauge transducers.

Applications of this kind first require a thin insulation layer to be provided on the metal substrate under stress, followed by depositing the thin-film gauge pattern, and finished by a protective layer. Because of the close contact between the thin-film sensor and the pressure-sensitive diaphragm, creep is virtually eliminated and heat transfer is improved, thus permitting higher current densities, i.e. higher outputs, than with cemented gauges. The use of thin-film semiconductor gauges has also been investigated, apparently with promising results [53].

4.2.3.4. *Strain-gauge transducers*

Soon after wire resistance strain gauges had been established as useful tools for the measurement of local strains and strain distributions on loaded structures, these gauges were also employed as strain sensors in transducers for the measurement of force, torque, acceleration, and pressure. Bonded wire gauges were later joined by unbonded types, by foil gauges and, in more recent times, by semiconductor strain sensors, and even by vacuum-deposited thin-film gauges.

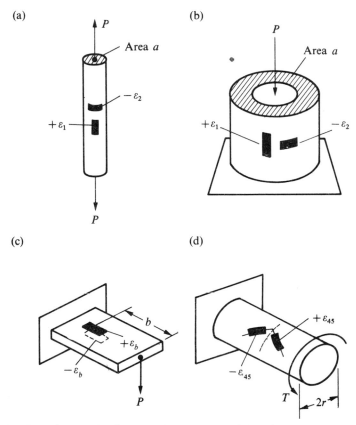

FIG. 4.2.22. Some force and torque transducers with bonded strain-gauge sensors, schematic diagram: (a) tension rod; (b) compression column; (c) bending cantilever; (d) torque shaft.

4.2.3.4.1. Force and torque transducers.   Force-measuring devices are generally known as dynamometers. Load cells measure weight [54]. Both the slender rod under tensile stress (Fig. 4.2.22(a)) and the squat cylinder under compressional stress (Fig. 4.2.22(b)) may carry one or more axially arranged gauges, and a similar number of circumferentially arranged gauges, the latter picking up (a somewhat lower) lateral strain of opposite sign, thus constituting push–pull pairs. Gauges should be distributed

round the circumference so that non-axial load components are compensated; tubes are preferred to solid cylinders for better stability.

The strain at any angle α to the axis is

$$\varepsilon_\alpha = \frac{\varepsilon_1}{2}\{(1-\nu)+(1+\nu)\cos 2\alpha\}, \tag{4.2.53}$$

where ε_1 is the principal strain along the axis and ν is Poisson's ratio. The
axial gauge therefore senses a strain of

$$\varepsilon_1 = \frac{P}{aE} \quad \text{for} \quad \alpha = 0, \tag{4.2.54 a}$$

and the circumferential gauge a strain of

$$\varepsilon_2 = \frac{P}{aE}(1-\nu) \quad \text{for} \quad \alpha = 90°, \tag{4.2.54 b}$$

where $P(N)$ is the force (load), $E(N\,m^{-2})$ is the Young's modulus, and a
(m^2) is the cross-sectional area.

A more efficient strain generator is the bending beam (cantilever) of
Fig. 4.2.22(c). The strain occurring at the gauge centre (distance b) is

$$\varepsilon_b = \frac{6Pb}{Ewt^2}, \tag{4.2.55}$$

where $w(m)$ and $t(m)$ are the beam width and beam thickness, respectively.

The basic arrangement for the measurement of torque is shown in Fig.
4.2.22(d). The strain experienced by a strain gauge mounted at an angle β
to the shaft axis is

$$\varepsilon_\beta = \frac{T}{\pi r^3 G}\sin 2\beta, \tag{4.2.56}$$

where $T(N\,m)$ is the torque, $r(m)$ the shaft radius, and $G(N\,m^{-2})$ the
modulus of rigidity. With $G = \frac{1}{2}E/(1+\nu)$ and $\beta = 45°$, we have

$$\varepsilon_{45} = \frac{2T}{\pi r^3 E}(1+\nu). \tag{4.2.57}$$

Both the bending beam and the torque shaft may carry (pairs of)
push–pull gauges.

Fig. 4.2.23 shows the elementary form of a ring dynamometer. If the
ring thickness t (m) is small compared with the radius r (m) the strain at
the indicated gauge locations can be estimated according to

$$\varepsilon = \frac{1\cdot08Pr}{Ewt^2} \quad \text{for} \quad t \ll r, \tag{4.2.58}$$

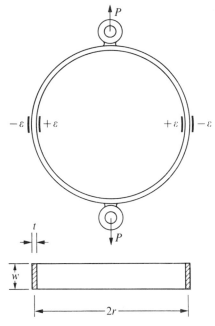

FIG. 4.2.23. Dynamometer ring.

though the stiffening effect of the loading hooks may reduce this value
appreciably. An accurate analysis is rather complex [55], particularly for
thick rings [56], and careful calibration will be essential in all cases.

Dynamometers and load cells with stressed elements producing tensile,
shear, compressional, or bending strains have been designed in a great
variety of shapes and sizes. They are highly developed devices with errors
less than 0·1 per cent of full-scale reading.

Up to this point we have considered strain measurements related to
individual forces and moments. There are, however, problems in modern
engineering where the simultaneous measurement of three orthogonal
forces and three similar moments, associated with a solid body suspended
in space, are necessary. This applies, for instance, to so-called sting
balances for the measurement of aerodynamic forces and moments acting
on models in wind-tunnels ([57]–[59]), as shown in Fig. 4.2.24(a). The
gauge locations are indicated in the side and plan views of Fig. 4.2.24(b),
relating to a so-called two-cage sting balance [60], while Fig. 4.2.24(c)
gives displacement patterns and bridge connections for the simultaneous
measurement of normal (N) and side (S) forces, of pitch (P), yaw (Y), and
roll (R) moments, and of axial forces (A). This last force represents
drag and is (it is hoped) small compared with the normal (lift) forces; it is
therefore difficult to measure accurately.

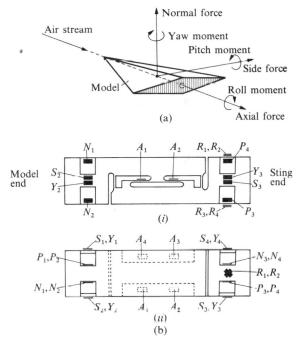

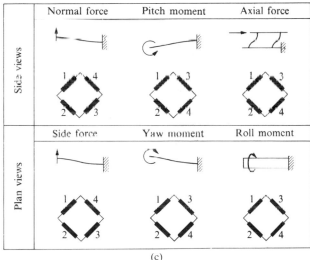

Fig. 4.2.24 (a) Aerodynamic forces and moments operating on wind-tunnel model. (b) (*i*) Side and (*ii*) plan view of a six-component strain-gauge sting balance. Gauge locations. (c) Displacement patterns and related bridge connections of a six-component strain-gauge sting balance (Royal Aircraft Establishment, Farnborough).

The body of the sting balance is machined from a solid piece of high-tensile stainless steel. In order to reduce interaction between forces and moments in a practical design the top and bottom beams must be left as strong as possible. The single flexures of Fig. 4.2.24(b) are usually replaced by double or triple flexures in order to improve axial sensitivity and de-coupling [61]. An alternative balance design is described in reference [62]. The use of semiconductor strain gauges for the measurement of the axial forces have also been tried with some degree of success [63].

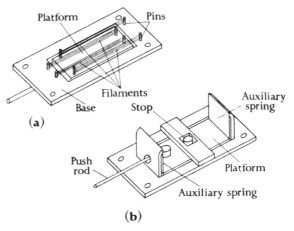

FIG. 4.2.25. Fine-wire extensiometer (unbonded strain-gauge-type transducer). General arrangement of early commercial type (Statham Laboratories, Oxnard, California). (a) Front view, (b) reverse view.

In contrast to bonded strain gauges the so-called unbonded strain gauge is, in fact, an extensiometer. An early version [64] is shown in Fig. 4.2.25. There may be up to twelve loops of high-tensile resistance wire (e.g. Alloy No. 479—see Table 4.2.3, p. 120) of a diameter as small as 0·001 mm wound between insulating pins let into both a movable platform and a stationary frame. A longitudinal platform movement results in push–pull resistance changes in the loops which make up a complete bridge circuit. When used as an extensiometer its range is set by stops to a maximum strain of ±0·15 per cent; the active loop length is 25 mm. When used as a dynamometer, the force range depends on wire material, wire diameter, and number of loop turns. In all cases the stiffness of the auxiliary guide springs should be small compared with the stiffness of the wire loops.

Another unbonded-wire force sensor employs a cross-shaped beam system which is welded to a solid outside ring of about 14 mm diameter [65]. Fine wire is wound over sapphire pins located at the points of inflection of the bending beams, i.e. at maximum angular displacements. Two wire loops on either side of the cross-shaped beam provide push–pull output. There are still other versions of unbonded strain sensing devices, all operating to the same basic principle and displaying similar characteristics [15].

Generally, the unbonded strain sensor provides a more definite anchorage of the gauge wires than bonded gauges, thus avoiding slip and eliminating faulty installation by inexperienced operators. On the other hand, danger of slack wires dictates low current densities, particularly since heat dissipation from freely suspended wires is worse than from bonded gauges. For the same reason transient errors at heat impact are usually larger.

Finally, we may mention that small cantilevers ($10\ \text{mm} \times 1\cdot 5\ \text{mm} \times 0\cdot 5\ \text{mm}$) are commercially available which measure loads of up to $40\ \text{g}$ force by means of an 'unbounded' semiconductor strain gauge mounted across a transverse slit in the cantilever [66].

4.2.3.4.2. Acceleration transducers. Accelerometers are force-measuring devices. According to Newton's second law of dynamics the force is

$$F = am \quad (\text{N}),\dagger \qquad (4.2.59)$$

where $a(\text{m s}^{-2})$ is the acceleration acting on the mass m (kg). If acceleration is quoted in multiples n of the acceleration due to gravity ($g = 9\cdot 81\ \text{m s}^{-2}$) then

$$F = ngm. \qquad (4.2.60)$$

An acceleration transducer consists of a mass supported by a compliant force-indicating element. In the present context the deformation of the latter is sensed by metal or semiconductor strain gauges. The design and material properties of springs have been treated in section 3.3.2 and the dynamic performance of vibratory systems has been discussed in section 3.1.2. Briefly, acceleration transducers have a 'flat' frequency response from zero frequency up to a cut-off frequency below its resonant frequency which depends on the magnitudes of mass, compliance, and damping. A high resonant frequency is generally desirable. The optimum frequency range occurs at about 70 per cent of critical damping, which is normally obtained by immersing the moving parts in (silicone) oil of suitable viscosity. However, these conditions do not necessarily lead to an 'optimum' transient response which would refer to minimum overshoot and/or time-delay values and which also depends greatly on the shape of the forcing function (see Table 3.1 and Fig. 3.8(a)–(f), p. 34.

† The newton (N) is, in fact, defined as the force required to accelerate a mass of 1 kg by 1 metre per second per second.

The so-called mushroom-type accelerometer would consist of a mass, carried by a (solid or hollow) strain-gauged cylinder, shown in Fig. 4.2.22(b). Because of its great stiffness this design is unsuitable for the measurement of low values of acceleration by strain-gauge sensors and invariably leads to large and heavy instruments.

A more congenial strain generator is the cantilever which provides the additional advantage of offering positive and negative strains of equal magnitude. The elementary form of this type consists of a mass, attached to the free end of a strain-gauged cantilever shown in Fig. 4.2.22(c). In practical transducers double cantilevers (operating in contre-flexure) are

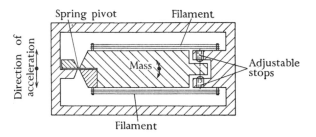

FIG. 4.2.26. Pivoted linear acceleration transducer with unbonded strain gauges, schematic diagram (Statham Laboratories, Oxnard, California.)

often preferred to single cantilevers in order to reduce transverse sensitivity. The effect of transferse forces on tip-deflection and root-strain in single and double cantilevers has been treated elsewhere [67].

A typical linear acceleration transducer using unbonded strain-gauge sensors is shown in Fig. 4.2.26. The seismic mass is pivoted by means of a spring hinge and restrained by fine-wire loops above and below. This design has a 'built-in' leverage resulting in rather short transducers in the direction of applied acceleration. They are, however, inherently sensitive to angular acceleration.

For the measurement of angular acceleration proper the arrangement of Fig. 4.2.27 has been devised. It has been shown in section 3.3.1 that under certain conditions the dynamic characteristics of the damping fluid contributes appreciably to the effective mass of an acceleration transducer. This normally undesirable effect has here been employed with advantage [68]. The angular accelerometer consists of a hollow paddle, pivoted at the centre and restrained by strain-sensitive filaments. The paddle is surrounded by a baffle. At angular acceleration applied about the pivot axis the oil will rush from one half of the housing to the other, thus exerting a force proportional to the angular acceleration. The hollow paddle facilitates balancing out the effects of linear acceleration and, because of its lightness, makes errors due to angular velocity negligible [69]. However, these instruments are rather large and heavy, especially in the lower ranges. For instance, the dimensions of a commercial transducer of the lowest range of ± 1.5 rad s^{-2} are 200 mm diameter and 75 mm height, and the weight is about 3.5 kg. Its natural frequency is 4 Hz. In fact, this instrument has been mentioned only as an interesting early attempt to measure low values of angular acceleration with good

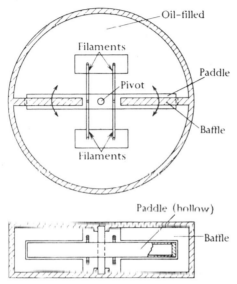

FIG. 4.2.27. Angular-acceleration transducer with liquid rotor and unbonded strain gauges, schematic diagram (Statham Laboratories, Oxnard, California).

accuracy; more recent methods will be described in Chapter 5 on force-balance transducers.

Finally, acceleration transducers with what might loosely be called 'unbonded' semiconductor strain gauges [66] (see section 4.2.3.4.1 above) have also been suggested for mushroom and cantilever types [70].

4.2.3.4.3. Pressure transducers.   Pressure transducers consist of two distinct parts; pressure-sensitive elements and, in the present context, strain-sensing components. A variety of common pressure-sensitive elements have been summarized in Fig. 3.21 (p. 56). The strain-sensing components may be metal strain gauges (bonded or unbonded wire gauges, foil gauges, or thin-film gauges), or semiconductor strain gauges.

Fig. 4.2.28 shows a number of pressure-summing elements in combination with strain-gauged cantilevers. Ideally, the cantilever stiffness should be high in comparison with the stiffness of the pressure-summing element in order to suppress the effects of instability and hysteresis inherent in the

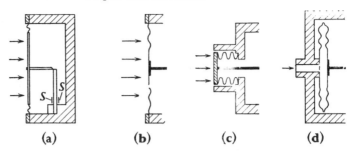

FIG. 4.2.28. Pressure transducers with bonded strain gauges on cantilever spring (S = strain gauge).

latter, though, at low ranges, both components may contribute to the total stiffness. Differential pressure types can be derived from the basic types of Fig. 4.2.28 by adding a second pressure chamber (diaphragm, bellows, or capsule) which is to oppose the movements of the first. In this arrangement the strain gauges need not be in contact with the pressure medium.

The single cantilever can be replaced by a double cantilever machined from one solid piece of high-tensile steel which is securely clamped at both its ends. The push-rod then acts at the centre of the double cantilever, both halves being 'decoupled' by transverse slits in order to reduce the longitudinal stresses. Pressure transducers of this type can be designed to have good thermal stability at both steady and transient environmental temperature variations by virtue of their symmetry and their excellent heat-conduction properties between the strain gauges on the cantilevers and the solid transducer body which carries the appropriate temperature-compensating elements [71].

If a tube under axial compression is substituted for the cantilever the instrument resembles a load cell (Fig. 4.2.23(b)). Because of the great stiffness of the tube this transducer type is useful only at high pressures. A special pressure summing component, the so-called catenary diaphragm [72], has been developed in order to combine the rigidity of a three-dimensional shell with the advantage at high temperature uses of thermal separation between diaphragm and strained element. A typical application is the engine indicator.

An apparently simple and commonsense arrangement for a pressure transducer is sticking strain gauges onto a clamped diagram. It is well known that such a diaphragm allows to form a complete bridge circuit by using (at least) two tensile gauges positioned in a radial direction near the circumference and two compressive gauges in a tangential direction near the centre of the diaphragm, thus taking advantage of push–pull strains at these locations (Fig. 4.2.29). In fact, etched foil gauges are commercially available which are designed to cover the whole area of a diaphragm by a printed foil pattern with (radial) meanders around the circumference and (tangential) spirals about the centre; but at close inspection the strain levels turn out to be disappointingly low if the strain–pressure relationship is to remain linear, and the diaphragm stresses to be kept within safe limits. A detailed analysis of this problem has been given elsewhere [73]. At small centre deflections, say less than half the thickness of a 'thin' clamped diaphragm of radius a (m) and thickness t (m), the maximum radial stress near the edges becomes

$$\sigma = \frac{3}{4} p \left(\frac{a}{t}\right)^2 \quad (\text{N m}^{-2}), \tag{4.2.61}$$

where p (N m^{-2}) is the pressure (see also Fig. 7.22, p. 58). At higher pressures the stress increases non-linearly with the load [74]. Diaphragm

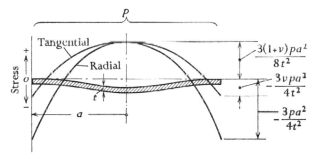

FIG. 4.2.29. Radial and tangential stress distribution in a circular diaphragm under small loads.

aspect ratios of $a/t \approx 100$ are most common; in thicker diaphragms ($a/t \to 10$) the sensitivity (strain versus pressure) declines, and in thinner diaphragms ($a/t \to 1000$) the measurable strain vanishes [73]. Materials with higher yield strength give a limited improvement; semiconductor strain gauges offer, of course, a better solution, if the designer is prepared to compensate for their shortcomings: non-linearity and thermal instability.

This reservation also applies to pressure-sensitive silicon diaphragms with diffused strain gauges [32]. The concept of integrating diaphragm and strain gauges with temperature-compensating and signal-conditioning elements on one piece of single-crystal silicon has recently been achieved in a pressure sensor intended for use in fuel metering and control systems [75]. In order to save cost in mass production, individual calibration has been abandoned and complete units are interchangeable within ±2 per cent of their specified output values.

Problems arising from imperfectly bonded gauges has also been eliminated in the thin-film strain sensor deposited on to a pressure-sensitive

metal diaphragm [52]. The gauges are in good thermal contact with the
diaphragm and, if made of suitable alloys, have low thermal instabilities.

Pressure transducers with unbonded strain sensors are now commercially available from various sources in a variety of designs [76]. Their
basic properties have been discussed in section 4.2.3.4.1 on force transducers. In the measurement of pressure the transducers are frequently
exposed to heat impact and temperature cycling, and it is in these
applications that the otherwise high accuracy of unbonded types suffers in
comparison with other types because of the low heat dissipation of the
freely suspended wires.

Apart from those strain-gauge pressure-transducer types discussed
above there are still a number of older designs which are now in less

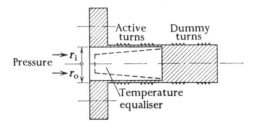

FIG. 4.2.30. Barrel-type pressure transducer.

common use than 10 years ago. The so-called barrel type shown in Fig.
3.21(j) (p. 56) and Fig. 4.2.70 operates by two active wire loops wound
on an open tube under internal pressure. Two more (dummy) loops
wound on a solid rod provide temperature compensation which is assisted
by a conical (copper) temperature equalizer. For thin tubes (wall thickness $s = r_0 - r_i \ll r_0$) the strain in the outer skin of the tube is

$$\varepsilon_0 = \frac{pr_i}{sE}\left(1 - \tfrac{1}{2}\nu\right), \tag{4.2.62}$$

where E (N m^{-2}) and ν are Young's modulus and Poisson's ratio of the
tube material. While barrel-type pressure transducers are useful at high
pressures, say 10 MN m^{-2} and above, tubes with elliptic cross-sections
[77] (Fig. 3.21(k), p. 56) can be employed at pressure ranges as low as
150 kN m^{-2}. They have also a better overload ratio since the tube will not
burst unless it is first deformed into a near-circular shape. Strain gauges
are bonded at positions of least curvature (tension) and of greatest
curvature (compression) of the elliptic cross-section, and a complete
bridge can thus be formed.

High (hydrostatic) pressures have been measured as early as 1856 [8]
by exposing a metal wire direct to the pressure medium. The resulting
fractural resistance change at a pressure p (N m^{-2}) can be shown

([16], [78]) to be

$$\frac{\Delta R}{R} = \frac{p}{E}(1-2\nu) + \frac{\Delta \rho}{\rho}, \tag{4.2.63}$$

where the first term represents the 'geometrical' contribution and the second the change in resistivity which is pressure dependent. The pressure sensitivity can thus be written

$$S_p = \frac{\Delta R/R}{p} = \frac{(1-2\nu)}{E} + k_p \quad (\text{m}^2 \, \text{N}^{-1}), \tag{4.2.64}$$

where $k_p = (\Delta \rho/\rho)/p$. Values of S_p for a number of relevant wire materials are listed in Table 4.2.7 [16].

TABLE 4.2.7

Some pressure-sensitive resistance materials

	Units	Constantan	Manganin	Bismuth
Resistivity ρ	$10^{-6} \, \Omega \, \text{m}$	0·48	0·43	1·17
Young's modulus E	$10^9 \, \text{N} \, \text{m}^{-2}$	150	135	23·5
Bulk modulus κ	$10^9 \, \text{N} \, \text{m}^{-2}$	125	132	30
Poisson's ratio ν	—	0·30	0·33	0·37
Pressure sensitivity S_p	$10^{-12} \, \text{m}^2 \, \text{N}^{-1}$	−7·0	+23·5	+158

Manganin is the most commonly used material [79], since it combines high pressure sensitivity with high temperature stability. It has also been employed in solid pressure media [80]. The very high pressure sensitivity of the semi-metal bismuth has been exploited in small sensors for the measurement of dynamic pressures [81]. Even carbon compound resistors (e.g. 100 Ω, 0·1 W), suspended by fine wires inside a cavity filled with grease, have been used to measure pressures with fair linearity and negligible hysteresis. However, the thermal noise, equivalent to about 30 kN m^{-2}, sets a lower limit to its useful pressure range [82]. A carbon resistor has also been employed to measure pressure waves [83]. The thermal stability of all carbon compound sensors is, of course, rather doubtful.

4.2.3.5. Resistance thermometers

Temperature measurements by means of conventional resistance thermometers is such a well-established technique that only the bare minimum of fundamentals will here be summarized. Detailed discussion of theoretical and practical aspects can be found elsewhere ([84], [85]).

4.2.3.5.1. Metal resistors.   The resistivity of pure metals and most alloys increases with increasing temperature, i.e. they have a positive temperature coefficient. For many practical purposes, and within limited ranges of temperature, the resistance–temperature relationship is linear, and thus the resistance R_2 of a conductor at temperature t_2 is given by

$$R_2 = R_1\{1+\alpha(t_2-t_1)\} = R_1(1+\alpha\,\Delta t), \qquad (4.2.65)$$

where R_1 is the resistance at temperature t_1, α the temperature coefficient at t_1, and $\Delta t = t_2-t_1$. The sensitivity of variable-resistance temperature transducers can thus be computed from eqn (4.2.65) and Table 4.2.8, which gives the temperature coefficients α of a number of pure metals and alloys over the temperature range 0–100 °C. Some non-metallic conductors have been included for comparison. Over wider temperature

TABLE 4.2.8

Temperature coefficients of resistance over a temperature range of 0–100 °C, per °C

Material	α	Material	α
Platinum	+0·00392	Iron (alloy)	+0·002 to +0·006
Gold	+0·0040	Manganin	±0·00002
Silver	+0·0041	Constantan	±0·00004
Nickel	+0·0068	Carbon	−0·0007
Copper	+0·0043	Thermistors	−0·015 to −0·06
Aluminium	+0·0045	Electrolytes	−0·02 to −0·09
Tungsten	+0·0048		

ranges, however, the resistance–temperature relationship may not be linear and a square term must be introduced,

$$R_2 = R_1(1+A\,\Delta t+B\,\Delta t^2). \qquad (4.2.66)$$

The best-known of all resistance-thermometer materials is pure platinum. Indeed the International Temperature Scale is based on the platinum resistance thermometer which covers the temperature range from $-182\cdot97$ °C (boiling point of liquid oxygen) to $+630\cdot5$ °C (above this temperature thermocouples and optical pyrometers are used.) Although the subtleties of temperature scales cannot be discussed here, Callendar's working equations for platinum resistance thermometers [86] are fundamental. At temperatures above 0 °C, eqn (4.2.66) can be written in the form

$$t-t_{Pt} = at(t-100)10^{-4}, \qquad (4.2.67)$$

where t_{Pt}, the 'platinum temperature', is obtained by assuming a linear resistance–temperature relationship, i.e.

$$t_{Pt} = \frac{100(R_t-R_0)}{R_{100}-R_0}, \qquad (4.2.68)$$

R_t, R_0, and R_{100} being the resistances at t °C, 0 °C, and 100 °C, respectively. The coefficient a for pure platinum has a value of 1·493, while the resistance ratio for the steam and the ice point is 1·3925. Below 0 °C further correction is required [87], which takes the form

$$t - t_{Pt} = at(t-100)10^{-4} + bt^3(t-100)10^{-8}, \qquad (4.2.69)$$

where b has the value of 0·11. At room temperature the platinum resistance thermometer permits the detection of temperature variations of the order of 10^{-4} °C. With great care accuracies of a similar order can be obtained. For practical purposes, however, 10^{-3}–5×10^{-3} °C are common

TABLE 4.2.9

Resistance factor of nickel wire for use in resistance thermometers

Temperature (0 °C)	Factor of resistance R_t/R_0
0	1·0000
100	1·6003
200	2·3994
300	3·3973

accuracies of measurement in the neighbourhood of room temperature. At 450 °C the reproducibility is down to about 10^{-2} °C and at around 1000 °C it will not be better than 10^{-1} °C.

Self-heating of resistance thermometers due to the measuring current is likely to cause errors. The heat conductivity of the medium around the element should be as high as possible. In doubtful cases two measurements at different currents i_1 and i_2 at the same bath temperature, say ice point, provides a correction factor. On the assumption that the resistance has changed, indicating an apparent temperature difference Δt, the measurements made at current i must be corrected by a factor

$$\Delta t i^2/(i_1^2 - i_2^2). \qquad (4.2.70)$$

Nickel wire is used because of its high sensitivity. However, its resistance variation with temperature is not as consistent as that of platinum. Table 4.2.9 gives the resistance values of nickel wire at temperatures up to 300 °C, its ultimate limit of application. Impurities are detrimental to high values of temperature coefficient of resistance. The mean value for a suitable nickel wire over the range 0–100 °C should not be less than 0·0058 per °C.

Copper is useful over temperature ranges of -150 °C to $+150$ °C. Its resistivity and temperature coefficient depend upon impurities and strain. Linearity is fairly good and reproducibility over the specified temperature range is about $\pm 0 \cdot 3$ °C. The main application of copper is probably in the measurement of coil temperatures in electrical machines, transformers, and similar equipment.

4.2.3.5.2. Thermistors [88]. In the practical application of thermistors reproducibility of characteristics is the most difficult problem. Since conductivity and temperature coefficient of semiconductors can be affected by less than one part in a million, only those compounds which are least sensitive to impurities are of any practical use. Some suitable materials are:

(a) Fe_3O_4 or corresponding spinels;

(b) sintered mixtures of NiO, Mn_2O_3, and Co_2O_3;

(c) sintered mixtures of TiO_2 and MgO.

Thermistors are normally made in the form of beads, discs, or flakes with small heat capacities, and are either enclosed in (gas-filled or evacuated) glass envelopes or in vitreous enamel.

With the effect of self-heating neglected, thermistors follow the approximate law

$$R = R_0 \exp\{b(1/T - 1/T_0)\}, \tag{4.2.71}$$

where

R = resistance at ambient (absolute) temperature $T(K)$,
R_0 = resistance at 'cold' (absolute) temperature $T_0(K)$,
b = material constant.

The temperature coefficient of a thermistor is defined as

$$\alpha = \frac{1}{R}\frac{dR}{dT} \quad \text{for} \quad R \to R_0, \tag{4.2.72}$$

and its value can be computed from the relationship

$$\alpha = -b/T^2. \tag{4.2.73}$$

It is seen from eqn (4.2.73) that α has a negative value which is large at low temperatures. It decreases rapidly with increasing temperature. Typical resistance–temperature curves of a miniature bead-type thermistor are shown in Fig. 4.2.31, and their current–voltage relationship in Fig. 4.2.32. With increasing current the voltage first increases according to Ohm's law, but with increasing self-heating the resistance drops and the voltage also starts decreasing. The maximum voltage E_{max} which can be developed across a thermistor depends on ambient temperature and on conditions of cooling. It is higher at lower temperatures and higher rates of cooling. In practice the input voltage must not be greater than E_{max} or a

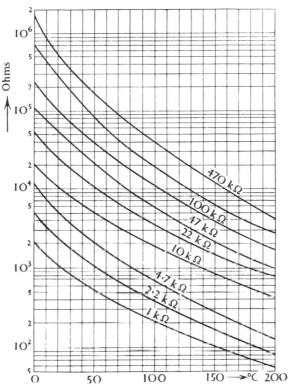

FIG. 4.2.31. Typical resistance–temperature curves of a miniature bead-type thermistor
(Mullard Ltd., London).

limiting resistor must be used for protection. Thermistors should not be
connected in parallel as small differences in their characteristics are
bound to cause unequal current-sharing.

4.2.3.6. *Temperature measurement in a fluid in motion*

In a streaming fluid, say air around an aircraft in flight, the measurement of the true fluid temperature is inherently an impossible task
because the flow of the medium is always disturbed by the insertion of a
temperature transducer. Inevitably a relative velocity exists between the
transducer and the fluid which causes the local temperature around the
transducer to change. The aerodynamic aspect of the transducer design
applies to both variable-resistance and thermocouple-type transducers,
although details of construction and the associated circuitry differ appreciably. The underlying theory and a discussion of basic aspects of the
performance of aircraft thermometers has been given by Clark [90]. In
the following we restrict ourselves to the transducer aspect supported by a
skeleton theory.

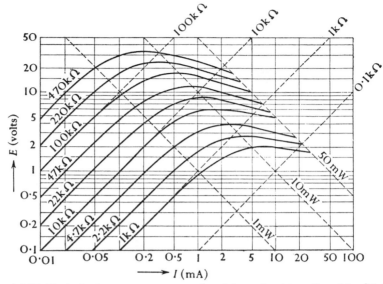

FIG. 4.2.32. Typical voltage–current curves of a miniature bead-type thermistor (Mullard Ltd., London).

Since it is impracticable to measure the true temperature of a fluid in motion we arrest the streaming medium in the neighbourhood of the temperature-sensitive resistance by means of a stagnation chamber (Fig. 4.2.33) and thus measure its stagnation temperature t_s. If no heat is lost or gained across the chamber boundaries and if external mechanical forces, such as gravitation or centrifugal forces, are negligible, it can be shown that in an ideal gas at adiabatic deceleration the temperature increase Δt over the true fluid temperature t_0 is

$$t_s - t_0 = \Delta t = v^2/2c_p, \qquad (4.2.74)$$

c_p being the specific heat of the fluid at constant pressure and v its velocity. However, the stagnation temperature t_s of eqn (4.2.74) and the actual temperature of the transducer element are not identical. Heat transfer through the boundary layer around the sensing element, heat conduction,

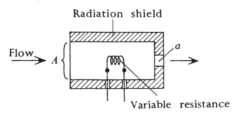

FIG. 4.2.33. Stagnation chamber, schematic diagram.

and heat radiation affect the magnitude of the latter. The stagnation chamber of Fig. 4.2.33 has a wide inlet of area A and a small leakage hole at the rear of area a. The leakage is essential to maintain 'ventilation' of the chamber, otherwise the heat losses and the time-lag of the transducer would be excessive. As a practical rule the ratio A/a should lie between 10 and 20. Under these conditions the element temperature is fairly independent of the fluid velocity v, even down to Mach numbers of $0 \cdot 1$–$0 \cdot 2$. The 'recovery factor' R of the transducer is defined as

$$R = \frac{t_{Ta}-t_s}{t_0-t_s},$$ (4.2.75)

t_{Ta} being the adiabatic temperature of the transducer in thermal equilibrium with no conductive or radiant heat transfer. The recovery factor is characteristic of a given geometry of the sensing element and determined by the heat transfer through the boundary layer. Eqn (4.2.74) can thus be written

$$\Delta t = Rv^2/2c_p.$$ (4.2.76)

The recovery factor must not be confused with the 'correction factor' k, which is the experimentally determined correction, including all heat losses or gains due to boundary-layer effects, heat conduction, and heat radiation. With a transducer temperature t_T at the sensing element, the correction factor is defined by

$$k = \frac{t_T-t_a}{t_0-t_s}.$$ (4.2.77)

While the recovery factor R in gases is always greater than zero and usually below unity, the correction factor k can assume any value, depending upon the various heat-transfer effects in a particular transducer under particular conditions of installation and test. In an adiabatic transducer, correction and recovery factor would be identical.

The value of the recovery factor of a sensing element depends upon its geometric shape and to some extent upon its surface roughness. In a well-designed stagnation chamber, temperature and velocity gradients are small because of the low velocity of the gas surrounding the sensing element, so that heat transfer and shear stresses are also low. Recovery factors very close to unity (about $0 \cdot 99$) have been achieved at Mach numbers $0 \cdot 2$–$3 \cdot 0$. As to the correction factor, calibration of the transducer under various conditions of use is essential. At room temperature range accuracies of around 1 °C can be secured. At higher temperatures greater errors are inevitable. In general a constant correction factor over a wider operating range, combined with low sensitivity to angle of yaw and low radiation, is more important than a correction factor of a value close to unity. If a short time-lag is called for, the platinum-wire resistance coil should be wound on a thin hollow metal cylinder. The shortest possible time-lag is probably attainable with an open-wound coil.

4.2.3.7. *Hot-wire anemometers*

Another specific application of the variable resistance principle is the hot-wire rate-of-flow, or fluid-velocity transducer. It consists basically of an electrically heated fine wire which is cooled by the streaming medium surrounding the wire (Fig. 4.2.34). The cooling effect is measured by way of the resistance change of the wire or, as an alternative, by way of

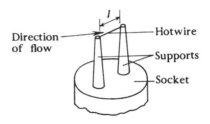

FIG. 4.2.34. Fast-responding hot-wire anemometer.

current change which is required to keep the wire at a constant temperature, i.e. at constant resistance. Although the instrument works on the same principle as the resistance thermometer, the wire temperature normally does not enter into calculation which, in the majority of cases, is in terms of stream velocity of the medium. The classical theory of the hot-wire anemometer, as developed by King [91], and the earlier applications in meteorology and general engineering have been treated by Burgers [92], Ower [93], and others, and need not be repeated here. However, more recent research into the turbulence of fluids demands the instantaneous measurement of flow patterns. These modern techniques as applied to transonic and supersonic air speeds have probably been best described by Lowell [94], who includes notes on the construction of miniature hot-wire anenometers, but discusses only the measurement of steady-state flow. A useful summary of the underlying theory and aspects of application in aerodynamics, including measurements of non-steady flow patterns, can be found in reference [95]. A more recent approach on the measurement of turbulence is given in reference [96].

Semiconductor (thermistor)-type anemometers have also been developed [97]. They have very high (negative) temperature coefficients but, because of their bulk, are useful only at slowly varying air flows.

Basically the hot-wire anemometer indicates the rate of heat transfer from the hot wire to the flowing stream. Heat transfer consists of

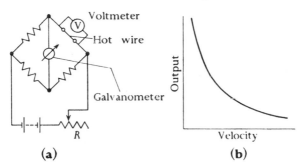

Fig. 4.2.35. (a) Circuit and (b) typical calibration curve of a hot-wire anemometer in 'constant-resistance' arrangement.

conduction, convection, and radiation. Simultaneous theoretical consideration of the three factors contributing to the cooling of the wire by the fluid stream would lead to expressions of excessive complexity. For forced convection, which is of paramount importance, the following basic design factors can be derived.

(a) *Constant-resistance method.* Under steady-flow conditions at velocity v the average heating current is

$$\iota = C_1 + C_2 v^{\frac{1}{4}}. \tag{4.2.78}$$

The sensitivity increases with heating current, i.e. with higher temperature differences between wire and medium, and decreases with increasing velocity (the constants C_1 and C_2 are determined by calibration). Fig. 4.2.35(a) and (b) show the circuit and a typical calibration curve. The bridge is operated by adjustment of resistance R, keeping the galvanometer deflection zero. Thus, the voltmeter indication is proportional to the wire current, since its resistance is being kept constant under balanced

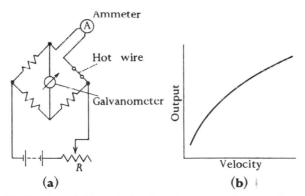

Fig. 4.2.36. (a) Circuit and (b) typical calibration curve of a hot-wire anemometer in 'constant-current' arrangement.

bridge conditions. In comparison with the constant-current calibration curve (Fig. 4.2.36(b)) a steeper slope indicates higher sensitivity at low velocity.

(b) *Constant-current method.* The wire resistance under steady-flow conditions is

$$R = \frac{R_a(C_1 + C_2 v^{\frac{1}{2}})}{C_1 + C_2 v^{\frac{1}{2}} - i^2},$$
                                                                          (4.2.79)

R_a being the wire resistance related to the adiabatic temperature of the wire. Circuit and typical calibration curves are shown in Fig. 4.2.36(a) and (b). The ammeter in series with the wire is used for the adjustment of constant current in the wire by means of resistance R, or replaced by a servo-loop. The readings are taken from the galvanometer.

The directional characteristics of hot-wire anomemeters, and two- or three-wire arrangements, can only be mentioned here.

References

1. K. KÜPFMÜLLER. *Einführung in die theoretische Elektrotechnik.* Springer, Berlin (1957).
2. S. BUTTERWORTH. On the alternating current resistance of solenoid coils. *Proc. R. Soc. A,* **107,** 693 (1925).
3. R. J. SULLIVAN. Resolution in precision wire-wound potentiometers. *Electron. Equip.,* Oct.–Nov. (1955).
4. G. W. A. DUMMER. *Fixed resistors,* pp. 47–51, Pitman, London (1956).
5. G. W. A. DUMMER. *Variable resistors,* p. 54, Pitman, London (1956).
6. R. PUTZ. Das lineare Potentiometer. *Arch. tech. Messen* No. 247, 191 (1956)
7. G. M. PARKER. The loading errors in linear potentiometers, *Electron. Engng.* **23,** 489 (1951).
8. W. THOMSON (Lord Kelvin). The electro-dynamic qualities of metals. *Phil. Trans. R. Soc.* **146,** 733 (1856).
9. P. W. BRIDGMAN. The effect of tension on the thermal and electrical conductivity of metals. *Proc. Am. Acad. Arts Sci.* **59,** 119 (1923–4).
10. C. C. PERRY and H. R. LISSNER. *The strain gage primer.* McGraw-Hill, New York (1962).
11. W. M. MURRAY and P. K. STEIN. *Strain gage techniques.* M.I.T. Press, Cambridge, Massachusetts (1956).
12. U. ZELBSTEIN. *Technique et utilisation des jaunes de contraintes.* Dunod, Paris (1956).
13. C. ROHRBACH and K. FINK. *Handbuch der Spannungs- und Dehnungsmessung.* V.D.I. Verlag, Düsseldorf (1958).
14. T. POTMA. *Strain gauges; theory and application.* Iliffe, London (1967).
15. H. K. P. NEUBERT. *Strain gauges; kinds and uses.* Macmillan, London (1967).
16. H. K. P. NEUBERT. The elasto-resistive effect in metal wires and films. R.A.E. Technical Report No. TR 69272 (1969).
17. E. JONES and K. R. MASLEN. The physical characteristics of wire resistance strain gauges. Aeronautical Research Council R & M No. 2661 (1948). H.M.S.O. London (1952).
18. P. JACKSON. The foil strain gauge. *Instrum. Pract.* **7,** 775–87 (1953).
19. J. SHIELDS. Strain gauge adhesives. *Metron* **2,** 163–171 (1970).
20. For example, S. G. STARLING. *Electricity and magnetism,* p. 101. Longmans Green, London (1920).
21. A. C. RUGE. The bonded wire gage torque meter. *Proc. Soc. exp. Stress Analysis* **1,** 68–72 (1943).

22. D. A. Drew. Slip-rings, a review. *Strain* **2**, No. 4 (1966).
23. R. Bamberger and F. Hines. Practical reduction formulas for use on bonded wire strain gages in two-dimensional stress fields. *Proc. Soc. exp. Stress Analysis* **2**, 113–27 (1944).
24. M. Hetényi (ed.). *Handbook of experimental stress analysis*, Chapter 9. Chapman & Hall, London (1950).
25. M. Cockrill. Two Fortran programs for rosette calculations. *Strain* **8**, 22–6 (1972).
26. H. K. P. Neubert. Resistance strain gauges, a critical review. *Metron* **2**, 123–30 (1970).
27. C. S. Smith. Piezoresistance effect in germanium and silicon. *Phys. Rev.* **94**, 42 (1954).
28. W. P. Mason and R. N. Thurston. Use of piezoresistance materials in the measurement of displacement, force and torque. *J. acoust. Soc. Am.* **29**, 1096 (1957).
29. G. R. Higson, Recent advances in strain gauges. *J. scient. Instrum.* **41**, 405–14 (1964).
30. H. K. P. Neubert. *Strain gauges; kinds and uses*, Chapter 3. Macmillan, London (1967).
31. W. G. Pfann and R. N. Thurston. Semiconducting stress transducers utilizing the transverse and shear piezoresistance effects. *J. appl. Phys.* **32**, 10 (1961).
32. N. Frantzis. A piezoresistive integral silicon pressure transducer. *J. Instrum. Soc. Am.* **4**, 334–48 (1965).
33. Anon. Absolute pressure transducer contains all transduction elements in a hybrid IC package. *IEEE Spectrum* **9**, 79 (1972).
34. M. Dean and R. D. Douglas (ed.). *Semiconductor and conventional strain gages*. Academic Press, New York (1962).
35. C. Rohrbach. *Handbuch für elektrisches Messen mechanischer Grössen*, pp. 134–9. V.D.I. Verlag, Düsseldorf (1967).
36. J. Dorsey. *Semiconductor strain gage handbook*, Sections I–VI. Baldwin–Lima–Hamilton, Waltham, Massachusetts (1963–5).
37. A. D. Kurtz and C. L. Gravel. Semiconductor transducers using transverse and shear piezoresistance. *J. Instrum. Soc. Am.* (1967). (Reprint No. P4-1-PHYMMID-6.)
38. J. C. Sanchez and W. V. Wright. Recent advances in flexible semiconductor strain gages. *J. Instrum. Soc. Am.* (1961). (Reprint No 46-LA-61.)
39. W. A. Leasure, N. Woodruff, and C. Gravel. Glass-bonding techniques for semiconductor strain gages. *Exp. Mech.* **11**, 235–40 (1971).
40. C. Rohrbach and N. Szaika. Über Funktion, Eigenschaften und Anwendung von Halbleitergebern. *Metallprüfung* **4**, 189–98 (1962).
41. F. J. Morin, T. H. Geballe, and C. Herring. Temperature dependence of the piezoresistance of high-purity silicon and germanium. *Phys. Rev.* **105**, 505 (1957).
42. J. Dorsey. n-type self compensating strain gages. *Exp. Mech.* **5**, 27A–38A (1965).
43. J. R. Parkins. Calibration and instrumentation of semiconductor strain gauges. *Strain* **4**, 10–18 (1968).
44. H. K. P. Neubert. *Strain gauges; kinds and uses*, p. 100. Macmillan, London (1967).
45. I. M. Ball. U.S.A. Patent No. 2556132 (1951).
46. R. L. Siddal and G. Smith. A thin-film resistor for measuring strain. *Vacuum* **9**, 144–146 (1959).
47. R. L. Parker and A. Krinsky. Electrical resistance–strain characteristics of thin evaporated metal films. *J. appl. Phys.* **34**, 2700–8 (1963).
48. M. J. Knight. Some fundamental properties of evaporated thin-film strain gauges. *Proceedings of the IEEE and IEE Joined Conference on the Applications of Thin Films in Electronics*, paper 21 (1966).
49. Z. M. Meiksin and R. A. Hudzinski. A theoretical study of the effect of elastic strain on the electrical resistance of thin metal films. *J. appl. Phys.* **38**, 4490–4494 (1967).
50. R. M. Hill. Electrical conduction in ultra-thin metal films. *Proc. R. Soc. A* **309**, 377–417 (1969).
51. L. G. Phillips. Self-adhesive strain gauges. R.A.E. Technical Report No. TR 67025 (1967).
52. P. R. Perino. Thin-film strain gage transducers. *Instrums Control Syst.* **38**, 119–21 (1965).

53. O. WATANABE, K. HATAKEYAMA, and T. SHIODA, Germanium-film strain gage with self-compensation of temperature. *Elect. Engng, Japan* **87**, 101–12 (1967).
54. V. L. TRAVATHAN and G. O. BEGEMAN. Weighing silos with strain gauges. *Instrum. Technol.* **17**, 45–50 (1970).
55. C. K. W. SCOTT and G. DROUBI. Prediction of zero strain vectors in dynamometer rings. *Strain* **6**, 156–61 (1970).
56. S. TIMOSHENKO. Strength of materials, Chaps 11 and 12, Van Nostrand, New York (1955).
57. J. R. ANDERSON. Strain gauge balances for windtunnels; an outline of practice in the United Kingdom. Advisory Group Aerospace R & D, Report No. 5 (1956).
58. R. M. HANSEN. Mechanical design and fabrication of strain gauge balances. Advisory Group Aerospace R & D Report No. 9 (1956).
59. K. H. McFARLAND and J. DINNEFF. Prolems involved in precision measurements with resistance strain gauges. Advisory Group Aerospace R & D Report No. 12 (1956).
60. G. F. MOSS and D. H. PAYNE. A compact design of a six-component internal strain gauge balance. R.A.E. Technical Note No. Aero 2764 (1961).
61. H. K. P. NEUBERT. *Strain gauges; kinds and uses*, pp. 127–31. Macmillan, London (1967).
62. H. R. MILLWARD. Miniature strain gauges employed on windtunnel balance at Warton Aerodrome. *Strain* **2**, 12–16 (1966).
63. C. KING. Semiconductor strain gauges applied to an axial force windtunnel balance. Unpublished R.A.E. report (1970).
64. R. D. MEYER. Applications of unbonded type resistance gages. *Instruments* **19**, 136–9 (1946).
65. Made by Bell & Howell Ltd., Basingstoke, Hants.
66. L. B. WILNER. Piezo-resistive force gages and their uses, *IEEE Trans. ind. Electron.* (*Contr. Instrum.*) (*Inst. elec. electron Eng.*) **16**, 40–3 (1969).
67. H. K. P. NEUBERT. *Strain gauges; kinds and uses*, pp. 146–8. Macmillan, London (1967).
68. G. N. ROSA. *Some design considerations for liquid rotor angular accelerometers.* Instrument Notes No. 26. Statham Laboratories, Oxnard, California (1954).
69. G. L. SMITH. Angular accelerometer errors arising for a bar-type design of seismic mass. *Rev. scient. Instrum.* **23**, 97 (1952).
70. F. E. DUFFIELD. Transducers with semiconductor strain gauges. Paper No. 1, *4th International Aerospace Instrumentation Symposium*, Cranfield, Bedford (1966).
71. Made by Electro Mechanisms Ltd., Slough, Berks.
72. C. S. DRAPER and Y. T. LI. New high-performance engine indicator of the strain gage type. *J. Aero/Space Sci.* **16**, 593–610 (1949).
73. H. K. P. NEUBERT. *Strain gauges; kinds and uses*, pp. 141–3. Macmillan, London.
74. S. WAY. Bending of circular plates with large deflections. *Trans. Am. Soc. mech. Engrs* **56**, Paper APM-56-12, 627–36 (1934).
75. A. R. ZIAS and W. F. J. HARE. Integration brings a generation of low-cost transducers. *Electronics* **45**, 83–88 (1972).
76. H. K. P. NEUBERT. *Strain gauges; kinds and uses*, pp. 133–7; 143–4. Macmillan, London (1967).
77. D. S. DEAN. British Patent Application No. 21479/53.
78. C. ROHRBACH. *Handbuch für elektrisches Messen mechanischer Grössen*, pp. 124–5. V.D.I. Verlag, Dusseldorf (1967).
79. H. LIPPMAN and M. RICHARD. Über das Verhalten von Manganin-Widerstandsmanometern im Druckbereich bis 6000 kp/cm². *Feingerätetechnik* **19**, 368–70 (1970).
80. J. LEES. Manganin gauges in solid pressure media. *High Temp. High Press.* **1**, 477–80 (1969).
81. Y. D. KLEBANOV and V. N. SUMAROKOV. A micro sonde for very high pressures. *Priborÿ Tekh. Éksp.* **15**, 224–225 (1970).
82. D. S. DEAN and K. BRIDGEWATER. Use of a 0·1 Watt resistor for pressure measurement. Unpublished Ministry of Aviation Report.

83. R. W. WATSON. Gauge for determining shock pressures. *Rev. scient. Instrum.* **38**, 978–980 (1967).
84. C. M. HERZFELD (ed.). *Temperature, its measurement and control in science and industry*, Vol. 3, Part 1: Basic concepts, standards and methods, §§ VI–VIII. Reinhold, New York (1962).
85. A. THULIN. High-precision thermometry using industrial resistance sensors. *J. scient. Instrum.* **4**, 764–68 (1971).
86. H. L. CALLENDAR. On the practical measurements of temperature. *Phil. Trans. R. Soc.* **178**, 160 (1887).
87. M. S. VAN DUSEN. Platin resistance thermometry at low temperatures. *J. Am. chem. Soc.* **47**, 326 (1925).
88. F. J. HYDE. *Thermistors.* Butterworth, London (1971).
89. D. A. WRIGHT. *Semi-conductors*, p. 48. Methuen, London (1950).
90. D. D. CLARK. An assessment of the probable causes of variation of speed correction coefficients of aircraft thermometers. Meteorological Research Communications M.R.P.677 S.C.I/58 (1951).
91. L. V. KING. On the convection of heat from small cylinders in a stream of fluid. *Phil. Trans. R. Soc.* A **214**, 373–432 (1914).
92. J. M. BURGERS. Hitzdrahtmessungen. In *Handbuch der Experimental Physik*, Vol. IV, Part 1, pp. 637–67. Akademische Verlagsgesellschaft, Leipzig (1931).
93. E. OWER. *The measurement of airflow.* Chapman & Hall, London (1949).
94. H. H. LOWELL. Design and application of hot-wire anemometers for steady-state measurements of transonic and supersonic air speeds. *Tech. Notes Natn Advis. Comm. Aeronaut., Wash.* **2117** (1950).
95. C. DEAN (ed.). *Aerodynamic measurements*, pp. 27–53 M.I.T., Massachusetts (1953).
96. P. BRADSHAW and R. F. JOHNSON. Turbulence measurement with hot-wire anemometers. *Notes appl. Sci.* **33** (1963).
97. K. KRAUS. Thermistoranemometer. *Arch. tech. Messen.* V 116–15, 41–4 (1972).
98. Made by Penny & Giles Transducers, Ltd., Mudeford, Hants.
99. A range of machinable ceramics and coatings are obtainable from Minnesota Mining & Manufacturing Co., St. Paul, Minnesota, and from Aremco Products, Briarcliff, New Jersey, U.S.A.

4.3. VARIABLE-INDUCTANCE TRANSDUCERS

THE physical quantity to be measured can be made to vary the inductance of a coil. As with a variable resistance, the varying inductance can be employed to change the parameters of an electric circuit which includes the inductance coil and an indicator. In this basic application the inductance is usually part of an a.c.-fed bridge circuit the output of which being proportioned to the input quantity. Transducers operating in this manner are commonly referred to as variable inductance types, and the greater part of this chapter will deal with their properties.

A variable inductance, however, can also be used to generate a voltage in the inductance coil if the coil is coupled to a magnetic circuit carrying a unidirectional flux generated, for instance, by a permanent magnet. This transducer type then produces an output voltage proportioned to the rate of change of flux. If the flux variations are produced by change of reluctance, it is called an electromagnetic transducer (see section 4.3.5), and if permeability variations are effective, a magnetostrictive (generator)-type transducer (see section 4.3.6.2).

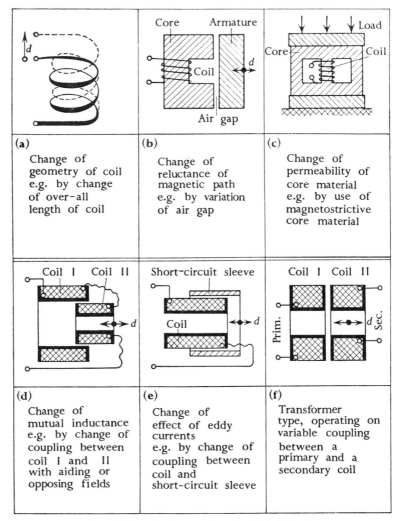

FIG. 4.3.1.   Basic examples of variable inductances.

Generally, a variation of the inductance of a coil can be achieved by changing:

(a) the geometry (Fig. 4.3.1(a)),

(b) the reluctance of the magnetic path (Fig. 4.3.1(b)),

(c) the permeability of the magnetic core material (Fig. 4.3.1(c)),

(d) the coupling of two or more elements (Fig. 4.3.1(d)) of a coil.

In practice case (a) has not been used in transducers, since useful changes of inductance can be obtained only with air-cored coils. Case (b) is the most common technique employed in variable-inductance transducers. A variation in permeability of the magnetic core material (case (c)) occurs in transducers using the magnetostrictive properties of some ferromagnetic materials. Transducers in which the coupling of two coils (d) or of a coil and a short-circuit sleeve (e) is altered by the input quantity have been made, but a more important application of variable coupling between coils is in the 'transformer' type transducer (f), which will also be dealt with in this chapter. However, single-phase and three-phase 'synchros', although based on variable coupling between coils, have produced a wide literature of their own [1]–[4]; they will not be treated here in order to avoid unnecessary duplication.

4.3.1. The equivalent circuit

As the variation of inductance is the basic property of such a transducer we require a coil arrangement which gives the largest possible

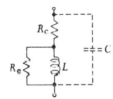

FIG. 4.3.2. Equivalent circuit of inductance coil L with 'copper' resistance R_c, eddy-current core loss resistance R_e, and self-capacitance C.

inductance change for a given input quantity, e.g. the displacement of an armature. An iron-cored coil will basically yield larger effects than an air-cored coil. The variable-air-gap type with ferromagnetic core and armature (Fig. 4.3.1(b)) is probably the most sensitive type if the fractional change of inductance for a given armature movement is considered. Inductances with ferromagnetic cores have also other advantages which we shall discuss later on, but it may be mentioned here that a closed magnetic circuit, as provided by a ferromagnetic core with only a small air gap, is least affected by external magnetic fields. Owing to its relatively low reluctance such coil needs fewer winding turns than an air-cored coil to achieve a required inductance level. This results in a smaller self-capacitance of the coil, which is of prime importance in certain applications. Therefore, variable inductance transducers have normally ferromagnetic cores, and thus our equivalent circuit must include the core characteristics. Inductance coils with ferromagnetic cores have been investigated thoroughly. A good survey of their properties and design is given in reference [5]; coils with ferrite cores are treated in reference [6].

Fig. 4.3.2 shows the pure inductance L of the transducer coil in series with a resistance R_c (representing the 'copper' losses of the coil) and in parallel with a resistance R_e (representing the eddy-current losses of the ferromagnetic core). A capacitance C in parallel with L and R_e represents the self-capacitance of the coil which is important particularly at high frequencies. Its effect on the transducer characteristics will be treated separately later on.

4.3.1.1. Inductance L

Let n be the number of turns of a uniformly wound toroidal coil with a magnetic core of length l (m). Then the magnetic field strength H (A m^{-1}) inside the core due to a coil current $I(A)$ is

$$H = nI/l.$$

With the permeability of empty space $\mu_0 = 4\pi \times 10^{-7}$ (H m^{-1}) and the relative permeability μ of the core material the flux density B (T, or Wb per m^2) is

$$B = \mu\mu_0 H = \mu\mu_0 nI/l,$$

and the total flux Φ (Wb) through a core cross-section A (m^2) is

$$\Phi = BA = \mu\mu_0 IA/l.$$

As the self-inductance of a coil in henries is defined as the number of flux linkages per unit current, its value is the product of the total flux and the number of turns divided by the current,

$$L = n\Phi/I = \mu\mu_0 n^2 A/l \quad \text{(H)}. \tag{4.3.1}$$

4.3.1.2. Copper loss resistance R_c

The resistance of a coil of n turns of wire of diameter d (m) and of resistivity ρ_c (Ω m) is

$$R_c = \frac{4\rho_c ns}{\pi d^2} \quad (\Omega), \tag{4.3.2}$$

where s (m) is the length of an average turn. R_c depends only upon the material and dimensions of the coil and is independent of frequency, if skin-effect and the effect of external shielding can be neglected. The dissipation factor D_c of a coil of inductance L (H) at a frequency $f = \omega/2\pi$ (Hz) with a series loss resistance R_c is

$$D_c = R_c/\omega L = c/f. \tag{4.3.3}$$

Thus the dissipation factor D_c of an inductance, caused by a series loss resistance R_c, is inversely proportional to frequency, where c is the factor of proportionality.

4.3.1.3. Eddy-current loss resistance R_e

Consider a ferromagnetic core built up of laminations of thickness t (m). The parallel loss resistance R_e, representing the eddy-current losses of such core, is given by

$$R_e = \frac{2p}{t} \cdot \frac{\cosh(t/p) - \cos(t/p)}{\sinh(t/p) - \sin(t/p)} \, \omega L, \qquad (4.3.4)$$

where p is the so-called 'depth of penetration' of the eddy currents and is defined as

$$p = \left\{ \frac{\rho_i}{\pi \mu \mu_0 f} \right\}^{\frac{1}{2}} \quad (m), \qquad (4.3.5)$$

ρ_i being the resistivity of the core material and μ its relative permeability. At fairly low frequencies f, i.e. for t/p values not greater than about 2,

TABLE 4.3.1

Highest frequencies in kHz for which the parallel eddy-current loss resistance R_e of laminated ferromagnetic cores is given by the simplified eqn (4.3.6)

Lamination thickness	0·05 mm	0·10 mm	0·15 mm
Core material	Frequencies (kHz)		
Mu-metal	23	5·7	2·7
Radiometal	104	26	12
Rho-metal	423	105	49
Stalloy	580	143	67

eqn (4.3.4) simplifies to

$$R \simeq \frac{6}{(t/p)^2} \, \omega L = \frac{12\rho_i A n^2}{lt^2} \quad (\Omega), \qquad (4.3.6)$$

where ωL is substituted from eqn (4.3.1). The maximum frequencies for which eqn (4.3.6) applies are shown in Table 4.3.1. These have been computed for various materials and lamination thickness 0·05 mm, 0·10 mm, and 0·15 mm, under the limiting condition $t/p = 2$ (i.e. the depth of penetration of eddy currents is half the lamination thickness). It is seen from Table 4.3.1 that audio frequencies below, say 10 kHz, are covered by laminations of 0·1 mm thickness, with the exception of Mu-metal. For higher frequencies Rho-metal laminations may be used with advantage.[†] Eqn (4.3.6) shows that the parallel loss resistance, representing the eddy-current losses of a laminated ferromagnetic core, is not only independent of frequency but also of the permeability of the core material. This is thus the most convenient representation of the

† Products of Telcon Metals Ltd. Their new range of 'Radiometals' cover equivalent properties, and so do soft magnetic materials from other sources (see also Table 4.3.3, p. 219).

eddy-current losses for the purpose of this section. However, we shall see later when investigating the performance of an inductance transducer in its associated circuit, the eddy-current losses are more suitably represented by an equivalent series resistance which can be readily lumped with the series copper resistance R_c.

From the eddy-current loss resistance R_e a dissipation factor D_e of a coil with eddy-current losses in its core can be computed. We have

$$D_e = \frac{\omega L}{R_e} = \frac{\pi \mu \mu_0 t^2 f}{6 \rho_i} = ef.$$ (4.3.7)

This dissipation factor is directly proportioned to frequency, e being the factor of proportionality.

4.3.1.4. Hysteresis losses

The representation of hysteresis core losses in our equivalent circuit is less straightforward than that of the copper and eddy-current core losses but, for low flux densities, i.e. in the region of initial permeability, Lord Rayleigh's theory permits a general treatment of hysteresis effects in ferromagnetic cores [7]. Rayleigh established by experiment that at very low flux densities, i.e. in the region in which the core permeability is not more than 10 per cent above its value at zero excitation, the magnetization curve follows the law

$$(B - B')/\mu_0 = \mu_i (H - H') + \tfrac{1}{2}\alpha (H - H')^2,$$ (4.3.8)

where $(B - B')$ is the change in flux density due to a change in field strength $(H - H')$. The initial permeability μ_i at $H = 0$, and Rayleigh's constant α are both functions of the magnetic material. Expression (4.3.8) is exact and does not contain terms of higher order. If $H' = B' = 0$, it becomes

$$B/\mu_0 = \mu_i H + \tfrac{1}{2}\alpha H^2.$$ (4.3.9)

B/μ_0 thus consists of a reversible term $\mu_i H$ and an irreversible term $\tfrac{1}{2}\alpha H^2$ as shown in Fig. 4.3.3(a). For a periodic change in H of $\pm H_1$, B oscillates between $+B_1$ and $-B_1$ (Fig. 4.3.3(b)), where

$$B_1/\mu_0 = \mu_1 H_1 + \alpha H_1^2.$$ (4.3.10)

For increasing H

$$B/\mu_0 = (\mu_i + \alpha H_1)H - \tfrac{1}{2}\alpha (H_1^2 - H^2),$$ (4.3.11)

and for H decreasing

$$B/\mu_0 = (\mu_i + \alpha H_1)H + \tfrac{1}{2}\alpha (H_1^2 - H^2).$$ (4.3.12)

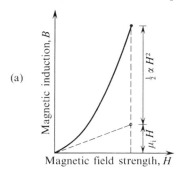

(a)

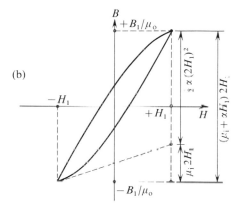

(b)

FIG. 4.3.3. Magnetization curves of ferromagnetic materials at low flux densities.

The area of the hysteresis curve is therefore

$$\int B \, dH - \tfrac{4}{3}\mu_0 \alpha H_1^3,$$

and the energy (W s) dissipated per unit core volume is

$$\tfrac{16}{3}\alpha H_1^3 \times 10^{-7}. \tag{4.3.13}$$

For a core of cross-section A (m^2) and length l (m) the total hysteresis losses are

$$\frac{16\pi}{3} A l \alpha H_1^3 \times 10^{-7} \quad \text{(W s per cycle)} \tag{4.3.14}$$

or

$$P_h = \frac{16\pi}{3} A l \alpha f H_1^3 \times 10^{-7} \quad \text{(W)}. \tag{4.3.15}$$

Since power $P = E^2/R$, the parallel resistance to represent the hysteresis losses becomes

$$R_h = E^2/P_h = 4\pi^2 f^2 L^2 I^2/P_h, \tag{4.3.16}$$

and as P_h is proportional to frequency (see eqn (4.3.15)), R_h is also proportional to frequency (in a similar way it can be shown that a series resistance would also depend on frequency.) Therefore, the hysteresis losses cannot be represented by a resistance in our equivalent of Fig. 4.3.2, but the dissipation factor D_h, representing the hysteresis losses, is invariant with frequency, as seen from $D_h = \omega L / R_h$. It can be shown that

$$D_h = h = \frac{2\alpha}{3\pi\mu_i} H_1, \qquad (4.3.17)$$

for small values of H_1, as assumed in this section.

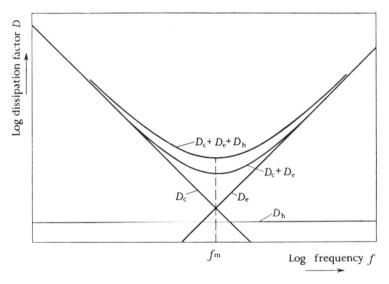

FIG. 4.3.4. The relationship between dissipation factor and frequency of coils with laminated ferromagnetic cores.

4.3.1.5. *Dissipation factor D and storage factor Q*

The total dissipation factor D of an inductance coil with a laminated ferromagnetic core is the sum of three separate dissipation factors

$$D = D_c + D_h + D_e = \frac{c}{f} + h + ef, \qquad (4.3.18)$$

f being the excitation frequency. This important result allows us to discuss the frequency characteristics of our equivalent circuit (Fig. 4.3.2), though still ignoring the self-capacitance C of the coil. Eqn (4.3.18) is symmetrical in frequency. When plotted on log–log paper (Fig. 4.3.4), the dissipation factor for the copper losses is represented by a straight line with a negative slope of unity and the dissipation-factor line for the eddy-current losses has a positive slope of unity. The dissipation factor for the hysteresis losses is represented by a horizontal straight line. At the intersection of D_c and D_e, which occurs at

$$f_m = \sqrt{(c/e)},\qquad\qquad(4.3.19)$$

the total dissipation factor D has a minimum, its value being†

$$D_m = h + 2\sqrt{(ce)}.\qquad\qquad(4.3.20)$$

The storage factor Q of a coil is the reciprocal of its dissipation factor, so from eqn (4.3.18),

$$Q = \frac{1}{D} = \frac{1}{c/f + h + ef},\qquad\qquad(4.3.21)$$

and for the maximum value of Q we have

$$Q_m = \frac{1}{h + 2\sqrt{(ce)}},\qquad\qquad(4.3.22)$$

which occurs at the same frequency f_m as the minimum value of D.

Before we discuss the parameters of a coil which establish Q_m and frequently f_m we can simplify our task by neglecting h in comparison with c and e at very low excitations, for, as it is seen from eqn (4.3.17), h approaches zero as H_1 decreases. With this simplification

$$Q_m = 1/2\sqrt{(ce)},\qquad\qquad(4.3.23)$$

and, together with eqn (4.3.19) for f_m and eqn (4.3.1) for L, the main design formulae for inductance coils with laminated ferromagnetic cores are thus established.

4.3.1.6. *Ferromagnetic cores with air gaps*

In transducer work we must consider magnetic cores with air gaps, the lengths of which vary with the physical quantities to be measured. The most convenient way to treat the effect of an air gap is by the introduction of an 'effective' permeability relating to the combination of a ferromagnetic core and an air gap. The reluctance of such a core with a gapped toroidal ring sample of (relative) permeability μ_s is

$$\frac{1}{A}\left(\frac{l-g}{g} + g\right) = \frac{l}{\mu A},$$

where l (m) is the length of the total path, g (m) the gap length, and μ the

---

† From experimental dissipation–frequency curves, when plotted on log–log paper, the numerical values of c, e, and h can be found by drawing asymptotes of slope unity to the curves. These asymptotes interesect the 1 Hz axis at values of c and e respectively. The value of h is the difference between the observed minimum and twice the value of the crossing-point of the two asymptotes.

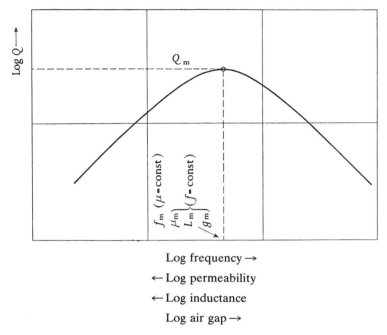

Fig. 4.3.5. Q-curve of coil with laminated ferromagnetic core at variable frequency, permeability, inductance, or air gap.

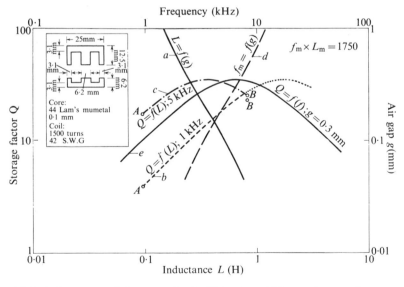

Fig. 4.3.6. Inductance- and Q-curves of coil with laminated Mu-metal core and armature at 1 kHz and 5 kHz, plotted against air gap.

'effective' permeability of the iron path and the air gap in series. Neglecting fringing, A (m^2) represents the cross-section of both the iron path and the air gap. This effective permeability μ, which takes its place in all above formulae if the core is gapped, reads

$$\mu = \frac{\mu_s}{1+(g/l)(\mu_s-1)} \tag{4.3.24}$$

or, since usually $\mu_s \gg 1$,

$$\mu = \frac{\mu_s}{1+(g/l)\mu_s}. \tag{4.3.25}$$

With the relatively short air gaps which occur in most inductance transducers, eqn (4.3.25) is sufficiently accurate. For larger air gaps the effect of fringing can be allowed for by introducing an 'effective' slightly-reduced air gap.

Another important relationship concerning inductances with ferromagnetic cores and air gaps can be deduced from the expressions for D_c (eqn (4.3.3)) and D_e (eqn (4.3.7)). It is seen that for a given structure D_c varies inversely and D_e directly with the product $f\mu$. Hence, at constant frequency f, D_c and D_e are functions of the effective permeability μ,

$$D_c = \frac{c'}{\mu} \quad \text{and} \quad D_e = e'\mu. \tag{4.3.26}$$

Therefore the Q-curve obtained for constant frequency and varying permeability will have the same shape as that obtained for constant permeability and varying frequency. Consequently, the Q-curve of an inductance can be obtained experimentally by varying the frequency at a fixed permeability or by varying the effective permeability (i.e. by varying the air gap) at a fixed frequency. This interesting relation is illustrated in Fig. 4.3.5, the arrows indicating the direction of increasing frequency, inductance (permeability), and air gap. The curve is useful for the interpretation of experimental results taken on prototype inductance coils with variable air gap or variable frequency, as may be convenient.

Consider the example illustrated in Fig. 4.3.6. Curve (a) represents the measured inductance L of a coil (x-axis) plotted as a function of the variable air gap g (y-axis). The measuring frequencies were 1 kHz and 5 kHz, and the inductance values obtained were identical within the accuracy of measurement for both frequencies. The core was made up of No. 450 Mu-metal laminations, 0·10 mm thick. A bar-shaped armature of similar laminations was positioned at a distance g opposite the three legs of the E-core. The dimensions of core and armature, and the coil-winding data are given in the insert to Fig. 4.3.6. For an ideal inverse relationship between inductance and air-gap length, the curve should be a straight line with a negative slope of 45°, but because of the reluctances of core and armature, and the uncertainty of the inductance value at 'butt joint', the experimental curve departs from such an ideal characteristic. At large air gaps, above 1 mm, curve (a) would level out because of fringing. The Q-values, measured simultaneously with inductances, are also plotted in Fig. 4.3.6 for 1 kHz (curve (b)) and for 5 kHz (curve (c)). They cover a range in terms of inductance between 'infinite' air gap (points A) and 'butt joint' (points B). At a frequency of 1 kHz, Q rises monotonically with increasing L, but at 5 kHz, Q reaches a maximum value of 35 at an inductance value of 0·35 H.

Analysis is facilitated by use of a template for drawing or checking the Q versus L curve [8]–[9]. Because of symmetry the general law $y = (x+1/x)^{-1}$ holds, and a template can thus easily be computed for any particular log–log paper. Using the appropriate template made from cardboard, curve (c) was checked and curve (b) was extended beyond its high-inductance end (small dots). A fictitious Q_m was thus obtained at $L_m = 1.7$ H, which was in good agreement with the theoretical value

$$L_{m1} = L_{m5}f_{m5}/f_{m1} = 1·75 \text{ H}. \qquad (4.3.27)$$

This procedure, as well as being a useful check on experimental results, also provides the designer with a means of drawing the appropriate Q-curve at any frequency considered for use, as long as the simplified theory of section 4.3.1 holds.

But the curves of Fig. 4.3.6 can be put to further use. Since Q_m occurs at a frequency f_m which is inversely proportional to the inductance L_m, the product $L_m f_m$ is a constant, and an inverse frequency scale can be plotted against L. It is more convenient, however, to plot a new curve (d) in Fig. 4.3.6 which represents the locus of f_m (x-axis) versus a given air gap g (y-axis). This curve is derived from curve (a) by first obtaining the product $L_m f_m$, i.e. $5000 \times 0·35 = 1750$. The air gap from curve (a), corresponding to 0·35 H, is 0·19 mm at 5000 Hz. To find a point on curve (d) corresponding to any point on curve (a), divide the abscissa of curve (a) into $(L_m f_m)$, i.e. 1750, from which will be obtained the abscissa of curve (d). Both curves have common ordinates.

Curve (d), which is the mirror image of curve (a), enables us to choose the appropriate air gap in order to get the maximum Q-value at any particular frequency. Take, for instance, an air-gap value of $g = 0·3$ mm, considered to be suitable for a particular transducer. From curve (d), f_m will be 6500 Hz, and the Q-curve with Q_m at 6500 Hz has been drawn in Fig. 4.3.6 by means of the template (curve (e)). If, for instance, it is required to know what frequency range under these conditions would yield Q-values not lower than 15, the answer obtained from curve (e) is 1·5–28 kHz. We also note that with increasing air gap the frequency range moves towards the higher-frequency end.

Core and coil combinations of a variety of shapes, dimensions, and materials have been investigated in this manner. The interested reader will find worked examples in references [8] and [9].

4.3.1.7. *Design parameters of inductance coils with ferromagnetic cores and small air gaps*

Consider the expressions for inductance L (eqn (4.3.1)), maximum storage factor Q_m (eqn (4.3.23)) and frequency f_m related to Q_m (eqn (4.3.19)). If we substitute the values for c and e we obtain the 'factorized' equations of Table 4.3.2, which allows us to appreciate at a glance the significance of the various design parameters. All symbols are defined

TABLE 4.3.2

Design formulae for L, Q_m, and f_m, factorized according to the physical meaning of the individual terms

	Dimensions of core	Material of core	Dimensions of coil	Material of coil
Inductance L	A/l	μ	n^2	—
Maximum storage factor Q_m	$\sqrt{(A/t^2 l)}$	$\sqrt{\rho_i}$	$\sqrt{(S/s)}$	$\sqrt{(1/\rho_c)}$
Frequency f_m where Q_m occurs	$\sqrt{(l/t^2 A)}$	$\sqrt{(\rho_i/\mu^2)}$	$\sqrt{(s/S)}$	$\sqrt{\rho_c}$

above except $S = \pi d^2 n/4$, the total copper cross-section (m²) of a coil in the core window. Some useful deduction, drawn from Table 4.3.2, are:

(a) the coil inductance is independent of coil dimensions, but increases with the square of the number of turns;

(b) Q_m and f_m increase with decreasing lamination thickness t and with increasing core resistivity ρ_i,

(c) Q_m is independent of μ and hence also of air gap;

(d) f_m decreases with increasing μ, hence it increases with increasing air gap;

(e) Q_m and f_m are independent of the number of turns n, if the total copper cross-section S and the average length of turn s are kept constant;

(f) proportional increase of all dimensions (except thickness of laminations t) increases Q_m and decreases f_m—this means that at low frequencies coils of high Q cannot be made below a certain size;

(g) a coil wound with resistance wire (for better temperature stability of resistance) has a lower Q_m which occurs at a higher frequency f_m.

4.3.1.8. *Higher harmonic distortions*

Referring to eqns (4.3.11) and (4.3.12), we have for a specimen in the Rayleigh region at a sinusoidal excitation $H = H_1 \cos \omega t$,

$$B/\mu_0 = (\mu_i H_1 + \alpha H_1^2)\cos \omega t \pm \tfrac{1}{2}\alpha H_1^2 \sin^2 \omega t, \qquad (4.3.28a)$$

which, when expanded in a Fourier series, becomes

$$B/\mu_0 = (\mu_i H_1 + \alpha H_1^2)\cos \omega t + \frac{4\alpha H_1^2}{3\pi} \sin \omega t -$$
$$- \frac{4\alpha H_1^2}{3\pi} (\tfrac{1}{5} \sin 3\omega t + \tfrac{1}{35} \sin 5\omega t + ...). \qquad (4.3.28b)$$

Thus, apart from producing an in-phase and a quadrature component of the magnetic induction, there will be components of higher, though only odd, orders. Now, since an e.m.f. induced in the coil is proportional to the rate of change of flux density, the voltage generated across the coil is proportional to

$$(\mu_i H_1 + \alpha H_1^2)\omega \cos(\omega t + \tfrac{1}{2}\pi) + \frac{4\alpha H_1^2 \omega}{3\pi} \cos \omega t -$$
$$- \frac{4\alpha H_1^2 \omega}{3\pi} (\tfrac{3}{5} \cos 3\omega t + \tfrac{1}{7} \cos 5\omega t + ...). \qquad (4.3.29)$$

For transducer work it is relevant that the third harmonic, which dominates the higher orders, is proportional to the square of the coil current (in H_1^2) and to Rayleigh's hysteresis factor α, which is an 'effective' α in case of an air-gapped core. It can be shown that with increasing air gap α decreases rapidly, so that coils having gapped ferromagnetic cores working at low flux densities have negligible hysteresis losses and hence produce negligible harmonic distortion. At flux densities above the Rayleigh region and at high frequencies the conditions for the generation of harmonic distortions are much more complex and cannot be discussed here. Even harmonics which are occasionally found in iron-cored coils can occur only if there is a lack of symmetry between the positive and negative lobe of the hysteresis loop. This may be caused, for instance, by an incompletely demagnetized core or by a d.c. component in the coil current.

4.3.1.9. *Inductance coil with shunt capacity*

Common to all inductance transducers is the fact that a capacitance C exists in parallel with the transducer coil, as indicated in Fig. 4.3.2. It arises from the unavoidable self-capacitance of the coil and the capacitance of the cable which connects the transducer to its associated circuit. If the impedance of a coil without capacitance is

$$Z = R + j\omega L, \qquad (4.3.30)$$

(*R* being a series resistance representing all coil and core losses), then the coil impedance with a capacitance connected across its terminals becomes

$$Z_s = \frac{R}{(1-\omega^2LC)^2+(\omega^2LC/Q)^2}+$$
$$+j\frac{\omega L\{(1-\omega^2LC)-(\omega^2LC/Q^2)\}}{(1-\omega^2LC)^2+(\omega^2LC/Q)^2}, \tag{4.3.31}$$

or, if $1/Q^2 \ll 1$, which applies to transducers with high values of $\omega L/R$,

$$Z_s = \frac{R}{(1-\omega^2LC)^2}+\frac{j\omega L}{1-\omega^2LC} = R_s+j\omega L_s. \tag{4.3.32}$$

It is seen from eqn (4.3.32) that with a capacity in parallel, the effective series loss resistance and the effective inductance are increased whilst the effective Q is decreased. It will be shown in the next two sections on sensitivity (section 4.3.2) and associated circuits (section 4.3.3) that variable inductance transducers with high Q-values have sensitivities proportional to the fractional change in inductance only. At small fractional changes dL/L the effective fractional change in the presence of a shunt capacitance then becomes

$$\frac{dL_e}{L_s} = \frac{1}{1-\omega^2LC}\cdot\frac{dL}{L}. \tag{4.3.33}$$

This result is of practical importance. It means that the sensitivity of an inductance transducer increases when a shunt capacitance is connected across it, according to the approach of resonance conditions in such a circuit. Thus an inductance transducer must be calibrated with the actual cable length as used in the measuring equipment (unless the cable capacity is eliminated by employing individually screened cables). With very large values of shunt capacitance or at extremely high frequencies, the transducer may enter, or even overstep, resonance conditions. In this case the L–C combination has ceased to be an 'inductance'. We shall return to this subject in section 4.3.3.1, where these conditions will be employed to obtain more linear calibration curves.

4.3.2. Sensitivity and linearity

4.3.2.1. *Inductance transducers with ferromagnetic cores and small variable air gaps*

The inductance L of a coil of n turns with a ferromagnetic core, length l (m) and cross-section A (m²), and a small air gap g (m) is given by eqn (4.3.1),

$$L = \mu\mu_0 n^2 A/l \quad (\text{H}),$$

where μ is here the effective permeability of the gapped core (eqn 4.3.25)),

$$\mu = \frac{\mu_s}{1+(g/l)\mu_s},$$

μ_s being the incremental permeability of a ring sample of the core material without air gap. Thus, for a given coil,

$$L = K\frac{1}{g+l/\mu_s}, \qquad (4.3.34)$$

where

$$K = \text{constant} = 4\pi An^2 10^{-7}. \qquad (4.3.35)$$

For a finite decrease in air gap δg we obtain a finite increase in inductance δL,

$$L+\delta L = K\frac{1}{g-\delta g+l/\mu_s}. \qquad (4.3.36)$$

Eliminating K from eqns (4.3.34) and (4.3.36) we have

$$1+\frac{\delta L}{L} = \frac{g+l/\mu_s}{g-\delta g+l/\mu_s}$$

and the fractional change in inductance is

$$\frac{\delta L}{L} = \frac{\delta g}{g-\delta g+l/\mu_s} = \frac{\delta g}{g}\cdot\frac{1}{1+(l/\mu_s)-(\delta g/g)}$$

$$= \frac{\delta g}{g}\cdot\frac{1}{1+l/g\mu_s}\cdot\frac{1}{1-(\delta g/g)\{1/(1+l/g\mu_s)\}}. \qquad (4.3.37)$$

Developing eqn (4.3.37) into a series for

$$\left|\frac{\delta g}{g}\cdot\frac{1}{1+l/g\mu_s}\right| \ll 1,$$

we have

$$\frac{\delta L}{L} = \frac{\delta g}{g}\cdot\frac{1}{1+l/g\mu_s}\left\{1+\frac{\delta g}{g}\cdot\frac{1}{1+l/g\mu_s}+\left(\frac{\delta g}{g}\cdot\frac{1}{1+l/g\mu_s}\right)^2+...\right\}. \qquad (4.3.38)$$

Similarly for an increase in air gap there is a decrease in inductance

$$L-\delta L = K\frac{1}{g+\delta g+l/\mu_s},$$

which becomes

$$\frac{\delta L}{L} = \frac{\delta g}{g}\cdot\frac{1}{1+l/g\mu_s}\cdot\frac{1}{1+(\delta g/g)\{1/(1+l/g\mu_s)\}}. \qquad (4.3.39)$$

The series for this expression is

$$\frac{\delta L}{L} = \frac{\delta g}{g}\cdot\frac{1}{1+l/g\mu_s}\left\{1-\frac{\delta g}{g}\cdot\frac{1}{1+l/g\mu_s}+\left(\frac{\delta g}{g}\cdot\frac{1}{1+l/g\mu_s}\right)^2-...\right\}. \qquad (4.3.40)$$

The two expressions (eqns (4.3.38) and (4.3.40)) for the sensitivity contain a linear term

$$\frac{\delta g}{g} \cdot \frac{1}{1+l/g\mu_s} \tag{4.3.41}$$

and terms of higher powers of $(\delta g/g)\{1/(1+l/g\mu_s)\}$ which represent the non-linearities of the relationship. It is seen from eqns (4.3.38) and

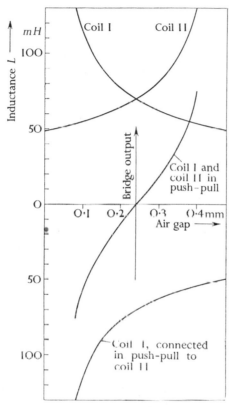

FIG. 4.3.7. Variation of inductance L of typical transducer coils I and II, as measured individually and as combined for push–pull performance.

(4.3.40) that, if a transducer is fitted with two variable-inductance coils in 'push–pull' (i.e. one coil inductance increases whilst the other decreases), the output of such a transducer is proportional to the sum of the fractional changes and thus the even–order terms disappear in the sensitivity expression. Push–pull transducers therefore have a more linear calibration curve than single-coil transducers. Fig. 4.3.7 shows the calibration curve of an actual variable-inductance transducer taken over a wider displacement range than would commonly be used. It indicates the individual coil-inductance variations and the combined effect of both coils in push–pull.

An infinitesimal variation in air-gap length causes a corresponding fractional change inductance

$$\frac{dL}{L} = -\frac{dg}{g} \cdot \frac{1}{1+l/g\mu_s},$$
(4.3.42)

as can be verified directly by logarithmic differentiation of eqn (4.3.34), and which is identical with the linear term in eqn (4.3.40). The negative sign indicates that dL/L increases with decreasing dg/g, and vice versa.

In general, it can be said that the sensitivity of an inductance transducer of this type is improved by reducing the term $l/g\mu_s$. This is achieved with a short core length l and a high incremental permeability μ_s. But if $l/g\mu_s$ is already small in comparison with unity the sensitivity for a given $\delta g/g$ cannot be improved any further. To obtain good linearity the fractional change of gap length must be kept small because only under these conditions are higher orders of $\delta g/g$ negligible. However, this leads to low sensitivities and it is the task of the designer to strike a compromise between the two conflicting requirements. In the case of pure push–pull inductances connected in a bridge circuit it can be shown [10] that linearity extends over a wider range than would be predicted from eqn (4.3.40). This effect is due to cancellation of non-linearities in transducer and bridge circuit. Normally the designer would aim at a fractional change in inductance of 0·1 to 0·2. In such a transducer, assuming push–pull action, the non-linearity will probably be 1–3 per cent at full-scale displacement, depending on the individual design.

Besides the change in inductance of a coil with a variable air gap, do we also have to expect a change in loss resistance? From eqns (4.3.2) and (4.3.6) it seems that since the effective permeability μ is missing in both R_c and R_e they are independent of air gap. Concerning first-order effects only, experiments have shown that it is so. But, unfortunately, the parallel eddy-current loss resistance R_e is inconvenient when investigating the characteristics of inductance transducers in their associated circuits, e.g. in a.c. bridges or tank circuits of oscillators. We must turn it into the equivalent series resistance

$$R_e' = R_e/(1+Q^2),$$
(4.3.43)

which can be easily lumped with the series copper loss resistance. For high values of Q ($Q^2 \gg 1$) and eqn (4.3.6)

$$R_e' = \frac{R_e}{Q^2} = \frac{\omega^2 L^2}{R_e} = \frac{16\pi^2 A t^2 n^2 \mu^2 f^2}{3 \times 10^{14} \rho_i l} \quad (\Omega),$$
(4.3.44)

which shows a variation with μ and thus with air gap.

The hysteresis loss resistance can be safely neglected if we keep inside or close to the Rayleigh region of excitation. Other so-called 'residual' losses are negligible in laminated cores compared with eddy-current losses, but may be of some importance in ferrite cores at higher frequencies.

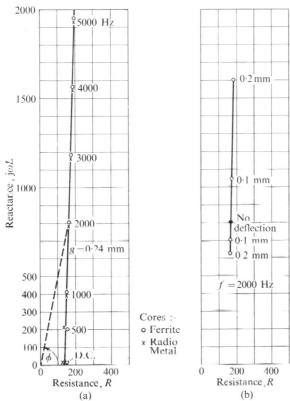

FIG. 4.3.8. Impedance diagram of typical transducer coil: (a) at variable frequency but constant air gap; (b) at variable air gap but constant frequency.

A convenient way of displaying inductance and loss-resistance changes is the impedance diagram. The coil characteristics shown in Fig. 4.3.8 are taken from a small acceleration transducer of which the inductance changes with air gap were shown in Fig. 4.3.7. In Fig. 4.3.8(a) reactance ωL and total series loss resistance R are plotted in a Gaussian plane at various frequencies, but for a fixed air gap. The storage factor Q at any frequency is represented by the angle ϕ according to

$$Q = \tan \phi = \omega L / R. \qquad (4.3.45)$$

The copper loss resistance R_c of the coil is given by the intersection with the real axis. Coil reactance increases approximately linearly with frequency, and since the plotted curve is a straight line within the limits of experimental errors the series loss resistance of this particular coil also seems to increase linearly with frequency. Its variation between d.c. and 500 Hz is, however, only 50 Ω, as compared with about 2000 Ω for ωL. The diagram of Fig. 4.3.8(a) is typical for small transducers with thinly laminated ferromagnetic cores, though ferrite cores have a similar performance, as the measured points in Fig. 4.3.8(a) indicate. For solid mild-steel cores, or at higher frequencies, a pronounced bend in the impedance locus may occur.

Now, if we keep frequency constant at 2000 Hz and vary the air gap we get the impedance locus of Fig. 4.3.8(b). A similar curve results which indicates the strongly non-linear variation of inductance and loss resistance at relatively large changes in air-gap length, and there is a corresponding change in Q-values. All these parameters, and their variations, will be significant when dealing with the associated circuits for inductance transducers in section 4.3.3.

4.3.2.2. *Inductance transducers with ferromagnetic plunger-type cores*

This type of inductance transducer operates on the principle of variation of reluctance in the leakage paths of the coil. It consists essentially of a coil, usually encased in a (slotted) ferromagnetic sleeve, and a (solid) ferromagnetic plunger-type core. The inductance of the coil depends on the length of that part of the core which has penetrated into the coil. The accurate theoretical treatment of such a magnetic system is more complex than that of an inductance coil with a small air gap in an otherwise-closed magnetic path. Its 'equivalent' air gap is large and its geometry is complicated.

With reference to Fig. 4.3.9(a) let

$$l = \text{length of coil (m)},$$
$$r = \text{radius of coil (m)},$$
$$n = \text{number of turns},$$
$$I = \text{current in coil (A)},$$

then the magnetic field strength H (A m^{-1}) along the axis is

$$H = \frac{In}{2l}\left[\frac{l+2x}{\{4r^2+(l+2x)^2\}^{\frac{1}{2}}} + \frac{l-2x}{\{4r^2+(l-2x)^2\}^{\frac{1}{2}}}\right], \qquad (4.3.46)$$

which has been plotted in Fig. 4.3.9(b). It is obvious from this curve that the contribution to a change of inductance will be much smaller if the plunger just penetrates or almost leaves the coil than if it is about half-way through the coil. A fairly linear relationship is expected only in the middle region of the coil.

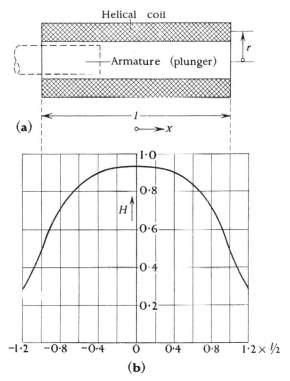

FIG. 4.3.9. Distribution of magnetic field strength along axis of single helical coil.

In a coil combination for push–pull action, as shown in Fig. 4.3.10(a), the field strength along the axis is given by

$$H = \frac{In}{2l}\left[\frac{l-2x}{\{4r^2+(l-2x)^2\}^{\frac{1}{2}}} - \frac{l+2x}{\{4r^2+(l+2x)^2\}^{\frac{1}{2}}} + \frac{2x}{(r^2+x^2)^{\frac{1}{2}}}\right], \quad (4.3.47)$$

which has been plotted in Fig. 4.3.10(b).

If, however, we neglect the non-uniform distribution of field strength in a coil of finite length, an approximate analysis of the plunger-type transducer can be obtained. In a single-layer coil of n turns, length of coil l (m), and radius r (m), the inductance is

$$L = 4\pi^2 n^2 r^2/10^7 l \quad (\text{H}). \quad (4.3.48)$$

If a ferromagnetic core of the same length as the coil and of radius r_c is introduced, the inductance will increase with the increased total flux to

$$4\pi^2 n^2\{r^2+(\mu_m-1)r_c^2\}/10^7 l,$$

where μ_m is an 'effective' permeability of the ferromagnetic core in this particular arrangement. With a core length l_c smaller than the coil length

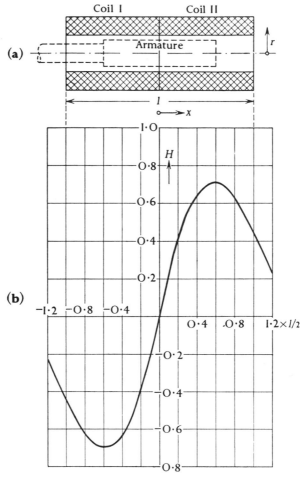

FIG. 4.3.10. Distribution of magnetic field strength along axis of pair of helical push–pull coils.

l, the inductance becomes

$$L = \left[\frac{4\pi^2 n^2}{l_c}\left(\frac{l_c}{l}\right)^2\{r^2+(\mu_m-1)r_c^2\}+\frac{4\pi^2 n^2}{l-l_c}\left(\frac{l-l_c}{l}\right)^2 r^2\right]10^{-7}$$

$$= 4\pi^2 n^2\{lr^2+(\mu_m-1)l_c r_c^2\}/10^7 l^2 \quad \text{(H)}. \tag{4.3.49}$$

It is seen from eqn (4.3.49) that for an increase in l_c by δl_c i.e. if the core in Fig. 4.3.11 is pushed farther into coil I by the amount δl_c, the inductance L increases by δL;

$$L+\delta L = 4\pi^2 n^2\{lr^2+(\mu_m-1)r_c^2(l_c+\delta l_c)\}/10^7 l^2, \tag{4.3.50}$$

and the change of inductance is

$$\delta L = 4\pi^2 n^2 r_c^2 (\mu_m - 1)\delta l_c / 10^7 l^2. \qquad (4.3.51)$$

The fractional change of inductance, which represents the sensitivity of a transducer of this type, becomes

$$\frac{\delta L}{L} = \frac{\delta l_c}{l_c} \cdot \frac{1}{1 + (l/l_c)(r/r_c)^2 \{1/(\mu_m - 1)\}}. \qquad (4.3.52)$$

In a similar way, coil II in Fig. 4.3.11 suffers an identical change of inductance of opposite sign.

It is seen from eqn (4.3.52) that the fractional change of inductance $\delta L/L$ is proportional to the fractional change in core length (i.e. core displacement) multiplied by a factor smaller than unity. To make the factor as large as possible, i.e. to obtain maximum sensitivity, the ratios l/l_c and r/r_c should approach unity and the effective permeability μ_m of the core should be as large as possible.

The plunger-type variable-inductance transducer has characteristics which may be summarized as follows.

(a) Because of the large air path, i.e. the high reluctance of the magnetic path, the sensitivity to a given mechanical movement is lower than that of the transverse-armature type.

(b) For the same reason a greater number of turns is required to attain a given inductance value. Thus a higher self-capacitance of the coil results which, at higher excitation frequencies, may bring the inductance–capacitance combination dangerously close to resonance conditions. This would cause the effective inductance to depend to a high degree upon the stability of its loss resistance.

(c) Since the diameter of the plunger has to be quite small it is usually made of solid mild steel. This results in rather large iron losses and consequently low Q-values of the coil.

(d) Because large portions of the magnetic path are in air it has an appreciable stray field and is therefore open to pick-up from external fields.

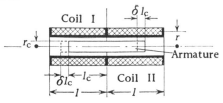

FIG. 4.3.11. Variable inductance transducer with two coils in push–pull and plunger-type armature.

(e) The shape of the coil or coil halves is important for the linearity
and stability of the calibration curve. Coil formers must be stable in
size and shape, while the transverse-armature type which does not
operate on leakage flux distribution, is greatly independent of coil
shape.

Another feature peculiar to the plunger-type transducer is the mutual
coupling between the two coils in a push–pull arrangement. (This is not
critical in transverse-armature types with small air gaps, save in exceptional cases, when the two coils share a common magnetic path.) The
sensitivity of a transducer with mutual coupling between the two push–
pull coils depends on whether the magnetic fields of the coil are in the

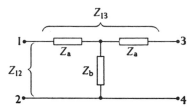

FIG. 4.3.12. Equivalent circuit of two magnetically coupled coils.

same or opposite directions. (In completely coupled coils there would be
mutual flux only and no output would be obtained when the armature was
displaced.)

Two magnetically coupled coils may be represented by the equivalent
circuit of Fig. 4.3.12 in which Z_a is the series impedance of each coil and
Z_b an impedance representing the coupling. Z_b is positive for fields in
opposite directions and negative for fields in the same direction.
$Z_{13} \neq 2Z_{12}$ and is larger or smaller depending on the direction of the
magnetic fields in the two coils. By measurement we may find Z_{12} and
Z_{13},

$$Z_{12} = Z_a + Z_b \quad \text{and} \quad Z_{13} = 2Z_a. \qquad (4.3.53)$$

The complex coupling factor k is defined by

$$k = \frac{Z_b}{Z_a + Z_b} = 1 - \frac{Z_{13}}{2Z_{12}}. \qquad (4.3.54)$$

It is positive for opposing fields and negative for aiding fields. If the load
impedance of the transducer is large, the real part of Z_b which is in series
with the load may be neglected. Its quadrature component produces a
phase shift given by

$$\psi \simeq \tan \psi = \frac{\text{imaginary component of } Z_b}{\text{load resistance}}, \qquad (4.3.55)$$

and thus the load impedance has acquired a factor $(1+j\psi)$. The significance of this phase shift will be appreciated later when the transducer is discussed in conjunction with its associated circuit.

The plunger-core-type transducer can be 'de-coupled' by providing independent magnetic paths for the two coils as shown in Fig. 4.3.13. The centre leg of the E-shaped core, which is inserted into the coil former, reduces the mutual flux considerably and thus avoids the disadvantages of coupling between the two coils.

The main feature of the plunger-type inductance is its longer linear range compared with the more sensitive transverse-armature type. This may be an overriding requirement when large movements are to be

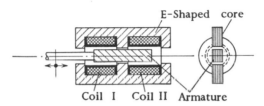

FIG. 4.3.13. Plunger-type inductance transducer with E-shaped laminated cores and armature giving reduced coupling between the coils.

measured. Also, at lower carrier frequencies, the disadvantages as quoted under (b) and (c) above are not relevant, though it may be useful to consider alternatives such as the transformer-type transducer with a plunger core, or the type as described in the next section.

4.3.2.3. Inductance transducers with short-circuit rings or sleeves

This transducer type consists basically of a coil with a short-circuit ring or sleeve of copper or silver which is movable in a direction parallel to the axis of the coil. Fig. 4.3.14(a) shows a single-coil arrangement and Fig. 4.3.14(b) a push-pull version. This is the most effective way of using the reaction of eddy currents generated by the coil current in a conductor positioned in the neighbourhood of the coil to cause a change of coil inductance; but any conductor, say a piece of copper sheet, penetrating the field of a coil will reduce its inductance as is well known from shielded coils, especially at higher frequencies.

Proximity transducers have been designed on this principle, using a conductive metal 'armature', similar to the ferromagnetic transverse armature. Here the effective change in coil inductance depends on distance, thickness, and conductivity of the armature plate, on coil dimensions, and on frequency. Inductance-proximity transducers, therefore, must be calibrated *in situ*, though some general performance characteristics can be predicted. At excitation frequencies in the kilohertz region

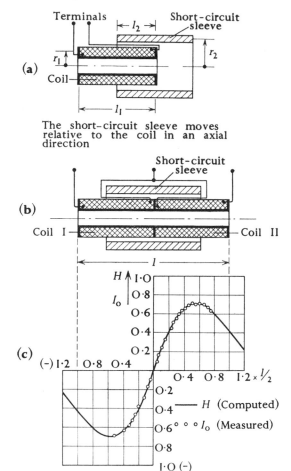

FIG. 4.3.14. Short-circuit ring or sleeve-type transducer: (a) single coil with short-circuit sleeve; (b) two coils in push–pull with short-circuit sleeve; (c) field strength H and output current I_o of (b).

the coil inductance decreases (non-linearly) with decreasing plate-distance. This effect is greatest with thick armature plates made of high-conductivity metals. Under these conditions the resistive component of the coil impedance increases only slightly. A quantitative analysis of coil–plate configurations with a variety of plate distances, thicknesses. and materials is given in reference [11].

Returning to the short-circuit ring-type transducer, the ring represents a secondary winding of a transformer (suffix 2 in Fig. 4.3.14(a)). The voltage across the coil (suffix 1) is

$$v = i_1(Z_1 + \omega^2 M^2 / Z_2), \tag{4.3.56}$$

where, neglecting the small secondary resistance R_2,

$$Z_1 = R_1 + j\omega L_1, \qquad Z_2 = j\omega L_2, \qquad M^2 = k^2 L_1 L_2, \qquad (4.3.57)$$

and thus

$$v = i_1\{R_1 + j\omega L_1(1 - k^2)\}, \qquad (4.3.58)$$

k being the coupling factor derived from the mutual inductance M of coil and ring. The 'effective' inductance L_1' of the coil with the ring coupled to it is

$$L_1' = L_1(1 - k^2). \qquad (4.3.59)$$

The coupling factor of two relatively long coils can be calculated from the partial coupling factors

$$k_{12} = \frac{B_1 A_1}{B_1 A_1} = 1, \qquad (4.3.60a)$$

$$k_{21} = \frac{B_2 A_1 l_2}{B_2 A_2 l_1} = \left(\frac{r_1}{r_2}\right)^2 \frac{l_2}{l_1}, \qquad (4.3.60b)$$

where

  A_1, A_2 = average cross-sectional areas of coil windings,

  B_1, B_2 = flux densities of coils, assumed to be uniform along axis,

  l_1, l_2 = lengths of coils,

  r_1, r_2 = average radii of coil windings.

Then the combined coupling factor is

$$k = \sqrt{(k_{12}k_{21})} = \frac{r_1}{r_2}\sqrt{\left(\frac{l_2}{l_1}\right)}, \qquad (4.3.61)$$

and the effective inductance of a coil with a ring or sleeve around it is

$$L_1' = L_1\left\{1 - \left(\frac{r_1}{r_2}\right)^2\left(\frac{l_2}{l_1}\right)\right\}. \qquad (4.3.62)$$

In a similar way the expression for the change in inductance can be computed to be

$$\delta L_1' = 4\pi^2 10^{-7}\left(\frac{r_1}{r_2}\right)^2\left(\frac{r_1}{l_1}\right)^2 n^2 \delta l_2 = L_1 \frac{\delta l_2}{l_2}k^2, \qquad (4.3.63)$$

and the fractional change in inductance is thus

$$\frac{\delta L_1'}{L_1'} = \frac{L_1(\delta l_2/l_2)k^2}{L_1(1 - k^2)} = \frac{\delta l_2}{l_2} \cdot \frac{k^2}{1 - k^2}. \qquad (4.3.64)$$

This interesting result shows that for perfect coupling between coil and ring ($k = +1$) the sensitivity would be infinite because the effective inductance which occurs in the denominator of eqn (4.8.64) would be zero.

However, L_1' must be finite and of a predetermined value. For a coupling factor $k = 1/\sqrt{2} = 71$ per cent, $\delta L_1'/L_1' = \delta l_2/l_2$, while for smaller k-values the fractional change of inductance will be smaller than the fractional change in sleeve position, and vice versa. These conditions apply strictly only to long coils and ideal coupling factors (eqn (4.3.61)), neglecting fringing. In practice $\delta L_1'/L_1'$ will always be smaller than $\delta l_2/l_2$. For short coils and rings, as they occur in practical transducers, the flux density B_1 of the coil is not constant along its axis but is given by eqns (4.3.46) and (4.3.47) of the previous section. Then the voltage induced in the ring, and hence the current, is proportional to the local field strength. The calibration curve is therefore of the same shape as the variation of field strength along the axis (as plotted in Fig. 4.3.14(c)) as computed from eqn (4.3.47). This curve has been verified experimentally on a small short-circuit ring push–pull transducer by plotting in Fig. 4.3.14(c) the reduced experimental values also. It is seen that they agree with the field-strength plot, even well beyond the region of linear relationship.

In general the short-circuit ring-type transducer has characteristics similar to those of the plunger-core type as described in the previous section, except that iron losses stemming from a (solid) plunger do not occur. Its linearity extends over an appreciable portion of its length, but its sensitivity is comparatively low. Maximum sensitivity can be expected in a push–pull transducer for a sleeve length of about 60 per cent of its over-all coil length.

4.3.3. Associated circuits

In preceding sections of this chapter we have dealt with various inductance-transducer types and their characteristics in terms of $\delta L/L$ as a function of $\delta l/l$, where δl represents in general the finite mechanical displacement of a (magnetic) core or armature. Now we intend to show how to make use of this variation of inductance to produce an electrical voltage or current which eventually will operate an indicator or recorder. There are several basic circuits known to be suitable for practical application, the choice of which depends upon a great variety of conditions.

4.3.3.1. Inductance transducers in a.c. potentiometer circuits

If an alternating voltage E_i is applied to the simple potentiometer circuit of Fig. 4.3.15 the voltage across the unloaded inductance coil will be

$$E_o = E_i \frac{Z}{R+Z}, \qquad (4.3.65)$$

Z being the impedance of the coil and R a series resistance. For a load resistance which is not large in comparison with the coil impedance the relationship would be non-linear, but for simplicity let the potentiometer circuit be followed by the high input impedance of an amplifier. Then the voltage variation δE_o for a given δZ can be shown to be

$$\delta E_o = E_i \frac{R/Z}{R/Z+1} \cdot \frac{\delta Z/Z}{R/Z+\delta Z/Z+1}. \qquad (4.3.66)$$

This is a non-linear relationship between δE_o and the fractional change of impedance $\delta Z/Z$, R/Z being the potentiometer ratio. If the series resistance R is large compared with Z, we have the linear relationship

$$\delta E_o = E_i \frac{Z}{R} \cdot \frac{\delta Z}{Z}. \qquad (4.3.67)$$

A compromise has to be found between good linearity (large R/Z ratio) and high sensitivity (small R/Z ratio). The fractional change in voltage

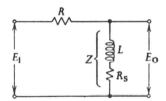

FIG. 4.3.15. Variable-inductance coil in an a.c. potentiometer circuit.

$\delta E_o/E_o$, incidentally, is

$$\frac{\delta E_o}{E_o} = \frac{R}{Z} \cdot \frac{\delta Z/Z}{R/Z+\delta Z/Z+1} \qquad (4.3.68)$$

which, for large potentiometer ratios R/Z, becomes

$$\delta E_o/E_o = \delta Z/Z, \qquad (4.3.69)$$

independent of R/Z.

Since the impedance of an inductance transducer coil can be represented by its series inductance L and its lumped series loss resistance R_s, we write

$$Z = R_s+j\omega L \quad \text{and} \quad \delta Z = \delta R_s+j\omega\,\delta L, \qquad (4.3.70)$$

and with the storage factor Q of the coil,

$$Q = \omega L/R_s, \qquad (4.3.71)$$

the fractional change in coil impedance $\delta Z/Z$, which represents the transducer's sensitivity, becomes

$$\delta Z/Z = \left(\frac{\delta R_s^2+\omega^2\,\delta L^2}{R_s+\omega^2 L}\right)^{\frac{1}{2}} = \left\{\frac{(\delta R_s/R_s)^2}{1+Q^2}+\frac{(\delta L/L)^2}{1+1/Q^2}\right\}^{\frac{1}{2}}, \qquad (4.3.72)$$

i.e. the fractional change of impedance is made up of the fractional change of inductance and the fractional change in loss resistance. The contribution of the loss resistance declines with increasing Q, whereas that of the inductance increases up to a maximum value, so that for $Q^2 \gg 1$ we have

$$\delta Z/Z = \delta L/L, \qquad (4.3.73)$$

even if the fractional change in loss resistance is comparable with $\delta L/L$. Since Z and δZ are complex, $\delta Z/Z$ is complex and its phase angle ϕ, derived from eqn (4.3.70), is

$$\tan \phi = \frac{R_s \omega \, \delta L - \omega L \, \delta R_s}{R_s \, \delta R_s + \omega L \, \delta L}$$

$$= \frac{\delta L/L - \delta R_s/R_s}{(1/Q)(\delta R_s/R_s) + Q \, \delta L/L} \qquad (4.3.74)$$

$$= \frac{(1/Q)(\delta L/L - \delta R_s/R_s)}{(1/Q^2)(\delta R_s/R_s) + \delta L/L} \qquad (4.3.74a)$$

$$= \frac{Q(\delta L/L - \delta R_s/R_s)}{\delta R_s/R_s + Q^2 \, \delta L/L}. \qquad (4.3.74b)$$

From this interesting result we find that the phase angle will be zero if the following is true.

(a) $\delta L/L = \delta R_s/R_s$—that is, if the two fractional changes are equal, i.e. if $\omega \, \delta L/\delta R_s = Q$. This will occur at a frequency $\omega = Q \, \delta R_s/\delta L$. In a vector diagram like that of Fig. 4.3.8(a) the condition $\omega \, \delta L/\delta R_s = \omega L/R_s$ means that a small incremental triangle is similar to the large impedance triangle and that the direction of the vector δZ lies in the direction of, and is an extension of, the vector Z. In many practical transducers of good design this condition is very nearly satisfied.

If $\delta R_s/R_s \neq \delta L/L$ the phase angle ϕ will also be zero if:

(b) Q is very large, i.e. if $\omega L \gg R_s$ (eqn (4.3.74a)),

(c) Q is very small, i.e. if $\omega L \ll R_s$ (eqn (4.3.74b)).

An inductance transducer that conforms to one of these conditions (a) to (c) has no phase angle associated with $\delta Z/Z$ and thus behaves with respect to the fractional voltage change $\delta E_o/E_o$ like a variable resistance with a fractional resistance change $\delta R/R$. The output voltage δE_o, however, will have a phase shift due to the complex factor Z/R, as seen from eqn (4.3.67).

Whilst the a.c. potentiometer circuit, as described in this section, employs the voltage drop across the transducer coil and which is normally fed into the high input impedance of an amplifier, the circuit of Fig. 4.3.16(a) may be used if a current output is required to operate a meter or a galvanometer recorder directly. Neglecting the loss resistance of the coil and disregarding for the moment the shunt capacity C, the indicator current is proportional to the variation of inductance if the indicator resistance is small compared with the coil reactance. For higher values of it the slope of the current curve decreases with increasing current. The inductance–air-gap relationship of the transverse-armature type transducer is also non-linear, and with increasing air gap we have also a

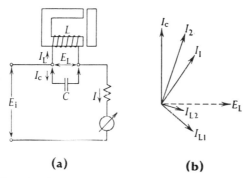

(a)   **(b)**

FIG. 4.3.16. (a) Basic series circuit of an inductance transducer and indicator. Capacitance C shunting transducer coil for compensation of non-linearity. (b) Vector diagram of current in circuit (a) at two different inductance settings.

decline in slope due to increasing leakage in the magnetic circuit, and thus the two non-linearities are additive. Now consider the capacitance C in Fig. 4.3.16(a), which is connected across the coil L. Its value is chosen to give the total current I a leading phase angle. In the vector diagram of Fig. 4.3.16(b) E_L is the voltage across the L–C combination and I_C the condenser current. At a large air gap the coil current I_{L1} will combine with I_C to I_1, but at a reduced air gap the smaller coil current I_{L2} will yield the larger total current I_2. Thus the 'compensating' capacitor C causes a reversal of the sign of the rate of change of the total current with respect to the air-gap change. By virtue of this inversion the two non-linearities— transducer and circuit non-linearity—are opposing and hence compensation is possible. This principle has been applied to inductance transducers in a variable impedance circuit as well as in a bridge circuit [12].

4.3.3.2. *Inductance transducers in a.c. bridge circuits*

The a.c. bridge is surely the circuit in which the majority of inductance transducers are used; no backing-off in its output is required, and it readily accepts push–pull type transducer. Further, if a push–pull transducer is connected in an a.c. bridge any equal changes in the two coils are cancelled in the output. This is an obvious advantage if the transducer is used at variable temperatures. There are monographs on the theory and application of various types of a.c. bridges [13]–[16]; they should be consulted if a more general understanding is required or if a special problem, say the theory of bridge balancing, is to be studied. For the purpose of transducer work we shall analyse here first a 'standard' a.c. bridge, and our main aim will be to correlate transducer and bridge characteristics to obtain workable equations for their over-all performance and design. Later on we shall introduce some modifications to this basic circuit to suit special requirements.

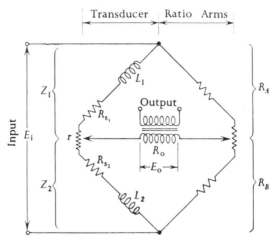

FIG. 4.3.17. Basic a.c. bridge for use with push–pull inductance transducers.

The circuit shown in Fig. 4.3.17 is fed from a (sinusoidal†) a.c. source of low impedance and suitable voltage, the frequency of which is high in comparison with the highest signal frequency to be 'carried'. A variable-inductance transducer of push–pull type has two coils, inductances L_1 and L_2 and series loss resistances R_{s1} and R_{s2}, connected to the ratio-arm resistances R_A and R_B. The centre of the ratio arms is adjustable for balancing purpose. At complete balance an a.c. bridge must satisfy two conditions simultaneously: the real and the imaginary part of the output voltage must be zero. This requires

$$\frac{L_1}{L_2} = \frac{R_{s1}}{R_{s2}} = \frac{R_A}{R_B}. \tag{4.3.75}$$

If only $L_1/L_2 = R_A/R_B$ the bridge will not be completely balanced, since we cannot expect the loss resistance ratio R_{s1}/R_{s2} to exactly equal the inductance ratio L_1/L_2. To correct this the small potentiometer r has been

† Non-sinusoidal excitation will be briefly discussed at the end of this section.

introduced. (A differential condenser across the ratio arms or a combination of a potentiometer and a condenser would be equally suitable [17].)
Although the equation of balance is independent of frequency, this is not
so in practice since the loss resistances R_{s1} and R_{s2} depend on frequency
and their ratio may vary with frequency. Therefore if the input voltage
contains harmonics, or if harmonics are generated in the transducer coils,
the bridge, although balanced for the fundamental, will not be exactly
balanced at higher frequencies, and a residual voltage occurs across the
output terminals.

Let the bridge be initially balanced and the pairs of parameters be
equal:

$$Z_1 = Z_2 = Z \quad \text{and} \quad R_A = R_B = R. \tag{4.3.76}$$

If now the armature of a push–pull transducer is displaced we have

$$Z_1 = Z + \delta Z \quad \text{and} \quad Z_2 = Z - \delta Z, \tag{4.3.77}$$

and the output voltage developed across the load resistance R_0 can be
shown to be

$$E_o = E_i \frac{\delta Z}{Z} \cdot \frac{R_0}{R + Z + 2R_0}, \tag{4.3.78}$$

neglecting a small squared term of $\delta Z/Z$ in the denominator. The
impedance Z of the transducer coil and its variation δZ can be written

$$Z = R_s + j\omega L \quad \text{and} \quad \delta Z = \delta R_s + j\omega \, \delta L, \tag{4.3.79}$$

and from eqns (4.3.78) and (4.3.79) we have, after some rearrangement,

$$E_0 = \frac{E_i R_0}{N_1^2 + N_2^2} \{(N_1 \, \delta R_s + N_2 \omega \, \delta L) + j(-N_2 \, \delta R_s + N_1 \omega \, \delta L)\}, \tag{4.3.80a}$$

where

$$N_1 = R_s(R_s + R + 2R_0) - \omega^2 L^2,$$
$$N_2 = \omega L.(2R_s + R + 2R_0). \tag{4.3.80b}$$

Before discussing this important result we may simplify it by assuming
that the bridge output is connected to a high impedance, e.g. the input
impedance of an amplifier. Then with $R_0 \to \infty$ and $Q = \omega L/R_s$,

$$E_o = \tfrac{1}{2}E_i \frac{R_s \, \delta R_s + \omega^2 L \, \delta L + j(R_s \omega \, \delta L - \omega L \, \delta R_s)}{R_s^2 + \omega^2 L^2}$$

$$= \frac{E_i}{2(1 + 1/Q^2)} \left\{ \frac{\delta L}{L} + \frac{1}{Q^2} \frac{\delta R_s}{R_s} + j \frac{1}{Q} \left(\frac{\delta L}{L} - \frac{\delta R_s}{R_s} \right) \right\}. \tag{4.3.81}$$

Evaluating eqn (4.3.81) leads to the following conclusions.

(a) The output voltage E_o depends on a complex factor consisting of an
in-phase and a quadrature component. Both components contain
the fractional changes of inductance and series loss resistance.

(b) In transducers of equal fractional changes in inductance and loss resistance, i.e. $\delta L/L = \delta R_s/R_s$, the quadrature component of the output voltage is zero. This is very nearly fulfilled in well-designed transducers, as discussed in the previous section. Note that the remaining in-phase component still contains the fractional change of series loss resistance as well as that of inductance.

(c) In transducers with coils of high Q-values the contribution of the fractional resistive change becomes negligible in comparison with that of the inductive change, and the quadrature component decreases with a factor $1/Q$. We thus have as a first approximation

$$E_o = \tfrac{1}{2}E_i(\delta L/L) \quad \text{for high } Q\text{-values.} \qquad (4.3.82)$$

(d) Multiplying numerator and denominator of eqn (4.3.81) by Q^2 reveals that for $Q \ll 1$ the fractional change of the series loss resistance predominates and the quadrature components disappears as in (c). To a first approximation the output voltage becomes here

$$E_o = \tfrac{1}{2}E_i\left(\frac{\delta R_s}{R_s}\right) \quad \text{for low } Q\text{-values.} \qquad (4.3.83)$$

Before the output voltage E_o is fed to an indicator it must be rectified and its carrier content removed. The latter is done by a suitable filter, unless the indicator has a (mechanical) frequency cut-off low enough to suppress the carrier efficiently. A phase-sensitive rectifier, or demodulator [18], [19], eliminates the quadrature component before the bridge output voltage is fed to the indicator but, since rectification normally occurs after amplification, the amplifier must carry both components. It is thus desirable to make the quadrature component, and residual harmonics, as small as possible in order to avoid overloading the amplifier. If Q cannot be made sufficiently large to suppress also the resistive contribution effectively (which may be the case at lower carrier frequencies or with miniature transducers having solid cores) the output will be affected by temperature variations and there may be resistive noise from cables or slip-ring units. However, the resistive contribution can be suppressed in a different way.

Consider eqn (4.3.80a) but with the quadrature component eliminated by phase-sensitive rectification. The resistive contribution in the in-phase component will disappear with $N_1 = 0$; thus from eqn (4.3.80b) we write

$$\omega L = \frac{R_s}{\omega L}(R_s + R + 2R_0) = \frac{Q}{Q^2 - 1}(R + 2R_0). \qquad (4.3.84)$$

Fig. 4.3.18 gives the factor $Q/(Q^2-1)$ for various values of Q. Using this graph and assuming that Q remains roughly constant over a reasonable

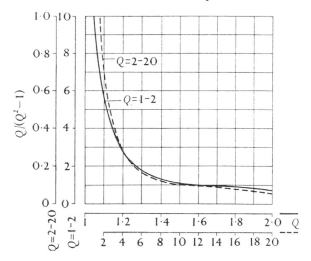

FIG. 4.3.18   The factor $Q/(Q^2-1)$ for various values of Q (see eqn (4.3.84)).

range of coil dimensions and number of turns, a suitable value of L can be computed from eqn (4.3.84).

Another important aspect is matching of transducer, ratio arms, and load. With phase-sensitive rectification the 'useful' power output derived from eqns (4.3.80a) and (4.3.80b) is

$$P_o = \frac{E_i^2 R_0 N_2^2 \omega^2}{(N_1^2+N_2^2)^2} \left\{ \frac{N_1}{\omega N_2} \delta R_s + \delta L \right\}^2. \qquad (4.3.85)$$

If we suppress the resistive contribution by making $N_1 = 0$, the power output is

$$P_o = \frac{E_i^2 R_0}{(2R_s+R+2R_0)^2} \left(\frac{\delta L}{L}\right)^2 \qquad \text{for} \quad N_1 = 0. \qquad (4.3.86)$$

At fixed values of E_i and $\delta L/L$ the maximum power output occurs for

$$R_0 = R_s + \tfrac{1}{2}R. \qquad (4.3.87)$$

Combining these results with those of the preceding section we have two conditions for an optimum design of an inductance transducer and its associated circuit:

(a) freedom from resistive noise, etc. $(N_1 = 0)$,

$$\omega^2 L^2 = R_s(R_s+R+2R_0), \qquad (4.3.88)$$

(b) maximum 'useful' power output (with phase-sensitive rectification),

$$R_s = R_0 - \tfrac{1}{2}R. \qquad (4.3.89)$$

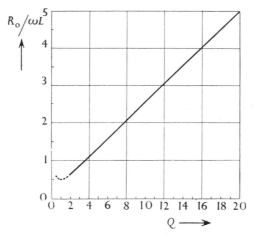

FIG. 4.3.19. Presentation of eqn (4.3.91).

The usual approach, however, is to design the transducer on the basis of other primary considerations, such as size, etc. and then to select the bridge parameters to satisfy the present conditions. Looking at the problem in this way the bridge parameters R and R_0 are given by

$$R = \tfrac{1}{2}\omega LQ(1-3/Q^2), \tag{4.3.90}$$

$$R_0 = \tfrac{1}{4}\omega LQ(1+1/Q^2). \tag{4.3.91}$$

The ratio R/R_0 depends only on Q and is

$$R/R_0 = \frac{2(Q^2-3)}{Q^2+1}. \tag{4.3.92}$$

In Fig. 4.3.19 the values of $R_0/\omega L$ and in Fig. 4.3.20 those of R/R_0 are plotted against Q. At large values of Q we have $R/R_0 \to 2$. At values of $Q < \sqrt{3}$ no values of R/R_0 exist.

Then there is the problem of coupling between the two inductance coils in a push–pull transducer. According to section 4.3.2.2 above the complex coupling factor k has a real component which can be neglected in comparison with the load resistance, and an imaginary component which produces a phase shift ψ. The load impedance thus has the form

$$Z_0 = R_0(1+j\psi). \tag{4.3.93}$$

The output from a partly coupled push–pull transducer in the bridge circuit of Fig. 4.3.17 is

$$E_o = \frac{E_i R_0}{(N_1')^2+(N_2')^2} [\{N_1'(\delta R_s + \psi\omega\,\delta L) + N_2'(\omega\,\delta L + \psi\,\delta R_s)\} +$$

$$+ j\{-N_2'(\delta R_s - \psi\omega\,\delta L) + N_1'(\omega\,\delta L + \psi\,\delta R_s)\}], \tag{4.3.94a}$$

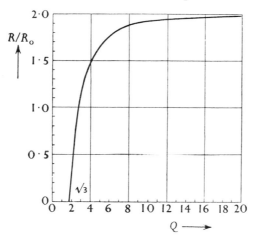

FIG. 4.3.20. Presentation of eqn (4.3.92).

where

$$N_1' = R_s(R_s+R+2R_0)-\omega^2L^2-2\omega LR\psi = N_1-2\omega LR_0\psi,$$
$$N_2' = \omega L(2R_s+R+2R_0)+2R_sR_0\psi = N_2+2R_sR_0\psi. \qquad (4.3.94b)$$

Evaluation of these equations shows that different sensitivities are obtained depending on whether the coils are wound to give aiding or opposing fields. The effect is greatest in transducers with plunger or short-circuit sleeves (unless the coils are decoupled as shown in Fig. 4.3.13) and is negligible in transverse-armature types. In a small push–pull inductance extensiometer with a short-circuit ring and 5 kHz carrier frequency the sensitivity at opposing fields was 70 per cent greater, while a transverse-armature accelerometer at 2 kHz showed an effect of only 2 per cent. Theory and experiment were in good agreement in both cases.

Now we wish to study modifications to the 'standard' a.c. bridge circuit shown in Fig. 4.3.17. An obvious alternative is the 'inverted' circuit of Fig. 4.3.21 in which the input and output terminals have been exchanged.

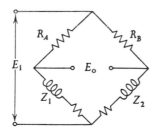

FIG. 4.3.21. Simplified circuit of Fig. 4.3.17 with the output and input terminals exchanged.

Analysis of this bridge circuit leads to an output equation similar to that given above (eqn (4.3.80)) though the parameters N_1 and N_2 are somewhat more complicated and thus less suitable for a general discussion. Under the simplified conditions ($R_0 \to \infty$; $Q \to \infty$; $|Z| = R$) the output is twice that of the circuit of Fig. 4.3.17.

Another alternative which has been used frequently is a bridge circuit with inductive ratio arms. These inductances can either be closely coupled or independent. An a.c. bridge with closely-coupled inductive ratio arms is known as a Blumlein bridge, and we shall discuss this circuit in section 4.4 on capacitance transducers, since its main advantage—freedom from the effect of stray capacitance—is most important in that type of transducer. Although it is also quite suitable for use with inductance transducers, its analysis will not be given here in order to avoid duplication. As to the independent inductive ratio arms it is seen immediately that no difference in sensitivity can be expected if the bridge is feeding into a high impedance, since eqn (4.3.78) yields for $R_0 \to \infty$

$$E_o = \tfrac{1}{2}E_i(\delta Z/Z). \tag{4.3.95}$$

which is independent of the ratio-arm impedance. If, however, R_0 is finite the power output from a bridge with reactive arms will be greater than that with resistive ones, since ideally no output power is dissipated in the reactances. The expression for output current is that given by eqn (4.3.80a) when divided by R_0, but the parameters N_1 and N_2 are here

$$N_1'' = R_s(R_s + 2R_0) - (\omega^2 L^2 + \omega^2 Ll),$$
$$N_2'' = 2\omega L(R_s + R_0) + R_s\omega l, \tag{4.3.96}$$

where l is the inductance of a ratio arm, its loss resistance being neglected. A general analysis is somewhat complex; in special cases, e.g. $l = L$, a numerical comparison can be made by using eqns (4.3.80a) and (4.3.96).

Now we consider the effect of shunt capacity across the input and output terminals of the bridge. Seen from the a.c. source the transducer bridge looks like an inductive impedance. With an appropriate capacitance shunting the bridge input the oscillator load can be made purely resistive and thus the loading characteristic of the generator can be improved. In a similar way a capacity across the output terminals to produce resonance condition improves the output current taken from the bridge by the load. However, in both cases the over-all accuracy of the circuit would depend critically upon frequency stability of the oscillator. Further, if a wide band of signal frequencies has to be transmitted the response of the resonant circuit would introduce an unwanted frequency characteristic, similar to that of a filter. The operation of a.c. bridges at resonance conditions is therefore usually to be avoided.

Finally, a.c. bridge circuits can successfully be operated with square-wave excitation. Apart from bridges for resistance transducers in which square-wave excitation has been employed to suppress the effect of small non-symmetrical capacities [20], miniature square-wave operated 'conditioning units' for use with inductance transducers have been developed recently [21]. The major advantage lies here in the possibility to design stable square-wave oscillators around commercial operational amplifier units which are small, reliable, and cheap. Such properties are of special attraction in aerospace applications where the signal conditioning unit must often be integrated with the transducer it is to serve. A particular system consists of a 3 kHz square-wave oscillator whose amplitude is controlled by a pair of back-to-back Zener diodes, feeding into a second operational amplifier which in turn energizes the transducer bridge circuit. This comprises a variable-inductance push–pull transducer and coupled inductive ratio arms. Bridge-balance is obtained by means of a secondary winding coupled with the ratio arms, in combination with an adjustable resistor in one of the arms. A third operational amplifier will serve as carrier amplifier. This is followed by a phase-lag network for the suppression of the resistive component of the bridge output. The demodulator employs a field-effect transistor switch which is driven by the square-wave oscillator. It provides half-wave phase-sensitive rectification and removal of the quadrature component of the bridge output. Finally, an active filter, also based on an operational amplifier unit, removes the carrier ripple and gives the necessary d.c. amplification for the required output level. A printed-circuit version of the complete conditional unit occupies an area of about 30 mm by 30 mm. It is energized from a centre-tapped 24 V d.c. source.

4.3.3.3. *Inductance transducers in tank circuits of oscillators*

An inductance transducer can be used to control the frequency of an oscillator. This is achieved by connecting the transducer to its tank circuit in which it represents the inductance. It has been found experimentally that the peak energy stored must be at least twice the energy dissipated in the load per cycle in order to obtain a good waveform of the oscillator current. A tank circuit can be represented either by an inductance L in parallel with a capacitance C and a series resistance R_s, or by a parallel arrangement of L, C, and R (Figs. 4.3.22(a) and (b)). In both cases the resistance represents the losses of the L–C configuration and, for the present purpose, also the load which is coupled to the tank circuit. To a first approximation the effective storage factor Q_e is

$$Q_e = \frac{\omega L}{R_s} = \frac{R_p}{\omega L} = \frac{1}{\omega C R_s} = \omega C R_p,$$

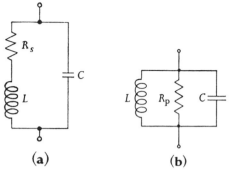

FIG. 4.3.22. Simplified equivalent circuit of a tank circuit of an oscillator.

or

$$R_s R_p = L/C. \tag{4.3.97}$$

Let the r.m.s. voltage across the L–C circuit be E and the r.m.s. current in L and C be I, then the peak energy per cycle stored in C is

$$P_s = E^2 C, \tag{4.3.98}$$

and the energy per cycle dissipated in R_s (where $R_s \ll \omega L$) is

$$P_d = I^2 R_s / f, \tag{4.3.99}$$

where f is the frequency. Thus the ratio of stored to dissipated energy is (Fig. 4.3.22(a)),

$$\frac{P_s}{P_d} = \frac{E^2 C f}{I^2 R_s}, \tag{4.3.100}$$

or, with $E = I/2\pi f C = 2\pi f L I$,

$$\frac{P_s}{P_d} = \frac{f L}{R_s}. \tag{4.3.101-}$$

Now since we require $P_s/P_d > 2$ we have

$$2\pi f L / R_s = Q > 4\pi. \tag{4.3.102}$$

The effective storage factor Q_e of a tank circuit must have a value of 4π if the stored energy is required to be twice the energy dissipated in the load. In our choice of L and C we must now satisfy the condition of matching the parallel resonant resistance R_p of the tank circuit to the oscillator characteristic. R_p is thus decided upon and we have the inductance value of the circuit in Fig. 4.3.22(b) from

$$\omega L = R_p / Q_e. \tag{4.3.103}$$

The required capacitance can be computed from the condition of resonance

$$\omega^2 = 1/LC. \tag{4.3.104}$$

The total wattage $P = E^2/R_p$ to be delivered to the tank circuit then is

$$P = E^2/\omega L Q_e = E^2 \omega C/Q_e. \tag{4.3.105}$$

While the product LC is fixed by the resonant condition (eqn (4.3.104)), the L/C ratio is

$$L/C = E^4/P^2 Q_e^2 \tag{4.3.106}$$

and thus depends on oscillator and tank-circuit design. At fixed voltage E and supply power P a low L/C ratio yields a high Q_e, and thus an improved waveform. On the other hand, for a given voltage E and fixed effective Q_e a small L/C ratio requires a greater power P to drive the tank circuit or, if P is limited, a smaller power output into the load will result. Practical tank-circuit design, then, is a compromise between these two conflicting conditions.

For small changes in L and Q the fractional change in frequency of the circuit of Fig. 4.3.22(a) is

$$\frac{\delta\omega}{\omega} - \frac{\delta Q/Q}{Q^2 - 1} \frac{1}{2}\frac{\delta L}{L}. \tag{4.3.107}$$

At higher values of Q the Q-contribution becomes negligible.

Inductance transducers for use in oscillator tank circuits are single-coil units. Therefore, their calibration curve is non-linear, but since frequency is proportional to $\sqrt{(1/LC)}$, it is of better linearity and may be satisfactory if only moderate changes of inductance are required. Also, in contrast to inductance transducers in bridge circuits, their minimum Q-value is dictated by eqn (4.3.102) based on the operational conditions of the oscillator.

4.3.4. Transformer-type transducers

This type is basically a transformer with variable coupling between two or more iron-cored inductance coils. It is excited on its primary side by a suitable voltage of convenient frequency and its secondary output varies with the coupling between primary and secondary coils (Fig. 4.3.23). With

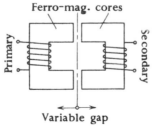

Ferro-mag. cores

Primary   Secondary

Variable gap

FIG. 4.3.23. Basic transformer-type transducer operating on variable coupling between primary and secondary iron-cored coils.

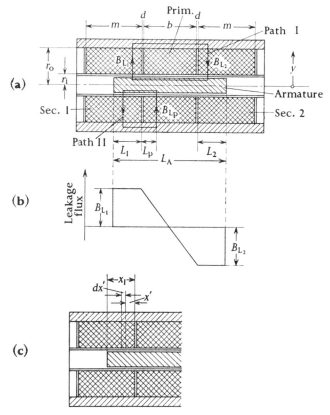

FIG. 4.3.24. Plunger-type differential transformer with primary and secondary coils: (a) cross-sectional view, general; (b) leakage flux distibution of (a); (c) cross-sectional view, particular.

only one secondary coil a compensating or dummy transducer is necessary to back off the transducer output in the no-signal position, unless a backing-off current arrangement at the indicator is preferred. The more usual, and superior, design, however, consists of two secondary coils working in push–pull. Such a transducer is known as a 'linear variable differential transformer' (LVDT). The two output windings are connected in opposition to produce zero output for the middle position of the armature, usually of the plunger type [22], [23].

The most common differential-transformer-type transducer consists (Fig. 4.3.24(a)) of a primary coil of length b and two identical halves of a secondary coil of length m [24]. The coils have an inside radius r_i and an outside radius r_o. The spacing between the coils is d. Inside the coils a ferromagnetic armature of length L_a and radius r_i (neglecting the bobbin thickness) can move in an axial direction. There is also a ferromagnetic stator around the coil assembly. Since the m.m.f. in the iron can be neglected in comparison with that in the air paths of the leakage flux we have for path I in Fig. 4.3.24(a),

$$\mu_o \oint H \, dl = \int_{r_i}^{r_o} (B_{L1} - B_{L2}) \frac{r_i}{y} \, dy = 4\pi A \, 10^{-7}$$

$$= r_i(B_{L1} - B_{L2})\ln(r_o/r_i) \qquad (4.3.108)$$

or

$$B_{L1} - B_{L2} = \frac{4\pi\Lambda}{10^7 r_i \ln(r_o/r_i)} \, (\text{Wb m}^{-2}), \qquad (4.3.108)$$

where $A = I_p n_p$ are the ampere turns of the primary coil and B the appropriate leakage flux densities. A magnetic path II in Fig. 4.3.24(a) at cross-section L_p inside the primary yields

$$B_{L_p} = B_{L1} - \frac{4\pi\Lambda L_p}{10^7 r_i b \, \ln(r_o/r_i)}. \qquad (4.3.109)$$

In general,

$$L_1 B_{L1} + \int_0^b B_{L_p} \, dL_p + L_2 B_{L2} = 0, \qquad (4.3.110)$$

and hence

$$B_{L1} = -B_{L2} \frac{2L_2 + b}{2L_1 + b}. \qquad (4.3.111)$$

Substituting from eqn (4.3.108) we have

$$B_{L1} = \frac{2L_2 + b}{L_a} \cdot \frac{2\pi A}{10^7 r_i \ln(r_o/r_i)}, \qquad (4.3.112a)$$

$$B_{L2} = -\frac{2L_1 + b}{L_a} \cdot \frac{2\pi A}{10^7 r_i \ln(r_o/r_i)}. \qquad (4.3.112b)$$

The distribution of the leakage flux in Fig. 4.3.24(b) has been idealized, for the non-uniformity of the field strength along the axis of the coil has been ignored. Also the flux from the end faces of the armature has been neglected. An armature with curved end faces and of a surface area equal to the cylinder surface of a cylindrical armature of a given length, however, will have only a small end-effect.

Now consider the induced voltage in the secondary coils. In Fig. 4.3.24(c) let $\phi_{x'}$, be the flux linking an elemental coil, width dx', distance x' from the primary. If the armature penetrates a distance x_1 and if the number of turns in the secondary is n_s, the total number of turns in the elemental coil is $n_s \, dx'/m$ and the flux linking it is $\phi_{x'} = 2\pi r_i x' B_{L1}$. Hence the total flux turns linked in the secondary are

$$\int_0^{x_1} \phi_{x'} \frac{n_s}{m} \, dx' = \frac{2\pi r_i B_{L1} n_s}{m} \int_0^{x_1} x' \, dx'$$

$$= \frac{\pi r_i B_{L1} n_s x_1^2}{m} \text{(Wb).}$$

(4.3.113)

With a primary sinusoidal excitation current I_p (r.m.s.) of frequency f, the r.m.s. voltage e_1 induced in the secondary coil 1 is

$$e_1 = \frac{4\pi^3}{10^7} \cdot \frac{fI_p n_p n_s}{\ln(r_o/r_i)} \cdot \frac{2L_2 + b}{mL_a} x_1^2,$$

(4.3.114)

and in coil 2

$$e_2 = \frac{4\pi^3}{10^7} \cdot \frac{fI_p n_p n_s}{\ln(r_o/r_i)} \cdot \frac{2L_1 + b}{mL_a} x_2^2.$$

(4.3.115)

The differential voltage $e = e_1 - e_2$ is thus

$$e = k_1 x (1 - k_2 x^2),$$

(4.3.116)

where $x = \frac{1}{2}(x_1 - x_2)$ is the armature displacement and k_1 the sensitivity e/x,

$$k_1 = \frac{16\pi f I_p n_p n_s (b + 2d + x_0) x_0}{10^7 \ln(r_o/r_i) mL_a} (\text{V m}^{-1})$$

(4.3.117)

with $x_0 = \frac{1}{2}(x_1 + x_2)$ and $k_2 = 1/(b + 2d + x_0)x_0$. k_2 is a factor of non-linearity in eqn (4.3.116), the non-linearity term ε being

$$\varepsilon = k_2 x^2.$$

(4.3.118)

Now it can be shown [24] that for a given accuracy and maximum displacement the over-all length of the transducer will be a minimum for $x_0 = b$, assuming that at maximum displacement the armature does not emerge from the secondary coils. Under these conditions eqns (4.3.116) and (4.3.117) combined read (neglecting $2d$ compared with b)

$$e = \frac{16\pi^3 f I_p n_p n_s}{10^7 \ln(r_o/r_i)} \cdot \frac{2b}{3m} \left(1 - \frac{x^2}{2b^2}\right).$$

(4.3.119)

Thus, to design a transducer of a given maximum displacement x_{max} and with a tolerated error of non-linearity ε, we obtain

$$b = x_{max}/\sqrt{(2\varepsilon)} \quad \text{(length of primary),}$$

(4.3.120)

$$m = b + x_{max} + \delta \quad \text{(length of secondary),}$$

(4.3.121)

δ being a small fraction of the armature diameter added to avoid emergence of the armature from the coils at maximum displacement.

With b thus settled we have

$$L_a = 3b + 2d \quad \text{(length of armature)}, \qquad (4.3.122)$$

and the output voltage can be computed from eqn (4.3.119). In practice r_i may be chosen as a reasonable fraction of the armature length, say $r_i/L_a = 0 \cdot 05$, and an r_o/r_i ratio of between 2 and 8. I_p and n_p of the primary coil are determined by the available winding space and by the primary voltage and frequency. The number of secondary turns should be as large as possible when feeding into a high impedance, or made to match the load impedance. At low frequencies, say below 500 Hz, the computed and measured sensitivities are in good agreement when transducers with solid iron cores are designed to this procedure. At higher frequencies the sensitivity drops appreciably, and at about 2000 Hz it may be little more than half of the theoretical value. This deficiency is obviously due to eddy-current losses in the core and stator, which have been ignored in the analysis. The iron-core losses are also important in another aspect, and so are harmonics: even if the two sides of the secondary coil are well balanced, the no-signal output will not be exactly zero, since the conditions of balance for the harmonics may differ from those for the fundamental. Differing coil capacities will have a similar effect. Proper matching of transducer and load is also advisable for the residual voltage will increase at frequencies below as well as above the flat part of the frequency-response curve of the 'transformer'.

4.3.5. Electromagnetic transducers

Electromagnetic transducers differ from 'conventional' inductance transducers by being bilateral in their mode of operation, i.e. they are capable of not only converting a mechanical input into an electrical output but also an electrical input into a mechanical output. The reader is referred to section 2.3.2, where the generalized theory of bilateral transducers has been treated in detail.

When used as an instrument transducer measuring a mechanical quantity by generating an analogue electrical signal, the electromagnetic transducer consists of an inductance coil wound on a ferromagnetic core with a variable air gap. A unidirectional flux can be produced by passing a constant magnetizing current through an auxiliary coil coupled to the magnetic path. In a practical transducer the magnetizing coil and current can be replaced by a permanent magnet.

Generally, the magnetic energy stored in the coil of n turns is

$$U_{mag} = (n\Phi)^2/2L \quad \text{(W s)} \qquad (4.3.123)$$

where Φ (Wb) is the magnetic coil flux and

$$L = \mu_0 a n^2/d \quad \text{(H)}, \qquad (4.3.124)$$

the coil inductance. $\mu_0 = 4\pi \times 10^{-7}$ (H m^{-1}) is again the permeability of empty space, a (m^2) is the cross-sectioned area of the magnetic core at the air gap, and

$$d = d_1 + \frac{\mu_1}{\mu_2} d_2, \qquad (4.3.125)$$

the 'effective' gap length, where d_1 (m) and d_2 (m) are the actual air and iron path length and μ_1 and μ_2 are the corresponding relative permeabilities. With $\mu_1 = 1$ for air and a current

$$I = n\Phi/L \quad \text{(A)} \qquad (4.3.126)$$

flowing in the coil, the magnetic energy becomes

$$U_{\text{mag}} = \mu_0 an^2 I^2 / 2d, \qquad (4.3.127)$$

and the force acting across the gap

$$f = -\frac{\partial U_{\text{mag}}}{\partial d} = \frac{\mu_0 an^2 I^2}{2d^2} \quad \text{(N)}. \qquad (4.3.128)$$

For simplicity let a constant magnetizing (bias) current I' (A) and a superimposed sinusoidal current of (smaller) amplitude i (A) be flowing in the same coil (Fig. 4.3.25). Then, eqn (4.3.128) can be written

$$f = \frac{\mu_0 an^2}{2d^2} (I'+i)^2$$

$$= \frac{\mu_0 an^2}{2d^2} (I'^2 + 2I'i + i^2), \qquad (4.3.129)$$

where the second term again is the sinusoidal transducer force

$$f_t = \frac{LI'}{d} i. \qquad (4.3.130)$$

Neglecting the coil resistance, we have

$$i = e/j\omega L, \qquad (4.3.131)$$

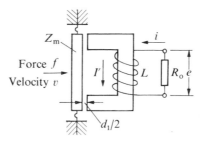

FIG. 4.3.25. Electromagnetic transducer, schematic diagram.

and eqn (4.3.130) becomes

$$f_t = \frac{I'}{j\omega d}\, e, \tag{4.3.132}$$

or, with the mechanical transducer impedance Z_m (see section 2.3.2),

$$f = Z_m v + \frac{I'}{j\omega d}\, e \quad \text{(N)}. \tag{4.3.133}$$

By comparison with eqns (2.22) and (2.28) (pp. 14 and 15) we have the complex transducer constant

$$N_{ef} = N = I'/j\omega d \quad \text{(N V}^{-1}\text{)}, \tag{4.3.134}$$

and since N is imaginary, we also have

$$N_{iv} = -N^* = I'/j\omega d \quad \text{(A s m}^{-1}\text{)}. \tag{4.3.135}$$

The electromagnetic transducer is therefore suitably represented by a current-source-type hybrid matrix (eqn (2.28), p. 15), namely,

$$\begin{bmatrix} f \\ i \end{bmatrix} = \begin{bmatrix} Z_m & I'/j\omega d \\ I'/j\omega d & 1/j\omega L \end{bmatrix}\begin{bmatrix} v \\ e \end{bmatrix}. \tag{4.3.136}$$

A 'negative stiffness'

$$N^2 Z_e = -\frac{I'^2 L}{j\omega d^2} \quad \text{(N s m}^{-1}\text{)}, \tag{4.3.137}$$

occurs, as can easily be verified by converting the hybrid matrix of eqn (4.3.136) into the equivalent impedance form. Its transducer constant would be real and frequency independent, but varying with L.

The characteristic transfer matrix equation of the electromagnetic transducer then follows from eqn (2.30) (p. 15) and eqns (4.3.134) and (4.3.135), thus,

$$\begin{bmatrix} f_t \\ v \end{bmatrix} = \begin{bmatrix} I'/j\omega d & 0 \\ -d/LI' & j\omega d/I' \end{bmatrix}\begin{bmatrix} e \\ i \end{bmatrix}, \tag{4.3.138}$$

which is shown in Fig. 2.3(d) (p. 20).

The transfer frunction, i.e. the dynamic sensitivity, of the electromagnetic instrument transducer can now be obtained from eqn (2.42) (p. 18) and eqns (4.3.134) and (4.3.135),

$$\frac{e}{f_0} = \frac{I'/j\omega d}{(I'^2/\omega^2 d^2) + \{c + j(\omega m - k/\omega)\}(Y_0 + Y + 1/j\omega L)} \quad \text{(V N}^{-1}\text{)}, \tag{4.3.139}$$

where d (m) is the (composite) initial air gap (see eqn (4.3.125)), L(H) is the initial coil inductance, $R_0 = 1/Y_0$ is the (large) indicator resistance, and

I' (A) the (large) magnetizing, or bias d.c. current. m, k, and c represent mass, stiffness and damping, respectively, i.e. the mechanical transducer impedance $Z_m = c + j(\omega m - k/\omega)$.

At large values of R_0 and small coil losses, $(Y_0 + Y) \ll 1/j\omega L$. The frequency dependence of the numerator of eqn (4.3.139), then, cancels that of $1/j\omega L$ in the denominator. The first transition usually occurs at resonance, the second at $\omega \simeq R_0/L$. The frequency response of the electromagnetic force, or pressure, sensor is therefore similar to that of the electrodynamic force sensor (see section 4.1.2), except that the second transition occurs here at a somewhat lower frequency, since the coil inductance of an electromagnetic transducer is usually larger than that of the electrodynamic type. The damping at resonance is complex, and the imaginary term represents a negative stiffness (eqn (4.3.137)) which may, or may not, be significant, depending on the relative values of the mechanical stiffness k and the electrical term $I'^2 L/d^2$.

Incidentally, the transfer function of an electromagnetic transducer operating as a 'sender' can be obtained from eqn (2.45) (p. 19).

$$\frac{-v}{i_0} = \frac{-I'/j\omega d}{\left(\dfrac{I'^2}{\omega^2 d^2}\right) + \{c + j(\omega m - k/\omega)\}(Y_0 + Y + 1/j\omega L)} \quad (\text{m A}^{-1}\text{s}^{-1}). \quad (4.3.140)$$

Like the electrodynamic sender, the response of the electromagnetic sender differs from that of the receiver or sensor, mainly by the relative values of (source) resistance R_0 and coil inductance L. Since, in the sender, R_0 is small, $Y_0 = 1/R_0$ dominates $1/j\omega L$, except at very low frequencies. The relative position of the two transition frequencies $\omega \simeq R_0/L$ and $\omega^2 = k/m$ (resonance) will then depend on the design parameters of a practical transducer. There is also a negative stiffness $I'^2 L/d$, if resonance occurs at the lower of the two transition frequencies.

4.3.6. Magnetostrictive transducers

Magnetostrictive transducers are bilateral in operation and, in acoustic and vibration work, they are used at the generating and receiving end of the systems. This type of transducer is normally designed to operate at mechanical resonance in order to obtain optimum energy transfer from sender to receiver. The design of these transducers is therefore concerned mainly with vibratory characteristics [25], which could be developed in the framework of the unified theory of bilateral transducers discussed in section 2.3.2 [26]. Instrument transducers are normally not operated at resonance and the design philosophy of these instruments follows different lines.

The main attraction of the magnetostrictive instrument of the transducer type is its high mechanical input impedance (i.e. negligible displacement under load), combined with its low electrical output impedance. It thus comes close to the ideal characteristics of a 'displacement-less' transducer with constant-voltage output. There are, however, a number of serious shortcomings in practical magnetostrictive transducers which will be discussed later.

Two materials frequently used in magnetostrictive transducers are pure nickel and a nickel iron alloy of 68 per cent nickel content, generally known as '68 Permalloy'. Some ferrite materials such as 'Ferroxcube B' also have quite sizeable magnetostrictive coefficients, but owing to their brittleness they are hardly suitable for transducer work. The hysteresis

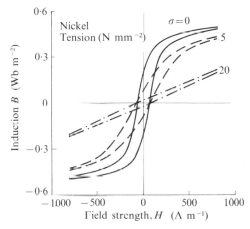

FIG. 4.3.26. Effect of tension on the hysteresis loop of nickel.

curves of nickel at various tensions σ are plotted in Fig. 4.3.26 [27]. With increasing tension in the nickel the slope of the hysteresis curve, i.e. the permeability μ, decreases and there is a large decrease in the remanence value B_r. These two effects can be used in magnetostrictive transducers. The decrease of B_r in nickel with increasing tension σ is known as negative magnetostriction. 68 Permalloy has positive magnetostriction. The curves of Fig. 4.3.27 [27] show a 'squaring' of the loop with increasing tension. The remanence B_r increases, and so does the permeability μ. Under compression the magnetostrictive effect changes its sign, so that alternating forces can be measured with the correct phase relationship. For a fuller understanding of magnetostrictive phenomena, the interested reader is referred to textbooks on ferromagnetism [27].

4.3.6.1. Variable-permeability types

Early magnetoelastic transducers employed the variable permeability of stressed nickel alloys [28]. If parts or the whole of a closed magnetic circuit consist of magnetostrictive material, variations of permeability produced by stresses in the structure will vary the inductance of

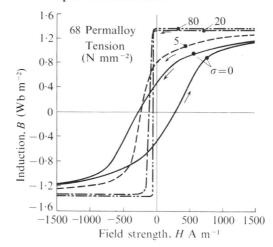

FIG. 4.3.27. Parts of hysteresis loops of 68 Permalloy under tension.

a coil which is linked with the magnetic circuit. Fig. 4.3.28 shows a schematic arrangement of a transducer of this type. Air gaps are not permissible since their reluctance would normally be large compared with the core reluctance and small gap variations would swamp the magnetoelastic effect. The core should be laminated in order to avoid excessive eddy-current losses, though some early transducers employed solid cores which were assembled by welding, thus facilitating the coil-winding process. Push–pull-type transducers are also possible if a coil pair can be arranged with tension and compression in separate cores under a given load, or as a differential transformer [29], [30].

The magnetizing current should be adjusted for maximum inductance. At this point of maximum permeability the inductance value is most sensitive to changes of stress, while at the same time it is virtually independent of current. Reliable data for the stress sensitivity (reversible

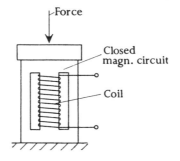

FIG. 4.3.28. Magnetostrictive transducer for the measurement of force, schematic diagram.

permeability variations as a function of stress) of nickel and 68 Permalloy cannot be given, but earlier curves obtained with an unspecified nickel–iron alloy [28] indicate permeability variations of about 10 per cent for stresses of 50 N mm^{-2} and about 24 per cent for 100 N mm^{-2}. In general the maximum stresses employed in instrument transducers are 20 N mm^{-2} for soft nickel and about 80 N mm^{-2} for Permalloy.

The inductance of the transducer also depends on the magnitude and frequency of the exciting current. The current error can be greatly reduced, as shown above, but the frequency must be stabilized. Zero-shift and variation of sensitivity with temperature depend largely upon the particular mechanical and electrical design of the transducer and can amount to 0·2–1 per cent per °C. Temperature-compensating resistances built into the transducer may reduce this error to 0·02–0·05 per cent per °C. The mechanical hysteresis, especially with laminated or built-up cores, can be quite high initially, though may be reduced to about 1 per cent by load and temperature cycling. The total error of practical transducers will be about 2–5 per cent of full scale.

4.3.6.2. *Variable-remanence types*

This type operates on the variation of magnetic induction at remanence. Generally speaking, any reasonable magnetic-induction level in magnetostrictive materials is sensitive to stress, as shown in Figs 4.3.29 and 4.3.30. Only at $B = 0$ and $B = B_s$ (saturation) is it zero. At small stresses the induction variation B_σ (Wb m^{-2}) for a given stress σ (N m^{-2}) can be expressed as a stress sensitivity

$$\Lambda = \left(\frac{B_\sigma}{\sigma}\right)_{\sigma \to 0} \quad \left(\frac{\text{Wb m}^{-2}}{\text{N m}^{-2}}\right). \tag{4.3.141}$$

For instance, the stress sensitivity of 45 Permalloy depends on the bias

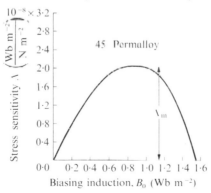

FIG. 4.3.29. Relationship between stress sensitivity Λ and bias induction B_0 for 45 Permalloy [27].

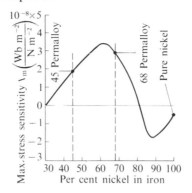

FIG. 4.3.30. The effect of nickel content on maximum stress sensitivity [27].

magnetic induction B_0 (Fig. 4.3.29) with a maximum at Λ_m. It can be shown theoretically [27] that this maximum value occurs at an induction value of $B_0 = (\sqrt{3}/3)B_s$. In Fig. 4.3.30 Λ_m has been plotted for various nickel–iron alloys, and it is seen that 68 Permalloy is close to the region of highest sensitivity (pure nickel occurs at the extreme right-hand side of the curve). In practice, however, for reasons of simplicity and even more so for magnetic stability, operation at the remanence point is preferred, i.e. the transducer core is magnetized to saturation and then the magnetizing current is removed. It is important that further accidental demagnetization be avoided. The Λ_r values at remanence are lower than the Λ_m values; with nickel, for instance, $\Lambda_r = 1 \cdot 5 \times 10^{-9}$ (Wb N^{-1}), as compared with $\Lambda_m = 5 \times 10^{-9}$.

Under compression the magnetic induction B_r in the magnetized nickel core of Fig. 4.3.28 increases. The open-circuit voltage across the coil is proportional to the rate of change of B_r and to the number of turns. Thus

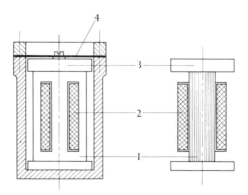

FIG. 4.3.31. Magnetostrictive acceleration transducer: 1, laminations; 2, coil; 3, seismic mass; 4, diaphragm.

an 'acceleration' transducer of this type produces in fact an output proportional to the first time derivative of acceleration. If acceleration is required the transducer output must be integrated with respect to time. The instrument shown in Fig. 4.3.31 may serve as an example for the design procedure [31]. The soft-nickel laminations have the dimensions given in Fig. 4.3.32. The compressive stress σ in the three uprights is proportional to acceleration a (m s^{-2}), thus

$$\sigma = c_\sigma a, \tag{4.3.142}$$

and the variation in magnetic induction $(B_r - B_{r0})$ at remanence is proportional to the stress

$$B_r - B_{r0} = c_B \sigma. \tag{4.3.143}$$

The voltage e_n generated in the coil of n turns is thus

$$e_n \doteq nc_n \frac{dB_r}{dt}. \tag{4.3.144}$$

In order to obtain an output voltage proportional to acceleration proper, e_n can be integrated by feeding it to a capacitance C (F) via a high resistance R (Ω). The voltage across C is

$$e_C = \frac{1}{RC} \int e_n \, dt \tag{4.3.145}$$

or

$$e_C - \frac{n}{RC} c_n c_B c_\sigma a. \tag{4.3.146}$$

The constants c_n, c_B, and c_σ may be worked out from the theory of ferromagnetism or obtained by calibration of a completed transducer.

The core of the transducer shown in Figs 4.3.31 and 4.3.32 was constructed from 29 laminations of pure and well-annealed nickel, 0·31 mm thick, which were bonded by Bakelite resin, thus constituting a

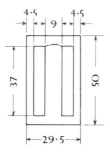

FIG. 4.3.32. Dimensions (mm) of nickel core laminations of transducer of Fig. 4.3.31. Lamination thickness 0·31 mm.

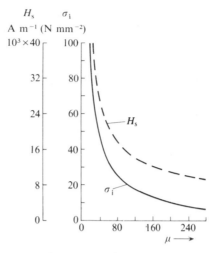

FIG. 4.3.33. Internal stress and saturation field strength plotted against permeability (nickel).

solid structure. Its total mass was 0·083 kg, half of which can be assumed to contribute to the inertia force. The core was loaded by a brass cylinder, the total inertia load amounting to 0·125 kg. At an acceleration a this mass produced a force F,

$$F = 0·125\, a \quad (N). \tag{4.3.147}$$

The stress in the three uprights was

$$\sigma = \frac{F}{A} = \frac{0·125\, a}{162} = 0·77 \times 10^{-3} \quad (N\, mm^{-2}), \tag{4.3.148}$$

where $A = 162\ mm^2$ was the total cross-section of the uprights. The coil consisted of 19 000 turns of copper wire of 0·07 mm diameter and was bonded in the same manner as the laminations in order to withstand large acceleration forces. The prepared laminations were fed into the coil and baked while compressed by means of a special jig. The brass cylinder was soldered on to the core and the complete unit was assembled in a steel housing shown in Fig. 4.3.31. The significance of the diaphragm will be discussed later.

The magnetic design procedure was based on the measurement of permeability of the nickel laminations which in this particular case was $\mu = 83·5$. From theory [27] this μ-value is related to an internal stress in the nickel (see Fig. 4.3.33) of

$$\sigma_i = 2 \times 10^3 / \mu = 24 \quad (N\, mm^{-2}) \tag{4.3.149}$$

which also constituted the upper limit of the useful range of the transducer. Hence, from eqns (4.3.148) and (4.3.149),

$$a_{max} = \frac{\sigma_i}{0.77 \times 10^3} = 31\ 200\ \text{m s}^{-2} = 3180\ \text{g},\qquad (4.3.150)$$

where $g = 9.81$ m s^{-2} is the acceleration due to the earth's gravity. Assuming that, for reasons of safety, the range is restricted to 1600 g, the external
stress is about half of σ_i, i.e. 12 N mm^{-2}. It is noted that the yield stress of
soft nickel is about 40 N mm^{-2}.

The magnetomotive force variation $(M_r - M_{r0})$ was computed from

$$M_r - M_{r0} = M_s \sigma / 4\sigma_i,\qquad (4.3.151)$$

and since the length of the active magnetic path was 74 mm, as compared
with the total length of the magnetic path of 108 mm, the variation of the
magnetic induction at remanence amounted to

$$D_r - D_{r0} = \frac{74}{108} \cdot \frac{B_s}{4\sigma_i} \sigma.\qquad (4.3.152)$$

The magnetic induction at saturation in nickel is approximately

$$B_s = 0.6\ \text{Wb m}^{-2},\qquad (4.3.153)$$

and the induction variation for a given stress σ becomes

$$B_r - B_{r0} = \frac{74 \times 0.6 \times \sigma}{108 \times 4 \times 24} = 4.28 \times 10^{-3}\sigma.\qquad (4.3.154)$$

With eqn (4.3.148) we thus have

$$B_r - B_{r0} = 3.3 \times 10^{-6} a \quad (\text{Wb m}^{-2}).\qquad (4.3.155)$$

These expressions are valid if the core is operating at the remanent
induction obtained from saturation. The magnetic field strength required
for saturation H_s can be read from Fig. 4.3.33 against the permeability μ.
At $\mu = 83.5$ the value of H_s is 17.4×10^3 A m^{-1}. With a coil resistance of
5500 Ω and a total length of the magnetic path of 108 mm an excitation of
1880 ampere-turns will be required, i.e. with 19 000 turns, the saturation
current becomes 99 mA and the coil voltage 540 V.

With a cross-section of $A_s = 80$ mm^2 of the centre core the voltage e_n
generated in the coil will be

$$e_n = nA_s \frac{dB_e}{dt} = \frac{0.264\ n}{10^9} \cdot \frac{da}{dt} \quad (\text{V}),\qquad (4.3.156)$$

and the integrated voltage across a capacitance C

$$e_C = \frac{1}{RC} \int e_n\ dt = \frac{0.264\ n}{10^9 RC} a \quad (\text{V}),\qquad (4.3.157)$$

if the time-constant RC is sufficiently large. The integrated voltage will then be proportional to the applied acceleration a (m s^{-2}), its magnitude also depending on the number of turns of the coil and upon the time-constant. A long time-constant improves the low-frequency response but reduces the voltage output. As an illustration of particular conditions of measurement let $RC = 0 \cdot 1$ s (cut-off frequency about 10 Hz). Then, with $n = 19\,000$ turns we have a sensitivity of

$$e_v \simeq 0 \cdot 5 \text{ mV g}^{-1}. \tag{4.3.158}$$

A recently magnetized transducer exhibits rather large negative over-shoot voltages when calibrated by a transient method. Mechanical ageing eventually reduces this error to a negligible magnitude. Other errors are caused by the leakage flux of the coil and core, though the steel housing of the transducer of Fig. 4.3.31 should provide an effective magnetic screen. However, when the transducer was tilted with reference to the magnetic field of the earth there was an error reading. A U-shaped core with astatically arranged coils would overcome this defect. A housing made of high-permeability material would also be an advantage.

Under transverse accelerations the nickel core experiences bending stresses which, although theoretically self-cancelling, produced an appreci-able error signal. When, however, the seismic mass was guided by a flexible diaphragm, as shown in Fig. 4.3.31, the transverse error was reduced to about 10 per cent of the longitudinal full-scale output.

The experimental sensitivity of the prototype transducer was 53 per cent of the theoretical sensitivity as computed from eqn (4.3.157). Con-sidering the uncertainty of some basic assumptions, and the loss in sensitivity by leakage and during ageing, the agreement between the theoretical and the practical values of sensitivity was reasonably good.

4.3.7. Construction of cores and coils

The inductance element of a variable-inductance transducer consists essentially of a wire-wound coil on a ferromagnetic core and a fer-romagnetic armature. The basic requirements for a good core material are high permeability and low losses, to obtain high sensitivity and high storage factor. But, as we have shown in section 4.3.1.7, the storage factor at a given carrier frequency depends on the over-all design, i.e. on the material and dimensions of core, coil, and air gap, and the design must therefore be balanced to yield a high Q-value without impairing the sensitivity. The core material must also permit operation at reasonably high flux densities without excessive generation of harmonics. Secondary requirements are high Curie temperatures, low cost, and good availability.

Variable-inductance transducers have an air path of diverse length, and very often the effective permeability of the core material is 'diluted' to such an extent that the reluctance of the 'iron' path becomes almost negligible in comparison with that of the air path. For the same reason the core losses are also reduced.

The results of section 4.3.1.7 are of importance in the design of transducers working under 'limit' conditions, e.g. at very high or very low carrier frequencies, and in circuits in which a high Q-value is essential.

TABLE 4.3.3

Properties of some soft magnetic nickel–iron alloys for use in inductance transducers

Type		1	2	3
Commercial name of alloy (Telcon Metals Ltd.)		Standard Mu-metal	Radiometal 50	Hyrho-radiometal
Initial permeability	—	60 000	6000	4000
Maximum permeability	—	240 000	40 000	65 000
Saturation induction	Wb m^{-2}	0·77	1·6	1·4
Remanence	Wb m^{-2}	0·45	1·0	1·0
Coercivity	A m^{-1}	1·0	10	10
Hysteresis loss	J m^{-3} per cycle	3·2	40	45
Curie temperature	°C	350	530	530
Resistivity	$\mu\Omega$ m	0·60	0·45	0·75
Tensile strength, annealed	MN m^{-2}	540	430	430
Young's modulus, annealed	GN m^{-2}	185	170	170
Specific gravity	—	8·8	8·3	8·3

Dust cores are of little use in transducer work because of the low effective permeability of the dust-core material itself which, in effect, introduces a large initial air gap, thus making the actual change of air gap in the transducer less effective. Dust cores are therefore not considered in the section to follow on soft magnetic materials.

4.3.7.1. *Soft magnetic materials*

Table 4.3.3 lists some magnetic, electric, and mechanical properties of three nickel–iron alloys [32], representing three types of soft magnetic materials for use in inductance transducers. A variety of other alloys with similar, or somewhat different, properties are available from commercial sources in this country and from abroad.

Type 1 comprises materials of the highest initial permeability, though they are difficult to handle since cold-working or strain spoils their pedigree characteristics appreciably. Type 2 materials are suitable for most inductance transducer applications, unless a.c. losses are an embarrassment. In this case, Type 3 alloys offer higher resistivities and thus reduced eddy-current losses. (For comparison, conventional 0·4 per cent silicon–iron has initial permeabilities of only about 200–300.)

There is also a range of nickel–iron alloys with well-defined permeability–temperature slopes available which can be used as magnetic shunts in systems needing temperature compensation.

An alternative core material for use in inductance transducers are the magnetically soft ferrites. They consist of mixed crystals of cubic ferrites

TABLE 4.3.4

Properties of some ferrite materials for use in inductance transducers

Ferroxcube (Mullard Ltd.)	Grade	A1	A2	A4	A5	A9	A13
Initial permeability	—	700	1000	1200	1150	3500	1850
Saturation induction	Wb m^{-2}	0·36	0·32	0·36	0·35	0·45	0·34
Power loss	kW m^{-3}	–	200	—	—	200	—
Curie temperature	°C	130	150	140	125	210	130
Resistivity	mΩ m	800	—	200	800	—	1000
Specific gravity	—	4·7	4·8	4·8	4·6	4·8	4·6

(thus the name 'Ferroxcube' [33]) of the general form MFe_2O_4, M representing a divalent metal. The three types in common use are:

Ferroxcube A: Manganese–zinc ferrites;

Ferroxcube B: Nickel–zinc ferrites;

Ferroxcube D: Magnesium–manganese ferrites.

They are homogeneous materials of a black ceramic appearance. Their specific gravities vary between 3·7 and 4·8 (according to grade), and they are as hard and brittle as porcelain. They have high permeabilities and their resistivities are about 10^6 times higher than those of metallic ferromagnetic materials; eddy-current losses are thus negligible. Hysteresis losses are small though the so-called residual losses may become prominent at higher frequencies [6], [34]. In transducer work the A grades are of main interest. Table 4.3.4 lists the major properties of six A grades in common use [33]. The B grades have lower values of initial permeability, flux density, and saturation flux density, higher coercivity and Curie temperature, and extremely high resistivity. Their main application is in the megahertz frequency range. They are also magnetostrictive. The D grades, rectangular $B–H$ loop materials, are used in switching and in memory devices.

With reference to Table 4.3.4, grade A1 is a material of medium permeability and relatively low losses suitable for use at low flux densities. Grade A2 can accommodate higher flux densities, but has somewhat higher losses. Grade A4 is a general purpose material of relatively high permeability and low losses for use in inductors and transformers. It is available in E-, I-, and U-cores and also as rods and toroids. Grade A5 is similar to A4 but has a higher resistivity and a somewhat lower Curie temperature. Low-frequency power applications are best served by the A9 grade permitting high flux densities at low core losses, even at elevated temperatures. Finally, grade A13 is a ferrite of very low losses, high permeability, and high temperature stability, which is available in the form of pot cores or in toroidal form. From this brief survey the designer should be capable of picking the most suitable ferrite material for his particular application, though he may wish to consult the manufacturer's literature for more detailed information on properties and available shapes and sizes of cores and armature suitable for inductance transducers.

4.3.7.2. Inductance coils

Large coils for use in inductance transducers are normally wound with enamelled copper wire on insulating coil formers which are offered by most core manufacturers in various shapes and sizes. In small transducers winding space is at a premium and self-bonding wire, requiring no former, would usually be preferred. 'Conybond' [35] wire with vinyl acetal insulation, and 'Conysolbond' [35] wire with polyurethane insulation fall into that category; similar products can be obtained from other sources in this country and from abroad. Rigid self-supporting coils are manufactured by passing a heating current through the coil, under controlled conditions, thus raising its temperature to about 150 °C, while it is still in its forming jig. These wires have also self-fluxing properties and may be soldered or dip-tinned.

Occasionally, coils of inductance transducers are wound with resistance wire with low or predetermined temperature coefficient, for the purpose of reducing or compensating temperature effects in the transducer. This is quite permissible, especially at higher carrier frequencies, when the 'copper' losses are small in comparison with the core losses. It is seen from Table 4.3.2. (p. 175) that an increasing ρ_c reduces the magnitude of Q_{max} but also shifts its occurrence towards higher frequencies. Thus, at carrier frequencies above f_{max} the Q-value of the coil will not change appreciably.

Earlier transducer terminals with organic insulation are now commonly replaced by miniature sintered glass or ceramic metal seals. These can either be obtained complete with an outer metal sleeve which is being soldered into the transducer body, or sintered *in situ* into the housing where lack of space forbids the use of even the smallest commercial types.

  The connecting cable of an inductance transducer, by virtue of the low coil impedance, can be used in great length and need be screened only on rare occasions. Anyhow, electrostatic screening is of little use at low impedances, but earth loops, caused by more than one earthing point along the line, must be avoided. The shunting capacity provided by the cable capacity must be considered when the transducer is calibrated with a cable different in length from that required for the actual measurement. At high carrier frequencies the cable length is limited as explained in section 4.3.1.9, and a cable of low capacity should be used. Screening may then be an advantage.

References

1. J. CARLSTEIN. Angle errors in resolvers used as synchros. *Control Engng* **10**, 79–80 (1963).
2. S. FARHI. Characteristic errors of a loaded synchro transmitter. *IEEE Trans. Appl. Ind.* (*Inst. elec. electron. Eng.*) **83**, 439–443 (1964).
3. C. LANG. Errors in common-wire synchros and resolver systems. *Electro-Technology* N.Y. **76**, 75–76 (1965).
4. E. HÖLSCHER. *Drehmelder-Systeme; Anwendung und Berechnung.* Oldenbourg, Munich (1968).
5. V. G. WELSBY. *The theory and design of inductance coils.* Macdonald, London (1950).
6. E. C. SNELLING. *Soft ferrites; properties and applications.* Butterworth, London (1969).
7. LORD RAYLEIGH. Notes on electricity and magnetism, III: On the behaviour of iron and steel under the operation of feeble magnetic forces. *Phil. Mag.*, Ser. 5, **23**, 225–45 (1887).
8. P. K. MCELROY. How good is an iron-cored coil? *Gen. Radio Expr*, March issue (1942).
9. P. K. MCELROY. Those iron-cored coils again. *Gen. Radio Expr*, December issue (1946) and January issue (1947).
10. P. E. BURRY. A 64 kc/s inductance bridge and demodulator for use with push–pull inductance transducers. R.A.E. Technical Note No. IR24 (1963).
11. F. FÖRSTER. Theoretische und experimentelle Grundlagen der zerstörungsfreien Werkstoffprüfung, I; Das Tastspulverfahren. *Z. Metallk.* **43**, 163–171 (1952).
12. U.S. Patent No. 2361173 (1943).
13. B. HAGUE. *Alternating current bridge methods.* Pitman, London (1946).
14. H. E. THOMAS and C. A. CLARKE (ed.). *Handbook of electrical instruments and measuring techniques.* Prentice-Hall, Englewood Cliffs, New Jersey (1967).
15. R. KAUTSCH. Induktive Messgrössenumformung. *Arch. tech. Messen* **J86-16** to **J86-19** November issue (1970)–February issue (1971).
16. H. HELKE. *Messbrücken und Kompensatoren für Wechselstrom.* Oldenbourg, Munich (1971).
17. H. K. P. NEUBERT. *Strain gauges; kinds and uses,* p. 63. Macmillan, London (1967).
18. D. G. TUCKER. *Modulators and frequency changers for amplitude modulated line and radio systems.* Macdonald, London (1953).
19. J. B. GRIMGLEBY and D. W. HARDING. A new high-performance phase-sensitive detector. *J. scient. Instrum.* **4**, 941–4 (1971).
20. C. ROHRBACH. *Handbuch für elektrisches Messen mechanischer Grössen,* p. 354 V.D.I. Verlag, Düsseldorf (1967).
21. W. R. MACDONALD. Unpublished R.A.E. report (1973).
22. H. SCHAEVITZ. The linear variable differential transformer. *Proc. Soc. exp. Stress Analysis* **4**, 79 (1946).

23. E. E. HERCEG. *Handbook for measurement and control.* Schaevitz Engineering, Pennsauken, New Jersey (1972).
24. P. D. ATKINSON and R. W. HYNES. Analysis of a linear differential transformer. *Elliott J.* **2**, 144–151 (1954).
25. T. F. HUETER and R. H. BOLT. *Sonics*, Chap. 5. Wiley, New York (1955).
26. L. CREMER and M. HECKL. *Körperschall*, p. 65. Springer, Berlin (1967).
27. R. M. BOZORTH. *Ferromagnetism.* Van Nostrand, New York (1951).
28. W. JANOWSKI. Über die magneto-elastische Messung von Druck-, Zug- und Torsionskräften. *Z. tech. Phys.* **14**, 466–72 (1933).
29. O. DAHLE. Der Pressduktor, eine neue Hochleistungsmessdose für die Schwerindustrie. *ASEA Z.* **5**, No. 1 (1960).
30. J. CHASS. The variable-mu pressure transducer. *I.S.A. Jl,* July issue, 71–4 (1963).
31. H. WILDE and E. EISLER. Ein Beschleunigungsmesser auf magnetostriktiver Grundlage. *Z. angew. Phys.* **1**, 359–66 (1949).
32. ANON. *Mumetal alloys; Radiometal alloys.* Telcon Metals, Ltd., Crawley, Sussex.
33. ANON. *Ferroxcube, PXE, permanent magnets.* Mullard Technical Handbook 3, Part 2. Mullard, Ltd., London (1971).
34. K. J. STANDLEY. *Oxide magnetic materials.* Clarendon Press, Oxford (1972).
35. Manufactured by Connolys (Blackley) Ltd., Liverpool.

4.4. VARIABLE-CAPACITANCE TRANSDUCERS

FIG. 4.4.1 shows examples of basic types of capacitance transducers. A variation in capacitance can be obtained by changing:

(a) the distance between two or more parallel electrodes (Fig. 4.4.1(a));

(b) the dielectric constant (relative permittivity) of the dielectric or parts of it (Fig. 4.4.1(b));

(c) the area of close proximity between electrodes constituting the capacitance proper (Fig. 4.4.1(c));

(d) the effective area of close proximity between serrated electrodes (Fig. 4.4.1(d)).

In practice case (a) is of major importance. The dielectric may be simply an air gap or may consist of several layers of different permittivities. Case (b) is a special case of (a) for the measurement of dielectric constant of, say, an insulation tape with the tape moving through the condenser. Case (c) illustrates one design of the variable-area type which comprises a great variety of similar arrangements, e.g. plunger in tube, transverse movement of two plates, etc. Case (d) is a special case of (c) suitable for short movements. A large number of capacitance transducers in practical use are of the variable air-gap type (case (a)) operating in the manner of a condenser microphone. The change of capacitance is obtained by means of a deflected membrane or thin diaphragm, rather than by a piston-like movement of a rigid plate. This design will be treated in sections 4.4.2.4 and 4.4.2.5 as far as it is relevant to the more general use of instrument transducers. The dynamic characteristics of d.c. polarized

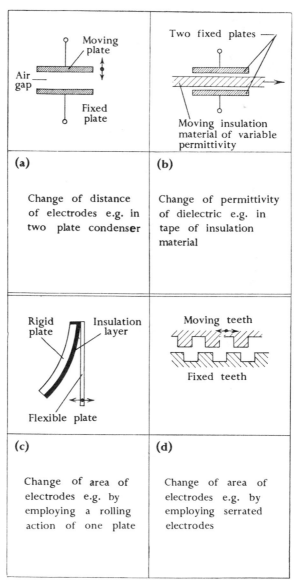

FIG. 4.4.1. Examples of variable capacitance for use in transducers.

'electrostatic' microphones will be discussed in section 4.4.3.2, though the reader interested in microphone technology is advised to consult the extensive literature on this subject [1], [2], [3].

4.4.1. The equivalent circuit

For most applications variable capacitance transducers can be represented by a pure capacitance C. Table 4.4.1 summarizes the design formulae of some basic condenser arrangements and Table 4.4.2 lists relative permittivity values of a variety of dielectrics and insulation materials.

Even at very high carrier frequencies (several megahertz) and with very small capacitance values, losses are normally small in capacitance transducers with the exception of transducers measuring moisture content of fabrics, etc., and in some high-temperature transducers with solid dielectrics. At low frequencies the losses of a condenser are represented by a parallel resistance R_p (Fig. 4.4.2) which comprises:

(a) d.c. conductance, usually negligible;

(b) dielectric losses in the insulating supports of the live electrodes;

(c) losses in gap dielectrics.

The dielectric losses (b) in the insulating structure constitute a conductance component of $1/R_p$ which increases with frequency, and the dissipation factor $(1/R_p\omega C)$ is thus roughly independent of frequency. The gap losses in an air condenser are normally negligible; but with solid dielectrics they clearly depend on the low-frequency dissipation of the dielectric material used. Losses caused by interfacial polarization usually decline with increasing frequency. At high frequencies the series resistance R_s in Fig. 4.4.2 represents the resistance of the leads, metal supports, and plates of the condenser. Its magnitude is significant only if eddy currents reduce the effective cross-section (skin effect) or in thin-film constructions. A more detailed discussion of the properties of dielectrics can be found in the literature [4], [5].

The total dissipation factor of a capacitance transducer is thus

$$D = \frac{1}{R_p\omega C} + R_s\omega C. \tag{4.4.1}$$

In general, the first term of eqn (4.4.1) has a slight negative slope with frequency. The second term has a pronounced positive slope. This means, as has been explained in detail in section 4.3.1.5 on the equivalent circuit of iron-cored inductances, that a frequency f_m exists at which the dissipation is a minimum, or the Q-value of the condenser a maximum.

TABLE 4.4.1

Design formulae for some basic condenser configurations

Geometrical arrangement	Design formulae of capacitance neglecting fringing (F)	Notations,etc.
	$\varepsilon \varepsilon_0 A/d$	A = Area of one electrode (m²) d = Distance of electrodes (m) ε = Relative permittivity of dielectric $\varepsilon_0 = 8 \cdot 854 \times 10^{-12}$ (F m⁻¹) = Permittivity of empty space
	$\dfrac{\varepsilon_0 A}{\left(\dfrac{d_1}{\varepsilon_1} + \dfrac{d_2}{\varepsilon_2} + \dfrac{d_3}{\varepsilon_3} \right)}$	Indices 1,2,3, indicate layers of different permittivity ε and thickness d
	$2\varepsilon \varepsilon_0 A/d$	A condenser of n alternately connected plates has $(n-1)$ times the capacitance of one pair of plates
	$2\pi \; \varepsilon \varepsilon_0 \, l/ln \; (R/r)$	l = length of cylinder (m)
	$\pi \; \varepsilon \varepsilon_0 \, l \, (R+r) \, / \, (R-r)$	For thin layers of dielectric

TABLE 4.4.2

Relative permittivity of some dielectrics

Material	Relative permittivity
Vacuum	1·00000
Air, dry	1·00054
PTFE	2·1
Polythene	2·3
Silicone oil	2·7
Araldite	3·3
Silica	3·8
PVC	4·0
Quartz	4·5
Glass	5·3–7·5
Porcelain	5·5–7
Mica	7
Alumina	8·5
Water	80
Barium titanate	1000–10 000

The series inductance L in Fig. 4.4.2 represents the total inductance of the current path between the terminals of the transducer. If a cable is attached to the transducer, L includes also the cable inductance. At any frequency $f = \omega/2\pi$ below resonance the effective capacitance C_e of the transducer is thus increased to

$$C_e = C/(1 - \omega^2 LC), \tag{4.4.2}$$

which is analogous to the effective inductance of section 4.3.1.9, where a capacitance was shunting the coil inductance. The effective fractional change in capacitance,

$$\frac{\delta C_e}{C_e} = \frac{\delta C/C}{1 - \omega^2 LC}, \tag{4.4.3}$$

is also increased.

4.4.2. Sensitivity and linearity

4.4.2.1. Capacitance transducer with solid dielectric and variable air gap between parallel plates

Neglecting fringing, the capacitance C (F) of a condenser with parallel plates, each of an area A (m²), and in layers of various dielectric

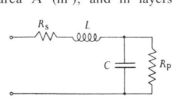

FIG. 4.4.2. Equivalent circuit of a variable-capacitance transducer.

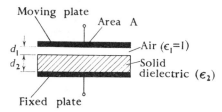

FIG. 4.4.3. Variable capacitance with solid dielectric and variable air gap between parallel plates.

materials, thicknesses $d_1, d_2,... d_n$, and relative permittivities $\varepsilon_1, \varepsilon_2, \varepsilon_3,... \varepsilon_n$ is (see Table 4.4.1)

$$C = \varepsilon_0 A \Big/ \left(\frac{d_1}{\varepsilon_1} + \frac{d_2}{\varepsilon_2} + \frac{d_3}{\varepsilon_3} + ... \frac{d_n}{\varepsilon_n}\right) \quad \text{(F)}. \tag{4.4.4}$$

$\varepsilon_0 = 1/(36\pi \times 10^9) = 8 \cdot 854 \times 10^{-12}$ (F m^{-1}) is the permittivity of empty space. A variable transducer may be designed to have two plates with a variable air gap ($\varepsilon_1 = 0$) of thickness d_1, and a layer of insulation material of thickness d_2 and permittivity ε_2 (Fig. 4.4.3). The capacity then is

$$C = \frac{\varepsilon_0 A}{d_1 + d_2/\varepsilon_2}. \tag{4.4.5}$$

If the air gap d_1 is decreased by δd_1, the capacity will increase by δC, hence

$$C + \delta C = \frac{\varepsilon_0 A}{d_1 - \delta d_1 + d_2/\varepsilon_2}, \tag{4.4.6}$$

and the fractional change in capacitance yields

$$\frac{\delta C}{C} = \frac{\delta d_1}{d_1 + d_2} \cdot \frac{1}{1/N_1 - \delta d_1/(d_1 + d_2)}, \tag{4.4.7}$$

where

$$N_1 = \frac{d_1 + d_2}{d_1 + d_2/\varepsilon_2} = \frac{1 + d_2/d_1}{1 + d_2/d_1\varepsilon_2}. \tag{4.4.8}$$

The total spacing $(d_1 + d_2)$, rather than d_1, has been used in eqn (4.4.7), because this is the actual distance between the plates. Rearranging eqn (4.4.7) we have

$$\frac{\delta C}{C} = \frac{\delta d_1}{d_1 + d_2} N_1 \frac{1}{1 - N_1 \delta d_1/(d_1 + d_2)}. \tag{4.4.9}$$

N_1 is the sensitivity factor; it depends on the ratio d_2/d_1 of the dielectric layer thicknesses and permittivity ε_2 of the solid dielectric. But N_1 is also the factor of non-linearity, as is clearly seen if eqn (4.4.9) is expanded into a series for $N_1 \delta d_1/(d_1 + d_2) \ll 1$, i.e. for small displacements,

$$\frac{\delta C}{C} = \frac{\delta d_1}{d_1 + d_2} N_1 \left\{1 + N_1 \frac{\delta d_1}{d_1 + d_2} + \left(N_1 \frac{\delta d_1}{d_1 + d_2}\right)^2 + \left(N_1 \frac{\delta d_1}{d_1 + d_2}\right)^3 + ...\right\}. \tag{4.4.10}$$

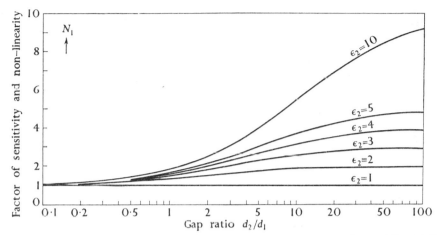

Fig. 4.4.4. Sensitivity and non-linearity factor N_1 of a variable capacitance (Fig. 4.4.3) plotted against the layer-thickness ratio d_2/d_1 at various permittivities ε [6].

Sensitivity and non-linearity thus increase with increasing permittivity ε_2 of the solid dielectric. For high values of d_2/d_1, i.e. for small air gaps, N_1 approaches ε_2. For values of $\varepsilon_2 > 1$, sensitivity and non-linearity also increase with d_2/d_1, but N_1 cannot fall below unity, even for very small values of d_2/d_1. The factor N_1 has been plotted in Fig. 4.4.4 against the thickness ratio d_2/d_1 for several values of ε_2. In a push–pull arrangement of two variable capacitances in a transducer the even harmonics in eqn (4.4.10) cancel, and linearity is improved appreciably.

To eliminate the effect of fringing (which has been neglected above) on the transducer capacitance, a constant electric field right up to the edges of the plates can be obtained by using a guard ring, as shown in Fig. 4.4.5. The guard ring is connected to the potential of the plate it surrounds. With no guard ring, laborious field mapping in the fringe areas would be necessary (see also section 4.4.2.3) if the exact capacitance value is required.

4.4.2.2. *Capacitance transducer with solid dielectric of variable permittivity or thickness and air gap between parallel plates*

Fig. 4.4.6 shows a capacitance C (F) of a condenser with guard ring and two parallel plates of active area A (m^2) at a distance a (m) apart. With a

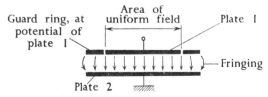

Fig. 4.4.5. Parallel-plate condenser with guard ring.

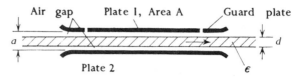

Fɪɢ. 4.4.6. Variable capacitance with solid dielectric of variable permittivity and air gap between parallel plates with guard plate.

solid dielectric material of constant thickness d (m), but variable permittivity ε, running through the gap, the capacitance is

$$C = \frac{\varepsilon_0 A}{a-d+d/\varepsilon} \quad \text{(F).} \tag{4.4.11}$$

An increase in permittivity of the solid dielectric by $\delta\varepsilon$, for instance due to a change in the moisture content of a fabric, will increase the capacitance by δC,

$$C+\delta C = \frac{\varepsilon_0 A}{a-d+d/(\varepsilon+\delta\varepsilon)}. \tag{4.4.12}$$

Hence the fractional change in capacity $\delta C/C$ can be shown to be

$$\frac{\delta C}{C} = \frac{\delta\varepsilon}{\varepsilon} N_2 \frac{1}{1+N_3(\delta\varepsilon/\varepsilon)}, \tag{4.4.13}$$

where the sensitivity factor

$$N_2 = \frac{1}{1+\varepsilon(a-d)/d} \tag{4.4.14}$$

and the non-linearity factor

$$N_3 = \frac{\varepsilon(a-d)/d}{1+\varepsilon(a-d)/d} = \frac{1}{1+\{d/\varepsilon(a-d)\}}, \tag{4.4.15}$$

and for small variations $N_3(\delta\varepsilon/\varepsilon) \ll 1$ we can write

$$\frac{\delta C}{C} = \frac{\delta\varepsilon}{\varepsilon} N_2 \left\{ 1 - N_3 \frac{\delta\varepsilon}{\varepsilon} + \left(N_3 \frac{\delta\varepsilon}{\varepsilon}\right)^2 - \left(N_3 \frac{\delta\varepsilon}{\varepsilon}\right)^3 + \ldots - \ldots \right\}. \tag{4.4.16}$$

The sensitivity factor N_2 and the non-linearity factor N_3 are plotted in Fig. 4.4.7(a) and (b) against the thickness ratio $d/(a-d)$ for various permittivities. Best results are obtained with high $d/(a-d)$ ratios, i.e. with thick layers of the test material and small air gaps. It is also seen that a material of low permittivity gives highest sensitivity and lowest non-linearity, particularly at medium values of $d/(a-d)$. For the measurement of the moisture content of fabrics, etc., the sensitivity must be found experimentally. Since the dielectric losses of moist fabrics are high there may be a contribution to the output due to the variation of the dielectric losses.

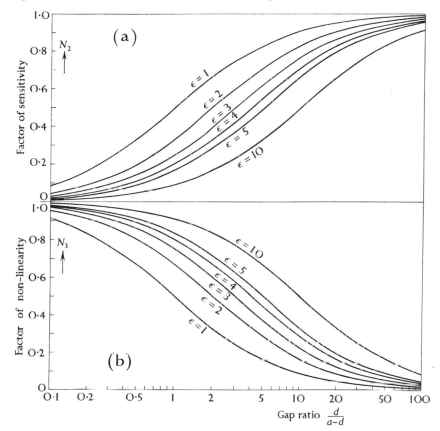

FIG. 4.4.7 (a) Sensitivity factor N_2 and (b) non-linearity factor N_3 of a variable capacitance (Fig. 4.4.6) plotted against the layer thickness ratio $d/(a-d)$ at various permittivities ε.

The arrangement of Fig. 4.4.6 can also be used to measure the thickness variation of solid test material of constant permittivity. With ε constant but d variable the fractional change in capacity can be shown to be

$$\frac{\delta C}{C} = \frac{\delta d}{d} N_4 \frac{1}{1-N_4(\delta d/d)},$$
(4.4.17)

where

$$N_4 = \frac{\varepsilon-1}{1+\varepsilon(a-d)/d}.$$
(4.4.18)

For $N_4(\delta d/d) \ll 1$, eqn (4.4.17) yields

$$\frac{\delta C}{C} = \frac{\delta d}{d} N_4 \left\{ 1 + N_4 \frac{\delta d}{d} + \left(N_4 \frac{\delta d}{d}\right)^2 + \left(N_4 \frac{\delta d}{d}\right)^3 + \dots \right\}.$$
(4.4.19)

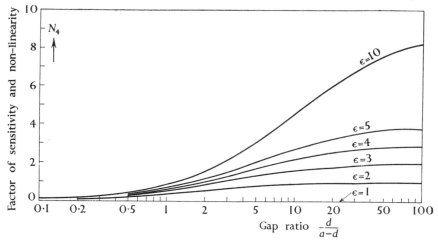

FIG. 4.4.8. Sensitivity and non-linearity factor N_4 of a variable capacitance (Fig. 4.4.6) plotted against the layer-thickness ratio $d/(a-d)$ at various permittivities ε.

The sensitivity and non-linearity factor N_4 is plotted in Fig. 4.4.8 against thickness ratio $d/(a-d)$. The character of the curves are similar to those of Fig. 4.4.4 and similar conclusions can be drawn for optimum design conditions.

4.4.2.3. Capacitance transducer with serrated plates

The serrated-plate condenser has been used for the measurement of small angular vibrations [6]. Consider a pair of flat serrated plates, as shown in Fig. 4.4.9. A small relative movement in the plane of the plates causes a change in capacitance. This movement must be kept small in comparison with the length of the teeth, otherwise the capacitance–displacement relationship yields ambiguous results. Neglecting for the moment fringing at the corners of the teeth the capacitance of the two plates in air is

$$C = \varepsilon_0 nlb/d \quad \text{(F)}, \tag{4.4.20}$$

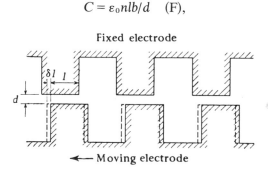

FIG. 4.4.9. Variable capacitance consisting of two serrated plates with n pairs of teeth.

where the dimensions (m) are

l = active length of tooth pair,
b = width of teeth, normal to plane of drawing,
d = distance between teeth in close proximity,
n = number of pairs of teeth.

The variation in capacitance δC due to a small displacement δl is

$$\delta C = \frac{\varepsilon_0 nb(l+\delta l)}{d} - C, \tag{4.4.21}$$

and hence the fractional change in capacity

$$\delta C/C = \delta l/l. \tag{4.4.22}$$

However, this result is only an approximation to the actual characteristics of a serrated-plate transducer. If the active length l of a pair of teeth is not large in comparison with the air gap d, the leakage flux in the fringing areas is an appreciable portion of the total flux. The actual flux distribution can be found by mapping the electric field between the teeth. This can be done by conformal representation [7], [8] with the object of reducing complicated field distributions to more basic ones. An electric field map for the gap and corner configuration, obtained in this way [8], is shown in Fig. 4.4.10. Good approximation of complex electrical field distributions can be obtained by an experimental analogy, the electrolytic-trough method, or by mapping 'curvilinear squares' of equal flux and equal permittance [8], [9].

Consider the shaded area in Fig. 4.4.10. It contains six parallel flux tubes stretching from one parallel face to the other. There are also six

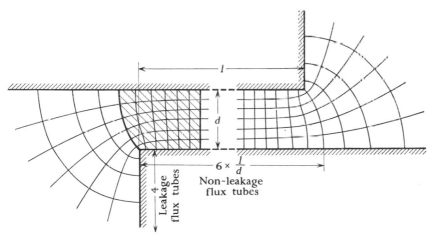

FIG. 4.4.10. Electric field distribution map of one pair of teeth of Fig. 4.4.9.

portions of equal permittance in each, constituting equipotential lines. If we now change l by δl, only the capacitance constituted by the $6l/d$ flux tubes between the parallel faces is appreciably increased. The eight leakage flux tubes issuing from the two end faces of the teeth remain substantially constant. The linear law of eqn (4.4.22) therefore requires a correction factor

$$N_s = \frac{\text{non-leakage flux}}{\text{total flux}} = \frac{6l/d}{6l/d+8};$$
(4.4.23)

hence

$$\frac{\delta C}{C} = \frac{\delta l}{l} \cdot \frac{1}{1+(4d/3l)}.$$
(4.4.22a)

Factor N_s has been plotted in Fig. 4.4.11 against the ratio d/l. For instance, for an active tooth length l, of only twice the air gap d, (i.e. $d/l = 0.5$) the 'efficiency' of the serrated condenser is only 60 per cent of its ideal sensitivity. It is also seen from eqn (4.4.22a) that the fringing causes a non-linear capacitance–displacement relation.

4.4.2.4. *Capacitance pressure transducer with a stretched diaphragm (membrane)*

Consider the change of capacitance which is caused by deflecting, under uniform pressure, a flexible diaphragm in proximity to a stationary rigid plate. There are two basic cases:

(a) deflection of a thin stretched diaphragm, or membrane, of negligible stiffness in bending;

(b) deflection of a clamped diaphragm thin in comparison with its diameter but of appreciable stiffness in bending.

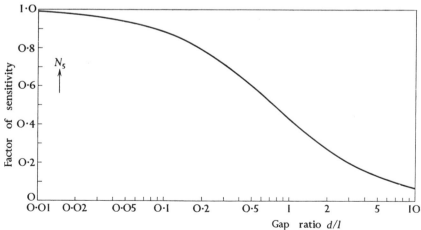

FIG. 4.4.11. Sensitivity factor N_s of a variable capacitance (Fig. 4.4.10) plotted against dimensional ratio d/l.

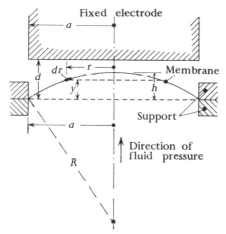

FIG. 4.4.12. Variable capacitance consisting of a stretched diaphragm (membrane), deflected by fluid pressure, and a solid plate.

A capacitance pressure transducer with stretched diaphragm or membrane is shown schematically in Fig. 4.4.12. The shape of a deflected circular membrane of diameter $2a$ (m) under one-sided fluid pressure is spherical and its deflection y (m) at any radius r (m) is

$$y = \frac{2S}{p}\left[\left\{1-\left(\frac{rp}{2S}\right)^2\right\}^{\frac{1}{2}} - \left\{1-\left(\frac{ap}{2S}\right)^2\right\}^{\frac{1}{2}}\right], \qquad (4.4.24)$$

where S is the tension in the diaphragm (N m^{-1}) and p is the fluid pressure (N m^{-2}). If eqn (4.4.24) is written in the form of a series, we have

$$y = \frac{p}{4S}\left\{(a^2-r^2)+\frac{1}{16}\left(\frac{p}{S}\right)^2(a^4-r^4)+\frac{1}{512}\left(\frac{p}{S}\right)^4(a^6-r^6)+...\right\}. \quad (4.4.25)$$

The first term in eqn (4.4.25), which represents the linear relationship between deflection and pressure, suffices if the deflections are small. Hence, for $(h/a)^2 \ll 1$,

$$y = \frac{p}{4S}(a^2-r^2). \qquad (4.4.26)$$

The partial capacitance dC (F) of a narrow annular zone on the sphere of width dr and length $2\pi r$, initially spaced d from the fixed plate, is

$$dC = \frac{\varepsilon_0 2\pi r\, dr}{d-y}. \qquad (4.4.27)$$

For small deflections $1/(d-y)$ can be approximated by

$$\frac{1}{d-y} = \frac{1}{d}\left(1+\frac{y}{d}\right) \quad \text{for} \quad y/d \ll 1, \qquad (4.4.28)$$

and the total capacitance of the spherical area of the deflected diaphragm, i.e. initial capacitance C plus incremental capacitance δC, is

$$C + \delta C = \int_0^a dC = \frac{2\pi\varepsilon_0}{d} \int_0^a \left(1 + \frac{y}{d}\right) r \, dr. \qquad (4.4.29)$$

Substituting for y from eqn (4.4.26) we have for small deflections

$$C + \delta C = \frac{2\pi\varepsilon_0}{d} \int_0^a \left\{1 + \frac{p}{4dS}(a^2 - r^2)\right\} r \, dr$$

$$= \frac{2\pi\varepsilon_0}{d} \left\{\frac{a^2}{2} + \frac{p}{4dS} \int_0^a (a^2 - r^2) r \, dr\right\}. \qquad (4.4.30)$$

The first term of eqn (4.4.30) is the initial capacitance C of the undeflected diaphragm; hence

$$\delta C = \frac{2\pi\varepsilon_0 p}{4d^2 S} \int_0^a (a^2 - r^2) r \, dr = \frac{\pi\varepsilon_0 a^4}{8d^2 S} p, \qquad (4.4.31)$$

and the fractional change in capacity is

$$\frac{\delta C}{C} = \frac{a^2}{8Sd} p. \qquad (4.4.32)$$

Eqn (4.4.32) represents the sensitivity $\delta C/C$ of a variable-capacitance transducer with a stretched diaphragm, or membrane, at small deflections. The centre-deflection y_{max} of the diaphragm, if substituting a piston-like movement for the actual shape of the deflected diaphragm, is (see Fig. 4.4.12)

$$y_{max} = h = \frac{pa^2}{4S} \qquad (4.4.33)$$

and

$$C = \frac{\pi\varepsilon_0 a^2}{d}. \qquad (4.4.34)$$

The initial plus the incremental capacitance for a small (piston-like) deflection h is

$$C + \delta C = \frac{\pi\varepsilon_0 a^2}{d}\left(1 + \frac{h}{d}\right), \qquad (4.4.35)$$

hence

$$\delta C = \frac{\pi\varepsilon_0 a^4}{4d^2 S} p. \qquad (4.4.36)$$

Comparison with the accurate result of eqn (4.4.31) reveals a factor 2, i.e. the actual sensitivity is only half that computed under the simplifying assumption of a piston-like movement of magnitude $y = h$.

The above results however, are applicable only to static deflections, since the cushioning effect of the thin layer of air behind the diaphragm has been neglected. This air cushion increases the (dynamic) stiffness and thus reduces the sensitivity to dynamic pressures. It can be shown [10], [11] that the sensitivity of a condenser-type transducer with a narrow air gap d is fairly independent of d if the external shunting capacity is negligible in comparison with the transducer capacity. This is brought about by the cancelling effect of increasing stiffness of the air cushion

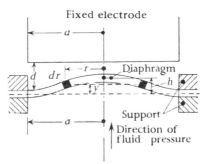

Fixed electrode
Diaphragm
Support
Direction of fluid pressure

FIG. 4.4.13. Variable capacitance, consisting of a clamped diaphragm deflected by fluid pressure, and a solid plate.

when the gap is decreasing, and vice versa. To overcome this limitation in the sensitivity of condenser microphones, the stationary electrode is usually perforated, thus reducing the cushioning effect to a negligible amount. This might improve the dynamic sensitivity up to 20 times. Another complication, especially at higher signal frequencies, is the inertia of the air layer next to the vibrating diaphragm and its effect on the dynamic sensitivity and frequency response. Its mass can easily be of the same order as that of the diaphragm itself.

4.4.2.5. *Capacitance pressure transducer with a clamped diaphragm*

In a pressure transducer of this type the diaphragm thickness is small compared with the diameter but there is no initial radial stress. Bending forces are actually the only forces considered. Stresses in the plane of the diaphragm, since they only count at large deflections, are neglected. For convenience a diaphragm, which for reason of low hysteresis etc. has been machined from the solid, will also be called a 'clamped' diaphragm.

The deflection y at any radius r of a clamped diaphragm is (Fig. 4.4.13)

$$y = \frac{3}{16} p \frac{1-\nu^2}{Et^3} (a^2 - r^2)^2,$$ (4.4.37)

where

p = fluid pressure (N m^{-2}),
a = radius of circular diaphragm (m),
t = thickness of diaphragm (m),
E = Young's modulus (N m^{-2}),
v = Poisson's ratio.

The initial capacitance C (F) plus the incremental capacitance δC of a deflected clamped diaphragm is derived by the same procedure as eqn (4.4.29) of the previous section and is given by

$$C + \delta C = \frac{2\pi\varepsilon_0}{d} \int_0^a \left(1 + \frac{y}{d}\right) r \, dr \quad \text{for} \quad y/d \ll 1. \tag{4.4.38}$$

Substituting for y from eqn (4.4.37) the fractional change in capacitance becomes

$$\frac{\delta C}{C} = \frac{(1-v^2)a^4}{16Edt^3} \, p. \tag{4.4.39}$$

Eqn (4.4.39) represents the sensitivity of a variable-capacitance pressure transducer with a clamped diaphragm at small deflections. If again compared with a fictitious sensitivity computed from a piston-like diaphragm movement of magnitude h, i.e. of the maximum centre deflection, the actual sensitivity is only one-third of the fictitious sensitivity. The discrepancy is here even greater than in the case of the stretched diaphragm of the previous section.

4.4.3. Associated circuits

The most common mode of operation of variable-capacitance transducers is amplitude modulation by means of a.c. bridge circuits, which will be treated in some detail in the section to follow. Other forms of modulation, such as frequency and pulse modulation [12], [13], have also been used; however, they will be ignored here in favor of discussing d.c. polarized circuits for use with 'electrostatic' transducers (see section 4.4.3.2).

4.4.3.1. Capacitance transducers in a.c. bridge circuits

Capacitance transducers can be used in a great variety of a.c. bridge circuits [14]–[16]. Here we shall discuss the so-called 'Blumlein' bridge circuit with tightly-coupled inductive ratio arms [17]–[19], which has particular advantages with variable-capacitance transducers.

Basically, a capacitance bridge is represented by the circuit of Fig. 4.4.14(a). It consists of two variable impedances Z, operating in push–pull

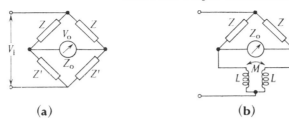

FIG. 4.4.14. Basic a.c. bridge circuit with variable impedance Z in push–pull fashion (a) in its most general form, (b) with coupled inductance ratio arms.

fashion, and two ratio arms Z'. The bridge is fed across the centre of the Z and Z' pairs by an a.c. voltage V_i, and the output voltage V_0 is developed across the other diagonal of the bridge. In comparison with the a.c. bridge discussed in section 4.3 on inductance transducers, the input and output terminals have been exchanged. Under certain conditions this bridge has a higher sensitivity than its alternative [20]. It also gives us an opportunity to discuss this bridge type in greater detail than was possible in section 4.3. The bridge output voltage can be shown to be

$$V_0 = \frac{\delta Z}{Z} V_i \frac{1}{1+\frac{1}{2}(Z'/Z+Z/Z')+ZZ'/Z_0}. \tag{4.4.40}$$

We omit here the obvious arrangement of the basic bridge circuit with resistive ratio arms ($Z = 1/j\omega C$ and $Z' = R$) which can easily be worked out by the reader, and turn to the Blumlein circuit of Fig. 4.4.14(b), with closely coupled inductive ratio arms. The coupled ratio arms have been redrawn in Fig. 4.4.15(a) in the form of a four-terminal network, while Fig. 4.4.15(b) shows the equivalent T-type network. Neglecting the ratio-arm resistances we have

and
$$\left. \begin{aligned} Z_s &= j\omega(L+M) \\ Z_p &= -j\omega M. \end{aligned} \right\} \tag{4.4.41}$$

Since
and
$$\left. \begin{aligned} Z_{12} &= Z_s+Z_p = j\omega L \\ Z_{13} &= 2Z_s, \end{aligned} \right\} \tag{4.4.42}$$

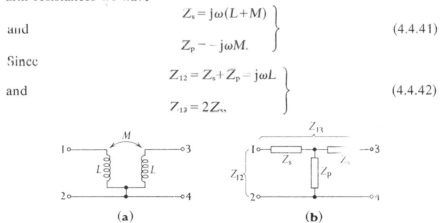

FIG. 4.4.15. Four-terminal network of (a) coupled inductances and (b) T-type network equivalent to (a).

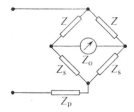

FIG. 4.4.16. A.c. bridge network with variable inductances Z in push–pull fashion. The coupled inductance ratio arms of Fig. 4.4.14(b) are replaced by the equivalent T-network of Fig. 4.4.15(b).

the coupling factor

$$k = \frac{Z_p}{Z_p + Z_s} = 1 - \frac{Z_{13}}{2Z_{12}} \tag{4.4.43}$$

and

$$Z_s = Z_{12}(1-k) = j\omega L(1-k). \tag{4.4.44}$$

Fig. 4.4.16 shows the bridge with the impedances of the T-network of Fig. 4.4.15(b) substituted. Without impedance Z_p the output voltage V_0' (eqn (4.4.40)) would be

$$V_0' = \frac{\delta Z}{Z} V_i \left[1 \Big/ \left\{ 1 + \frac{1}{2}\left(\frac{Z_{12}(1-k)}{Z} + \frac{Z}{Z_{12}(1-k)} \right) + \frac{Z + Z_{12}(1-k)}{Z_0} \right\} \right], \tag{4.4.45}$$

but with Z_p in series with the bridge impedance Z_B the output voltage V_0 becomes

$$V_0 = V_0' \frac{Z_B}{Z_B + Z_p} = V_0' \frac{Z + Z_{12}(1-k)}{Z + Z_{12}(1+k)}, \tag{4.4.46}$$

or finally

$$V_0 = \frac{\delta Z}{Z} V_i \frac{\left\{ 1 + \dfrac{Z_{12}(1-k)}{Z} \right\} \Big/ \left\{ 1 + \dfrac{Z_{12}(1+k)}{Z} \right\}}{1 + \dfrac{1}{2}\left\{ \dfrac{Z_{12}(1-k)}{Z} + \dfrac{Z}{Z_{12}(1-k)} \right\} + \dfrac{Z + Z_{12}(1-k)}{Z_0}}. \tag{4.4.47}$$

Eqn (4.4.47) is the general expression for the output voltage of the bridge circuit of Fig. 4.4.16 with coupled inductance ratio arms and impedances Z variable in push–pull fashion.

Now consider a capacitance transducer (Fig. 4.4.17), with

$$Z = 1/j\omega C \quad \text{and} \quad \delta Z = 1/j\omega\, \delta C, \tag{4.4.48}$$

ratio arms $Z_{12} = j\omega L$, feeding into a high load impedance, $Z_0 \to \infty$. The winding sense of the ratio arms may be arranged so that k is positive for current flowing from the source through the ratio arms when the bridge is

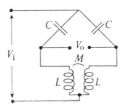

FIG. 4.4.17. Variable push–pull capacitance transducer in a.c. bridge with coupled inductance ratio arms.

balanced. Then the ratio-arm impedances

$$Z_s = Z_{12}(1-k) \tag{4.4.49}$$

are reduced by the factor $(1-k)$ and will become zero for $k = +1$; i.e. for
100 per cent coupling. This means that a voltage drop cannot develop
across the ratio arms and any stray or cable capacitance which would be
in parallel to the ratio arms is ineffective. This important feature of the
Blumlein bridge circuit simplifies screening and earthing problems in
capacitance bridges of this type, i.e. zero-stability of the circuit is greatly
improved. The circuit is now widely used in capacitance-transducer work,
and it is also useful in a.c. bridges for inductance and resistance transducers, especially at higher carrier frequencies.

However, with respect to the out-of-balance current the coupling factor
has its sign reversed, because this current flows 'round' the bridge, i.e.
from terminal (1) to (3) in Fig. 4.4.15(a), or reverse. This means a
coupling factor $k = -1$ in the case of 100 per cent coupling, and

$$Z_s = Z_{12}(1-k) \rightarrow 2Z_{12}. \tag{4.4.50}$$

If, as before, $Z_0 \rightarrow \infty$ and $Z = 1/j\omega C$, $\delta Z = 1/j\omega\,\delta C$, $Z_{12} = j\omega L$, and for 100
per cent coupling, eqn (4.4.47) becomes

$$V_0 = \frac{\delta C}{C}\, V_i\, \frac{1-2\omega^2 LC}{1-\omega^2 LC-(1/4\omega^2 LC)}$$

$$= \frac{\delta C}{C}\, V_i\, \frac{4\omega^2 LC}{2\omega^2 LC-1} \qquad \text{for} \quad k = -1;\; Z_0 \rightarrow \infty. \tag{4.4.51}$$

Eqn (4.4.51) is the output voltage of an a.c. bridge with two variable
capacitances in push–pull and tightly-coupled inductive ratio arms, feeding into a high impedance. The factor $4\omega^2 LC/(2\omega^2 LC-1)$, which represents the circuit sensitivity, has been plotted in Fig. 4.4.18 against $\omega^2 LC$.
Resonance occurs at $\omega^2 LC = \frac{1}{2}$, i.e. for $\omega L = 1/2\omega C$. Below resonance and
for $\omega^2 LC \ll 1$, sensitivity is proportional to $\omega^2 LC$. Above resonance and

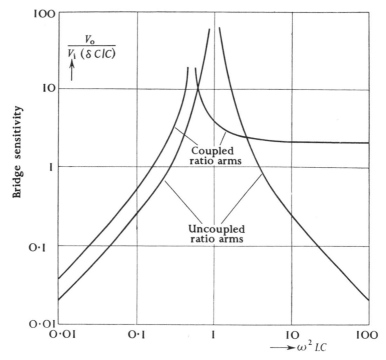

Fig. 4.4.18. Sensitivity of the circuit of Fig. 4.4.17 with tightly coupled, and with un-coupled, ratio arms.

for $\omega^2 LC \gg 1$, the sensitivity factor tends asymptotically to 2. For good stability, therefore, $\omega^2 LC$ should be made greater than, say, 2, so that variations of sensitivity with varying frequency or inductance may be avoided.

The inductance value L in the above formulae is that of the 'effective' uncoupled inductance of one ratio arm. If a cable capacitance C' shunts the ratio arms

$$L = \frac{L'}{1 - \omega^2 LC'},\qquad(4.4.52)$$

L' being the uncoupled inductance value of a ratio-arm coil. At small values of transducer capacitance C, and at moderate frequencies ω, it is often not practicable to satisfy $\omega^2 LC \gg 1$. Here a large cable capacitance C' improves stability. Indeed, it may be advantageous to shunt the ratio arms with a large constant capacity, thus improving stability at the expense of sensitivity.

For the purpose of comparison we shall also derive the output voltage of a bridge circuit with uncoupled inductive ratio arms ($k = 0$). From eqn

(4.4.47),

$$V_0 = \frac{\delta C}{C} V_i \frac{1}{1-\frac{1}{2}\{\omega^2 LC+(1/\omega^2 LC)\}}$$

$$= \frac{\delta C}{C} V_i \frac{-2\omega^2 LC}{(\omega^2 LC-1)^2} \qquad \text{for} \quad k=0; Z_0 \to \infty. \qquad (4.4.53)$$

Eqn (4.4.53) has also been plotted in Fig. 4.4.18. It is seen that for small values of $\omega^2 LC$ the bridge circuit with tightly coupled ratio arms has twice the sensitivity of a bridge with uncoupled inductive ratio arms. For higher values of $\omega^2 LC$ the uncoupled bridge has no horizontal characteristic, and hence no regions where sensitivity does not vary with frequency or inductance.

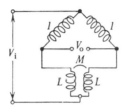

FIG 4.4.19. Variable push–pull inductance transducer in a.c. bridge with coupled inductance ratio arms.

It is convenient to discuss here also the sensitivity of a variable-inductance transducer in a Blumlein bridge network. Let

$$Z = j\omega l \quad \text{and} \quad \delta Z = j\omega \, \delta l \qquad (4.4.54)$$

be the transducer impedance and the impedance variation, respectively. Assume a push–pull inductance transducer, as shown in Fig. 4.4.19, of high Q-value $(R_s \to 0)$. Then, from eqn (4.4.47) we have, for $Z_0 \to \infty$ and with tightly coupled ratio arms,

$$V_0 = \frac{\delta l}{l} V_i \frac{1+2L/l}{1+L/l+L/4l}$$

$$= \frac{\delta l}{l} V_i \frac{4L/l}{2L/l+1} \qquad \text{for} \quad k=-1; Z_0 \to \infty. \qquad (4.4.55)$$

Eqn (4.4.55) has been plotted in Fig. 4.4.20. For comparison the output voltage for uncoupled ratio arms $(k = 0)$ has also been plotted, namely,

$$V_0 = \frac{\delta l}{l} V_i \frac{1}{1+\frac{1}{2}\{L/l+1/(2L/l)\}}$$

$$= \frac{\delta l}{l} V_i \frac{2L/l}{(L/l+1)^2} \qquad \text{for} \quad k=0; Z_0 \to \infty. \qquad (4.4.56)$$

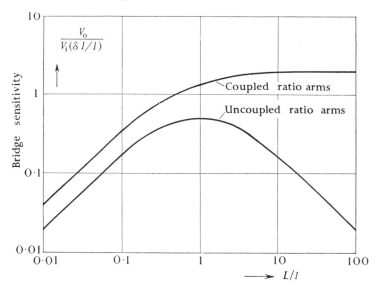

F<small>IG</small>. 4.4.20. Sensitivity of the circuit of Fig. 4.4.19 with tightly coupled, and with uncoupled, ratio arms.

The bridge with coupled ratio arms has a higher sensitivity throughout.
At low values of L/l this factor is 2, at high L/l values the improvement in
sensitivity increases with L/l, but the more important characteristic is the
horizontal portion of the curve for the coupled arms bridge at L/l values
above, say, 2. In this region the bridge sensitivity is independent of
ratio-arm inductance variations.

Summarizing, it can be said that the Blumlein bridge circuit has a
higher stability than a bridge with uncoupled inductance ratio arms or
with resistance ratio arms. The above results apply strictly only to purely
reactive impedances in transducer and ratio arms; they remain a good
approximation as long as the Q-values are not too low. The more general
case of the loaded bridge ($Z_0 \neq \infty$) cannot be discussed here. The interested reader is invited to study this case with the help of the general
eqn (4.4.47).

From the great number of capacitance transducers used with a.c. bridge
networks, a miniature pressure transducer for use in wind-tunnel models
may be of particular interest [21]. A cross-sectional view is shown in Fig.
4.4.21. Its over-all size is about 3 mm×6 mm×9 mm. The transducer
consists of two pressure chambers separated by a diaphragm. The deflection of the diaphragm is sensed by variation of capacitance between it and
two insulated electrodes in the end-pieces. The transducer is made of
aluminium alloy and insulation between the individual parts is effected by
anodization. The diaphragm and the two electrodes are connected to

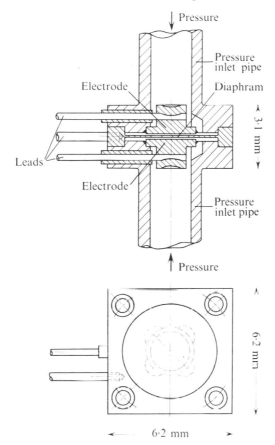

FIG. 4.4.21. Miniature differential pressure transducer for use in wind-tunnel models
(Royal Aircraft Establishment, Farnborough).

three individually screened leads by way of a connector block attached to
the side of the transducer. The nominal pressure range is ± 30 kN m^{-2}. It
is used in conjunction with a Blumlein bridge circuit, followed by a carrier
amplifier and a phase-sensitive demodulator. At a carrier frequency of
20 kHz and an input voltage of 10 V, the full-scale output is $+4\cdot5$ V. With
air gaps of $0\cdot05$ mm and initial capacities of $1\cdot2$ pF, the full-scale change
in capacitance is $0\cdot1$ pF. The non-linearity of calibration is less than $\pm1\cdot5$
per cent of full-scale output, and zero-drift and change of sensitivity with
temperature are less than $0\cdot15$ per cent per °C. The error caused by
acceleration is less than $0\cdot02$ per cent per g. With an inlet tube trimmed
to a length of 3 mm the rise time of the transducer is less than 40 μs.
Therefore, an amplifier with a carrier frequency of 400 kHz has also been
developed for use in shock tubes and for other fast-transient work.

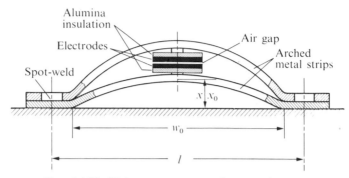

FIG. 4.4.22. High-temperature capacitance strain gauge.

The variable-capacitance method should be particularly suitable for measuring small displacements at high environmental temperatures [22] since the dielectric constant of air is virtually independent of temperature. Fig. 4.4.22 shows, in fact, a capacitance strain gauge [23] which has recently been developed by the Central Electricity Research Laboratory and G.V. Planer Ltd. Its operational temperature range is $-269\,°C$ to $+650\,°C$, though the most useful span is $20\,°C$ to $600\,°C$. The gauge consists of two superimposed arched metal strips which carry the insulated electrodes. Contact is made by ribbon leads, and mineral-insulated screened cables connect the transducer to the ratio-arm bridge circuit.

Strain in the substrate changes the differential height of the arches and thus the capacitance between the electrodes. The height variation of an individual arch can be computed from (see Fig. 4.4.22)

$$x^2 - x_0^2 = -\tfrac{15}{64}\varepsilon l(\varepsilon l + 2w_0),\tag{4.4.57}$$

where

ε = strain (m per m),
x = instantaneous height of arch (m),
x_0 = initial height of arch (m),
w_0 = unstrained width of arch (m),
l = gauge length (m).

The manufacturer's provisional performance specification quotes as follows

nickel-alloy based body (Type N),

gauge length: 20 mm,

strain range: $\pm 5000\ \mu$m per m,

gauge factor: 100, approximately,

initial capacitance: $0\cdot5$–1 pF,

drift at steady temperatures:
   low strain at 20 °C: <5 μm per m per month,
   any strain at 600 °C: <5 μm per m per day,

drift on temperature cycling (20–600 °C):
   low strain (1st cycle): <20 μm per m per cycle,
   any strain (after 1st cycle): <5 μm per m per cycle.

Other types of capacitance strain gauges for use at high temperatures
have been described elsewhere [24]–[26].

4.4.3.2. *Capacitance transducers with d.c. polarization (electrostatic transducers)*

The capacitance transducer when used in circuits providing d.c. polarization has bilateral characteristics, i.e. it can be operated as a sender as

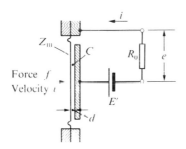

FIG. 4.4.23. Electrostatic transducer, schematic diagram.

well as a receiver. The reader is referred to section 2.3.2, where bilateral
transducers have been analysed generally.

The so-called electrostatic-transducer type (Fig. 4.4.23) consists essen
tially of two flat and parallel plates of area a (m²), separated by a distance
d (m) (if one plate is a flexible diaphragm, as in Fig. 4.4.23, results
obtained here for piston-like movements are an approximation—see
sections 4.4.2.4 and 4.4.2.5 above). There is a (large) constant bias
voltage E' (V) and, in an electromechanical sender, also a (small) sinusoidal voltage e (V) superimposed on E' which produces a sinusoidal force
f (N). In the receiver configuration the plate distance d (m) is varied
sinusoidally, for instance, by a sound pressure if the transducer is used as
an 'electrostatic' microphone. Then, the bias voltage E' causes a sinusoidal current i (A) to flow in the circuit which, in turn, produces a
sinusoidal output voltage e (V).

Generally, the electric energy stored in a condenser is

$$U_{el} = Q^2/2C \quad \text{(W s)}, \tag{4.4.58}$$

where Q (C) is the charge and

$$C = \varepsilon\varepsilon_0 a/d \quad \text{(F)} \tag{4.4.59}$$

the capacity, ε the relative permittivity of the dielectric ($\varepsilon = 1$ for air), and ε_0 again the permittivity of empty space. Then, at a potential difference

$$E = Q/C \quad \text{(V)}, \tag{4.4.60}$$

eqn (4.4.58) becomes

$$U_{el} = \varepsilon\varepsilon_0 aE^2/2d, \tag{4.4.61}$$

and the force acting between the plates

$$f = -\frac{\partial U_{el}}{\partial d} = \frac{\varepsilon\varepsilon_0 aE^2}{2d^2} \quad \text{(N)}. \tag{4.4.62}$$

For the electrostatic transducer with a bias voltage E' (V) and a superimposed sinusoidal voltage of amplitude e (V), eqn (4.4.62) can be written

$$f = \frac{\varepsilon\varepsilon_0 a}{2d^2}(E'+e)^2 = \frac{\varepsilon\varepsilon_0 a}{2d^2}(E'^2 + 2E'e + e^2). \tag{4.4.63}$$

The second term on the right-hand side is the sinusoidal transducer force

$$f_t = \frac{CE'}{d} e, \tag{4.4.64}$$

while the first term represents a constant force produced by the constant voltage E', and the third term a small-of-second-order sinusoidal force of twice the frequency. Neglecting leakage,

$$e = i/j\omega C, \tag{4.4.65}$$

and eqn (4.4.64) becomes

$$f = \frac{E'}{j\omega d} i \quad \text{(N)}. \tag{4.4.66}$$

Force equilibrium at the transducer then requires

$$f = Z_m v + \frac{E'}{j\omega d} i, \tag{4.4.67}$$

where Z_m is the mechanical transducer impedance as defined in section 2.3.2.

Comparing eqn (4.4.67) with eqn (2.25a) (p. 14) yields

$$M_{fi} = M = E'/j\omega d \quad \text{(N A}^{-1}), \tag{4.4.68}$$

and with eqns (2.17) and (2.18) (p. 13), and since M is imaginary,

$$M_{ev} = -M^* = E'/j\omega d \quad \text{(V s m}^{-1}\text{)}, \tag{4.4.69}$$

i.e. the transducer constant does not change sign.

The electrostatic transducer can therefore be represented by a voltage-source-type impedance matrix equation (see section 2.3.2),

$$\begin{bmatrix} f \\ e \end{bmatrix} = \begin{bmatrix} Z_m & E'/j\omega d \\ E'/j\omega d & Z_e \end{bmatrix} \begin{bmatrix} v \\ i \end{bmatrix}, \tag{4.4.70}$$

where

$$Z_e = 1/j\omega C \tag{4.4.71}$$

is the electrical transducer impedance.

It can easily be seen that the equivalent hybrid matrix eqn (2.27) (p. 15), which applies also to the electrostatic transducer, displays an extra mechanical impedance term

$$M^2/Z_e = -E'^2 C/j\omega d^2 \quad \text{(N s m}^{-1}\text{)}. \tag{4.4.72}$$

The physical significance of eqn (4.4.72) is the existence of a 'negative stiffness' which constitutes an instability inherent in the electrostatic transducer, unless the attraction force $(-E'^2 C/d^2)$ is swamped by a large mechanical restoring force k (N m^{-1}) of, for instance, a spring or a diaphragm. The extra term does not occur in the impedance matrix of eqn (4.4.70) since this was derived under the assumption of limited available energy (eqn (4.4.58)), while in the hybrid matrix the transducer force is a function of voltage which, in principle, can feed any amount of energy into the negative spring. We also note that in the hybrid form the transducer constant is real and independent of frequency (eqn (4.4.64)), but since it depends on C, it would become meaningless for the ideal transducer with $Z_e \to 0$ or $C \to \infty$. With eqn (2.29) (p. 15) the transfer matrix equation of the electrostatic transducer takes the form

$$\begin{bmatrix} f \\ v \end{bmatrix} = \begin{bmatrix} E'/j\omega d - Z_m d/E'C & j\omega d Z_m/E' \\ d/CE' & j\omega d/E' \end{bmatrix} \begin{bmatrix} i \\ e \end{bmatrix}, \tag{4.4.73}$$

and with $f_t = f - Z_m v$ the characteristic transfer matrix equation of the transducer becomes

$$\begin{bmatrix} f_t \\ v \end{bmatrix} = \begin{bmatrix} E'/j\omega d & 0 \\ -d/CE' & j\omega d/E' \end{bmatrix} \begin{bmatrix} i \\ e \end{bmatrix}, \tag{4.4.74}$$

which is shown in Fig. 2.3(c) (p. 20).

The transducer constant of the electrostatic force receiver is

$$M = E'/j\omega d,$$

and its (dynamic) transfer function from eqn (2.42) (p. 18) becomes

$$\frac{e}{f_0} = \frac{R_0 E'/j\omega d}{\dfrac{E'^2}{\omega^2 d^2} + \{c + j(\omega m - k/\omega)\}(R_0 + 1/j\omega C)} \quad (\text{V N}^{-1}), \quad (4.4.75)$$

where d (m) is the initial air gap, C (F) the initial transducer capacity, E' (V) the (large) polarizing, or bias, d.c. voltage, R_0 (Ω) the (large) indicator resistance, and ω the circular frequency. m, k, and c represent mass, stiffness, and damping, respectively, i.e. the mechanical transducer impedance $Z_m = c + j(\omega m - k/\omega)$.

In a practical electrostatic force, or pressure, receiver the first transition point on the frequency-response curve occurs at $\omega = 1/CR_0$, well below resonance. Above that frequency, R_0 dominates $1/\omega C$, and since also the frequency dependence of the numerator cancels that of the stiffness term in the denominator in eqn (4.4.75), the response is constant up to resonance, where $\omega m = k/\omega$. In this flat region the electrostatic transducer is an ideal force receiver. R_0 should be made as large as possible in order to boost both output voltage and useful frequency range. The damping at resonance, $\{c + E'^2/(\omega^2 d^2 R_0)\}$, depends mainly on its mechanical term, since the electrostatic interaction forces are weak.

The electrostatic sender, though of little use because of the feeble electromechanical coupling forces, has the transfer function (eqn (2.44), p. 19)

$$\frac{-v}{e_0} = \frac{E'^2/j\omega d}{(E'^2/\omega^2 d^2) + \{c + j(\omega m - k/\omega)\}/j\omega C} \quad (\text{m V}^{-1}\text{ s}^{-1}). \quad (4.4.76)$$

Since, in the sender's case, the source resistance R_0 is small compared with $1/\omega C$, it has been omitted in eqn (4.4.76), and so has the transducer resistance R (no losses). We note that the frequency dependence of the numerator cancels that of $1/j\omega C$ in the denominator in eqn (4.4.76), and the frequency response of the electrostatic sender (like that of the piezoelectric sender) is a simple resonance curve. The 'damping' factor $(c - E'^2 C/j\omega d^2)$ at resonance is small; the imaginary (electrical) component, though insignificant in practice, represents a negative stiffness, as already predicted by eqn (4.4.72). (In the electrostatic receiver a (negligible) negative stiffness would occur only at frequencies well below the useful frequency range.)

4.4.4. Construction of variable capacitances

When discussing the construction of variable capacitances used in transducers it seems more difficult to divorce the basic capacitance unit from the specific transducer design than in the previous chapter on variable-inductance transducers, where coil, core, and armature were distinct elements. Essentially, we require conductive electrodes, two in the case of a single-ended capacitor, or three for a push–pull type, and the appropriate one or two variable air gaps. Since the electrodes, though insulated from each other, must be connected mechanically, solid supports made of insulation material must be provided. The choice of a suitable insulation material is of great importance. It must be of sufficient mechanical strength and, even more important, of an extremely high form-stability. Its thermal coefficient should be as low and as predictable as possible; in many cases it must match the coefficient of expansion of other structural parts of the transducer, usually made of metal, so as to achieve improved zero-stability by way of compensation. Ceramic insulation materials are generally a better choice than plastics or organic materials. A number of machinable and castable ceramic materials, even for use at high environmental temperatures, are now available. For the more difficult ceramics ultrasonic machining has been used widely in the production of prototype insulators without the need for expensive moulding or sintering tools.

The metal parts, such as diaphragms, electrodes, supports, and transducer housings, also require a high degree of form stability. Low-expansion and high-temperature nickel–iron alloys can be considered, though some are difficult to machine. The faces of the variable condenser plates inside the transducer cannot normally be cleaned in use. The air gap must therefore be protected from humidity and condensation, as well as from corrosion and dust. Stainless-steel construction or rhodium plating has proved useful and is a necessity for electrodes exposed to humidity, unless they can be deposited in the form of thin metal film (e.g. gold) on the insulator faces. Metal and alloy combinations used in the design of the transducer must be chosen to avoid electrolytic corrosion, particularly so if the transducer cannot be effectively sealed against the atmosphere. In pressure transducers the metal surfaces must be compatible with the pressure media. The transducer housing must be made absolutely rigid to avoid any distortion when mounted on uneven surfaces, or the sensitive condenser unit inside the housing must be mechanically insulated from the housing.

Connections are usually made by miniature glass or ceramic terminals. Electrostatic screening of capacitance transducers leads is essential and should be carried right down to the transducer housing without leaving an unscreened gap. Transducer-circuit configurations requiring double screening should be avoided, especially for use at high environmental temperatures. A convenient length of cable is often made an integral part of the transducer.

References
1. A. E. ROBERTSON. *Microphones.* Iliffe, London (1951).
2. M. L. GAYFORD. High-quality microphones. *Proc. Instn. elect. Engrs* **106B**, 501–16 (1959).
3. M. L. GAYFORD. *Electroacoustics; microphones, earphones and loudspeakers.* Butterworth, London (1970).
4. A. R. VON HIPPEL. *Dielectric materials and application.* Wiley, New York (1954).
5. H. FRÖHLICH. *Theory of dielectrics.* (2nd edn). Clarendon Press, Oxford (1968).
6. B. C. CARTER, J. R. FORSHAW, and J. F. SHANNON. Electric capacitance gauges, in: *Handbook of experimental stress analysis* (ed. M. Hetényi), Chap. 7. Chapmann & Hall, London (1950).
7. C. CARATHEODORY. *Conformal representation.* Cambridge University Press, Cambridge (1941).
8. E. WEBER. *Electromagnetic fields; theory and application,* I. Mapping of fields, p. 301. Wiley, New York (1950).
9. S. S. ATTWOOD. *Electric and magnetic fields.* Wiley, New York (1949).
10. E. C. WENTE. The sensitivity and precision of the electrostatic transmitter for measuring sound intensities. *Phys. Rev.* **19**, 448 (1922).
11. D. GOECKE. Schwingmembranmanometer für Druckmessungen von 10^{-3} bis 10^3 Torr. *Arch. tech. Messen* **V1342-6**, 49–54 (1972).
12. S. Y. LEE, Y. T. LI, and H. L. PASTAN. Signal transduction with differential pulse width modulation. *IEEE Trans. ind. Electron. (Contr. Instrum.)* *(Inst. elec. electron. Eng.)* **17**, 39–43 (1970).
13. Y. T. LI, S. Y. LEE, and H. L. PASTAN. Air-damped capacitance accelerometers and velocimeters, *IEEE Trans. ind. Electron. (Contr. Instrum.)* *(Inst. elec. electron. Eng.)* **17**, 44–8 (1970).
14. B. HAGUE. *Alternating bridge methods.* Pitman, London (1946).
15. H. E. THOMAS and C. A. CLARKE (ed.). *Handbook of electrical instruments and measuring techniques.* Prentice-Hall, Englewood Cliffs, New Jersey (1967).
16. H. HELKE. *Messbrücken und Kompensatoren für Wechselstrom.* Oldenbourg, Munich (1971).
17. R. WALSH. Inductance ratio arms in alternating current bridge circuits, *Phil. Mag.* **10**, 49–70 (1930).
18. A. T. STARR. A note on impedance measurement. *Wireless Engr. exp. Wireless* **9**, 615–617 (1932).
19. H. A. M. CLARK and P. B. VANDERLYN. Double-ratio a.c. bridges with inductively coupled ratio arms. *Proc. Inst. elect. Engrs* **96**, 365 (1949).
20. A. C. SELETZKY and L. A. ZURCHER. Sensitivity of four-arm bridge, *Electl Engng, N.Y.* **58**, 723–8 (1939).
21. H. K. P. NEUBERT, W. R. MACDONALD, and P. W. COLE. Sub-miniature pressure and acceleration transducers, *Control* **4**, 104–106 (1961).
22. H. K. P. NEUBERT, *Strain gauges; kinds and uses,* Macdonald, London, pp. 19–20 (1967).
23. B. E. NOLTINGK, D. F. A. McLACHLAN, C. K. V. OWEN, and P. C. O'NEILL. High-stability capacitance strain gauge for use at extreme temperatures, *Proc. Inst. elect. Engrs* **119**, 897–903 (1972).
24. O. L. GILLETTE and L. E. VAUGHN. Research and development of a high-temperature capacitance strain gauge, *Tech. Rep. Nr. AFFDL-TR-68-27*, Airforce Flight Dynamics Lab., Wright-Patterson Airforce Base, Ohio (1968).
25. O. L. GILLETTE and J. L. MULLINEAUX. Development of high-temperature capacitance strain gauges, *Instr. Soc. Amer. Trans.* **8**, 52 (1969).
26. ANON. Capacitance strain gauge operates at 1750 °F, *Electronic Design (USA)* **20**, 34 (1972).

4.5. PIEZOELECTRIC TRANSDUCERS

PIEZOELECTRIC transducers are energy converters and are therefore generator-type transducers. However, in contrast to the electrodynamic-generator type of section 4.1, the generated quantity is an electric charge.

The indicated voltage thus depends on the capacitance of the transducer and indicator circuit. This fact, together with the high impedances normally prevailing in these circuits, determines the approach in the design of piezoelectric transducers and their associated electronic equipment. Owing to the finite insulation resistance of the transducer circuit and the shunting effect of the load resistance, the generated charge gradually leaks away and there is therefore ño steady-state response.

Piezoelectric transducers are force-sensitive devices (the prefix *piezo* from the Greek meaning to press) and are therefore employed for the measurement of physical quantities which can be reduced to forces, such as pressure, stress, or acceleration. Piezoelectricity occurs in crystals of certain configurations when exposed to compression or tension. It is fundamentally different from electrostriction whose polarization does not change sign with the strain. Electrostriction occurs to different degrees in all solid dielectrics but is smaller in magnitude than piezoelectricity in the presence of which it can normally be neglected. Pyroelectricity (Greek *pyro*: fire) is an effect observed in some crystals when heated uniformly. The physical significance of these and other related effects are discussed in greater detail in references [1] and [2]. The pyroelectricity of tourmaline was known in Europe as early as about 1700, while the piezoelectricity of quartz was discovered by the brothers Curie in 1880. In the design and use of piezoelectric transducers the pyroelectric effect is unwanted.

The main attraction of piezoelectric transducers for the measurement of force, pressure, or acceleration is their high mechanical input impedance (stiffness). They are genuine force-measuring instruments of negligible deformation under load. In the language of transducer performance they are instruments with high natural frequencies. The output side normally represents a high electrical impedance which requires special precautions and has led to a new type of output circuit (charge amplifier). Another advantage of piezoelectric transducers is the possibility of designing instruments of very small dimensions.

The main disadvantages of piezoelectric transducers are their lack of steady-state response and their high electrical output impedance, coupled with the need for low-noise cables of low capacitance value. The maximum working temperatures of most piezoelectric materials are in the neighbourhood of 200-250 °C. Their Curie temperatures are higher (e.g. quartz at 576 °C), but loss of insulation resistance above, say, 200 °C sets a practical limit.

4.5.1. The piezoelectric effect

Cady [1] defines piezoelectricity as an electric polarization produced by mechanical strain in crystals belonging to certain classes, the polarization being proportional to the strain and changing sign with the strain (direct piezoelectric effect). The inverse effect, i.e. an electrical polarization producing a strain (variation in dimension) in the crystal, represents the same fundamental property of the crystal. Piezoelectricity is thus classed as a reversible effect, such as, for instance, Faraday's law of electromagnetism.

The magnitude of the piezoelectric effect can best be represented by the vector of polarization,

$$\bar{P} = P_{xx} + P_{yy} + P_{zz}, \tag{4.5.1}$$

where x, y, z refer to a conventional orthogonal system related to the crystal axes. In terms of axial stresses σ and shear stresses τ we can write

$$P_{xx} = d_{11}\sigma_{xx} + d_{12}\sigma_{yy} + d_{13}\sigma_{zz} + d_{14}\tau_{yz} + d_{15}\tau_{zx} + d_{16}\tau_{xy},$$

$$P_{yy} = d_{21}\sigma_{xx} + d_{22}\sigma_{yy} + d_{23}\sigma_{zz} + d_{24}\tau_{yz} + d_{25}\tau_{zx} + d_{26}\tau_{xy}, \tag{4.5.2}$$

$$P_{zz} = d_{31}\sigma_{xx} + d_{32}\sigma_{yy} + d_{33}\sigma_{zz} + d_{34}\tau_{yz} + d_{35}\tau_{zx} + d_{36}\tau_{xy},$$

where the constants d_{mn} are the piezoelectric coefficients. The subscripts 1–6 are related to the three crystal axes x, y, z according to Fig. 4.5.1 [3].

The physical significance of the d-coefficients of eqn (4.5.2) is indicated by their dimensional notation. For the direct piezoelectric effect under short-circuit conditions we have

$$d_{m,n} = \frac{\text{coulomb}}{\text{newton}} \quad \text{or} \quad \frac{\text{C m}^{-2}}{\text{N m}^{-2}} \text{ (direct effect)} \tag{4.5.3}$$

i.e. charge output per unit force input (or charge density per unit pressure), and for the inverse piezoelectric effect under no-load conditions,

$$d_{m,n} = \frac{\text{metre}}{\text{volt}} \quad \text{or} \quad \frac{\text{m per m}}{\text{V per m}} \text{ (inverse effect),} \tag{4.5.4}$$

i.e. strain per unit applied field strength. Note that the dimensions of eqns (4.5.3) and (4.5.4) are identical and, in SI units, equal to (A s^3 kg^{-1} m^{-1}).

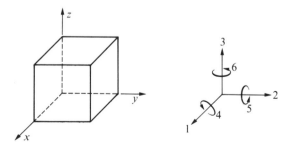

FIG. 4.5.1. Axial notation of piezoelectric coefficients.

For convenient computation two more coefficients have been defined. The g-coefficients are obtained by deviding the d-coefficients by the (absolute) dielectric constant $\varepsilon\varepsilon_0 = \varepsilon_{m,n} \times 8\cdot854 \times 10^{-12}$ (F m^{-1}), thus yielding

$$g_{m,n} = \text{V m}^{-1}/\text{N m}^{-2}, \tag{4.5.5}$$

which represents the voltage gradient in the crystal per unit pressure input. The h-coefficients are convenient for computing the voltage gradient per unit strain,

$$h_{m,n} = \text{V m}^{-1}/\text{m per m}. \tag{4.5.6}$$

They are obtained by multiplying the g-coefficients by the Young's moduli valid for the appropriate crystal orientation.

Another common notation is the coupling coefficient k. It represents the square-root of the mechanical energy converted to electrical energy, to the input mechanical energy, and vice versa, and—at frequencies well below the (lowest) mechanical resonant frequency—is thus a measure of the efficiency of the crystal as an energy converter. Numerically, its value is

$$k_{m,n} = (d_{m,n} \times h_{m,n})^{\frac{1}{2}}. \tag{4.5.7}$$

In instrument transducer work the coupling coefficient is probably of less importance than in acoustic and ultrasonic applications, but since it is linked with the equivalent-circuit concept of piezoelectric transducers we shall return to it in section 4.5.3.1.

4.5.2. Piezoelectric materials

Piezoelectric materials for use in instrument transducers can be divided into two groups: natural piezoelectric crystals, such as quartz, and polarized piezoelectric ceramics, such as barium titanate and its derivatives.

4.5.2.1. Natural crystals

4.5.2.1.1. Quartz.
To a first approximation quartz can be described [4] as a helix with one silicon and two oxygen atoms alternating along the helix, thus forming a hexagonal plan view of the crystal cell. The sign of rotation of the helix represents right- or left-handed quartz. A simplified though instructive model [5] of this arrangement is shown in Fig. 4.5.2(a), looking down on a crystal cell in the direction of the optical (z) axis. Quartz is chemically SiO_2. In one cell there are three silicon atoms and six oxygen atoms, the latter being lumped in pairs, thus constituting the hexagonal shape. It is known that Si-atoms carry four positive charges and the O-atoms two negative charges. In the unstressed arrangement of Fig. 4.5.2(a) all charges are compensated and there is no external effect.

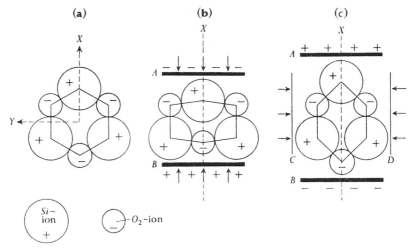

FIG. 4.5.2. (a) Simplified crystal structure of quartz, showing arrangement of ions in neutral cell. (b) Quartz cell loaded in direction of x-axis. (c) Quartz cell loaded in direction of y-axis.

If, however, a force is applied to the crystal in the direction of the x-axis (Fig. 4.5.2(b)) the balance is disturbed and the cell becomes polarized, generating an electric charge on the two faces A and B (longitudinal effect). A force in the direction of the y-axis (Fig. 4.5.2(c)) causes a distortion and thus a polarization of the crystal cell which leads to a charge whose polarity is opposite to that of Fig. 4.5.2(b), on the same crystal faces A and B (transverse effect). It is easily seen from Fig. 4.5.2(b) and (c) that reversal of the forces (tension instead of compression) produces charges of opposite sign. The model of Fig. 4.5.2 can also be used to explain the inverse piezoelectric effect when assuming that an electric charge is applied to the faces A and B thus causing contraction or expansion of the crystal cell. Since there is symmetry in the z-direction no polarization occurs with forces applied in the direction of the (optical) z-axis.

The magnitude of the charge generated from longitudinal compression is obviously proportional to the degree of distortion of the crystal cell and thus depends on the pressure applied. The charge generated on a given area of the crystal face is also proportional to the area affected by the pressure. The total charge output is thus proportional to the total force applied to a quartz crystal, independent of the area of the crystal. It is also independent of the crystal thickness in the x-direction, since only the surface layers A and B produce free charges. Inside the crystal, polarization is cancelled out.

With forces applied in the transverse (y) direction conditions are different. Since the collecting faces A and B are not identical with those under pressure (C and D), a force applied to a crystal which is longer in the y-direction than in the x-direction yields a charge which is larger by the area ratio A/C, i.e. the length ratio l_y/l_x. Therefore, for the measurement of small forces, a (slender) length-expander crystal is more sensitive than a thickness-expander crystal of similar volume. In practice, however, a slender crystal prism under longitudinal compression may be liable to buckle and length-expander quartz crystals are therefore only occasionally used in instrument transducers.

The piezoelectric d-coefficients of quartz, according to eqn (4.5.2), are

$$
\begin{matrix}
d_{11} & -d_{11} & 0 & d_{14} & 0 & 0 \\
0 & 0 & 0 & 0 & -d_{14} & -2d_{11} \\
0 & 0 & 0 & 0 & 0 & 0
\end{matrix}
\tag{4.5.8}
$$

Their numerical values are

$$
d_{11} = 2 \cdot 3 \times 10^{-12}\,\mathrm{C\,N^{-1}}; \qquad d_{14} = -0 \cdot 67 \times 10^{-12}\,\mathrm{C\,N^{-1}}. \tag{4.5.9}
$$

Now consider a quartz disc cut normal to the x-axis. The diameter of the two (metallized) faces is d (m) and the disc thickness is t (m). A force F_x (N) is applied in the x-direction. Then the generated charge is

$$
q_x = d_{11}F_x \quad (\mathrm{C}), \tag{4.5.10}
$$

and the generated voltage across the metallized faces becomes

$$
v_x = q_x/C_x = d_{11}F_x/C_x \quad (\mathrm{V}), \tag{4.5.11}
$$

where C_x is the capacity of the disc,

$$
C_x = \varepsilon\varepsilon_0 \pi d^2/4t \quad (\mathrm{F}). \tag{4.5.12}
$$

If, however, a voltage v_x (V) is applied across the same (unloaded) disc, the disc will contract (or expand), according to the inverse piezoelectric effect, by

$$
\Delta l_x = d_{11}v_x \quad (\mathrm{m}), \tag{4.5.13}
$$

or, in terms of strain,

$$
\frac{\Delta l_x}{l_x} = d_{11}\frac{v_x}{l_x} = d_{11}e_x, \tag{4.5.14}
$$

where e_x (V m^{-1}) is the electric field strength.

If a force F_y is applied to a block of quartz of dimensions l_x, l_y, and l_z in the transverse (y) direction the charge generated in the x-direction becomes

$$
q_x = d_{12}\frac{l_y}{l_x}F_y \quad (\mathrm{C}), \tag{4.5.15}
$$

TABLE 4.5.1

Properties of some piezoelectric crystals [6]

Material	Crystal cut	Basic plate action	Free dielectric constant	d-coeffi-cient	Volume resistivity	Density	Young's modulus	Maximum safe stress	Maximum safe tem-perature	Safe humidity range	Remarks
Units	—	—	—	$C\,N^{-1}$	$\Omega\,m$	$kg\,m^{-3}$	$N\,m^{-2}$	$N\,m^{-2}$	$°C$	per cent	
Multiply by	—	—	1	10^{-12}	1	10^{3}	10^{9}	10^{6}	1	1	
Quartz	0°x	TE	4·5	2·3	10^{12}	2·65	80	98	550	0–100	
	0°x	LE	4·5	2·3	10^{12}	2·65	80	98	550	0–100	
Tourmaline	0°z	TE	6·6	1·9	10^{11}	3·10	160	—	1000	0–100	
	0°z	VE	6·6	2·4	10^{11}	3·10	—	—	1000	0–100	Hydraulic pressure
Rochelle salt (30 °C)	0°x	FS	350	550	10^{10}	1·77	—	14·7	45	40–70	Twister bimorph
	45°x	LE	350	275	10^{10}	1·77	19·3	14·7	45	40–70	Bender bimorph
	0°y	FS	9·4	54	10^{10}	1·77	—	14·7	45	40–70	Twister bimorph
	45°y	LE	9·4	27	10^{10}	1·77	10·7	14·7	45	40–70	Bender bimorph
Ammonium dihydrogen phosphate	0°z	FS	15·3	48	10^{8}	1·80	—	20·6	125	0–94	Twister bimorph
	45°z	LE	15·3	24	10^{8}	1·80	19·3	20·6	125	0–94	Bender bimorph
Lithium sulphate	0°y	TE	10·3	16	10^{10}	2·06	46	—	75	0–95	
	0°y	VE	10·3	13·5	10^{10}	2·06	—	—	75	0–95	Hydraulic pressure

and since $d_{12} = -d_{11}$, according to eqn (4.5.8),

$$q_x = -d_{11} \frac{l_y}{l_x} F_x, \qquad (4.5.16)$$

or the voltage

$$v_x = q_x/C_x = -d_{11}F_x/C_x \quad (V), \qquad (4.5.17)$$

where now $C_x = \varepsilon\varepsilon_0 l_y l_z/l_x$ (F). The inverse effect yields here

$$\Delta l_y = -d_{11} \frac{l_y}{l_x} v_x \quad (m) \qquad (4.5.18)$$

and the strain

$$\frac{\Delta l_y}{l_y} = -d_{11} \frac{l_y}{l_x} e_x. \qquad (4.5.19)$$

Table 4.5.1 summarizes the basic properties of some piezoelectric crystals for different modes of deformation [6] which are explained in Fig. 4.5.3.

The piezoelectric sensitivity of quartz varies with temperature. Fig. 4.5.4 shows the dependence of d_{11} in particular [7]. Between 20 °C and 200 °C the temperature coefficient of d_{11} is about $-0\cdot016$ per cent per °C. At lower temperatures the variation is less pronounced. On heating quartz up to $+576$ °C the piezoelectric α-phase changes into the inactive β-phase, the α-phase being re-established on returning below $+576$ °C. If, however, taken above the melting-point, quartz loses its piezoelectric character permanently. Variations of relative dielectric constant and volume resistivity at temperatures up to the Curie point are shown in Fig. 4.5.5. While the dielectric constant changes very little with temperature, there is an immediate and continuously increasing drop of volume resistivity, amounting to about 6 orders between room and Curie temperature.

The main application of quartz is in transducers for the measurement of high-level, fast-transient pressures and accelerations. Ambient temperature variations, especially if only of a transient nature, are of little consequence with quartz, since sensitivity and permittivity are hardly affected and the pyroelectric effect is negligible. However, because of the very low dielectric constant, shunt capacitance, e.g. in the cable, reduces the output voltage appreciably. Without shunt capacitance the voltage output can be sizeable. Quartz has no useful shear mode and cannot therefore be employed in twister elements. It has also no volume expander mode. Artificially grown quartz is normally preferred since it is purer and free from twinning.

4.5.2.1.2. Tourmaline. Because of its high cost (semi-precious stone) and low sensitivity it is little used in practical transducers. The piezoelectric coefficients of tourmaline, according to eqn (4.5.2), are

$$\begin{matrix} 0 & 0 & 0 & 0 & d_{15} & -2d_{22} \\ -d_{22} & d_{22} & 0 & d_{15} & 0 & 0 \\ d_{31} & d_{31} & d_{33} & 0 & 0 & 0. \end{matrix} \qquad (4.5.20)$$

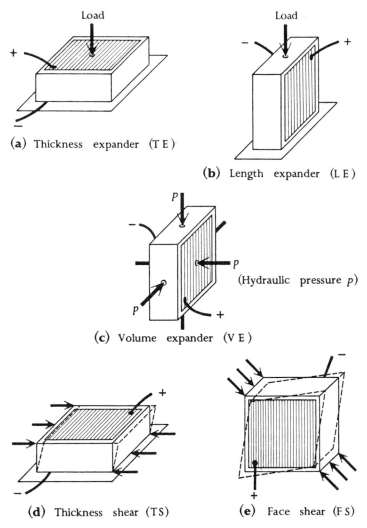

(**a**) Thickness expander (T E)

(**b**) Length expander (L E)

(Hydraulic pressure p)

(**c**) Volume expander (V E)

(**d**) Thickness shear (T S)   (**e**) Face shear (F S)

FIG. 4.5.3. Deformation modes of single piezoelectric elements.

It is mentioned here as the only natural material exhibiting a large volume-expander mode ('hydrostatic' coefficient d_h) related to the z-axis,

$$d_h = d_{33} + 2d_{31} = 1 \cdot 26 d_{33}, \qquad (4.5.21)$$

since $d_{31} = 0 \cdot 31 d_{33}$. Also, its useful temperature range is very wide.

4.5.2.1.3. Rochelle salt. It is the only piezoelectric material that is being grown on an industrial scale since it still plays a leading role in applications such as gramophone pick-ups and microphones, where its high shear

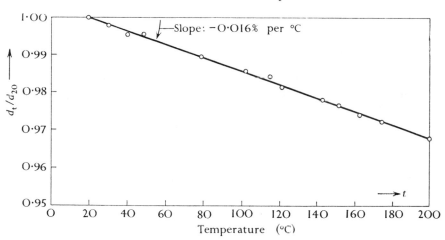

FIG. 4.5.4. Variation of the d_{11} coefficient of quartz with temperature, referred to d_{11} at 20 °C.

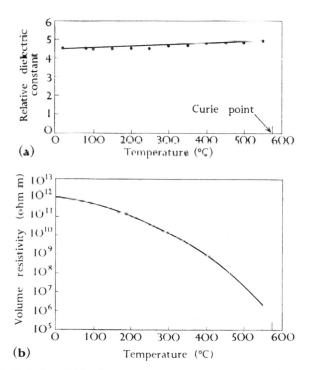

FIG. 4.5.5. Variation of (a) relative dielectric constant and (b) volume resistivity of quartz at higher temperatures.

sensitivity,

$$P_{xx} = d_{14}\tau_{yz}, \tag{4.5.22}$$

and high permittivity are of prime importance. However, its shortcomings—low mechanical strength, hysteresis, fatigue and limited temperature and humidity range, and a large change of permittivity between 0 °C and 40 °C—make Rochelle salt unsuitable for instrument transducers.

4.5.2.1.4. Ammonium dihydrogen phosphate (ADP). This artificial crystal is in many respects similar to Rochelle salt, though its permittivity is lower. Its basic plate action is face shear (FS) (see Fig. 4.5.3). Length-expanders are cut from FS-plates in a diagonal direction. Although ADP stands temperatures up to about 120 °C and fairly high values of humidity its insulation resistance is relatively low and decreases by about an order for every 20 °C temperature increase. ADP is probably best as a twister bimorph (see below) but length-expander elements have been used in acceleration and pressure transducers.

4.5.2.1.5. Lithium sulphate (LH). This material has an outstandingly high sensitivity to 'hydrostatic' pressure, but it can also be used as a thickness-expander, if the low permittivity is no embarrassment. Like Rochelle salt and ADP, LH is grown artificially.

4.5.2.2. Piezoelectric ceramics

The earliest of these piezoelectric materials is barium titanate, a polycrystalline ceramic substance of composition BaTiO₃. Because of its high dielectric constant (1000–5000) it was first used in small capacitors,

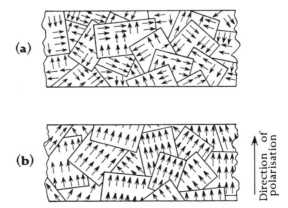

FIG. 4.5.6. Schematic arrangement of zones in barium titanate: (a) unpolarized; (b) polarized.

but soon its ferroelectric properties (so named by analogy to the properties of ferromagnetic materials) were realized and the possibility of
permanently polarizing the material opened up new applications. We may
visualize the material as consisting of zones with spontaneous polarization
(similar to Weiss zones in ferromagnetic materials) which can be partially
oriented by the application of an external electric field. Fig. 4.5.6 shows
schematically the material in a virgin state (a), and after polarization (b).
If electric induction is plotted against variable electric field strength, a
complete hysteresis loop is described (Fig. 4.5.7), and minor loops with
appreciably smaller slopes and negligible hysteresis can be produced in a

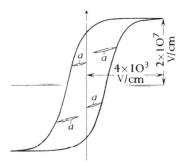

FIG. 4.5.7. Hysteresis loop of barium titanate (a = minor loops).

fashion familiar to the worker in ferromagnetics. Likewise the weak field
variations cause little or no domain motion, and the dielectric constant of
the minor loops is about equal to that at saturation where all domains are
completely aligned. The spontaneous polarization of the individual domains can be explained by the fact that below 120 °C the crystal cells of
barium titanate are tetragonal in shape with a length ratio 1·01 of the
crystal axes. This gives the small titanium atom at the centre of the cell
(Fig. 4.5.8) a preferred direction of movement between the oxygen atoms,
and thus polarizes the cell in this direction. Above 120 °C barium titanate
enters a cubic phase which exhibits no ferroelectric properties. For a
fuller understanding of the physics of ferroelectric ceramics, especially to
the explanation of secondary transition points of barium titanate at about
+10 °C and −80 °C, the reader should consult the appropriate textbooks
[2], [8], [10]. They also provide an introduction to other theories of
ferroelectricity which differ from the theory (that of Mason and Matthias)
followed here.

The stability of the remanent polarization of ferroelectric ceramics
relies on the coercive force of the dipoles displayed in the hysteresis loop.
In 'straight' barium titanate there are indications that this polarization

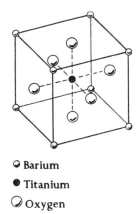

○ Barium

● Titanium

○ Oxygen

FIG. 4.5.8. Crystal structure of barium titanate.

may decrease with time as does the remanent polarization of soft magnetic materials. In order to improve the stability of polarization impurities
have been introduced in the basic material with the idea that the
polarization may be 'locked' into position. Lead titanate, calcium titanate,
and yttria have been tried among others. Their general effect on the
temperature behaviour of the compound is a lowering of the secondary
transition points and a smoothing of the sensitivity and capacitance versus
temperature curves. This desirable effect is normally coupled with a
lowering of sensitivity and capacitance levels. Commercially available
barium-titanate compounds are carefully balanced mixtures which are
tailored for particular applications.

A typical ceramic body of this kind consists of a solid solution of lead
zirconate and 10–60 mole per cent of lead titanate [11], with a performance peak at about 45 mole per cent (pure lead zirconate is not
ferroelectric and cannot be polarized). Its main advantage over barium
titanate is its higher Curie point at 320–350 °C. The piezoelectric coefficients of commercial lead zirconate–titanate compounds differ widely.
Typical examples are listed in Table 4.5.2. Positive polarization is in the
direction of the z-axis.

Piezoelectric ceramics have three active d-coefficients, d_{31} (length-
expander mode), d_{33} (thickness-expander mode), and d_{15} (shear mode),
according to the scheme

$$
\begin{matrix}
0 & 0 & 0 & 0 & d_{15} & 0 \\
0 & 0 & 0 & d_{15} & 0 & 0 \\
d_{31} & d_{31} & d_{33} & 0 & 0 & 0.
\end{matrix}
\qquad (4.5.23)
$$

Lead zirconate–titanate is the most common material for use in instrument transducers. Lead metaniobate which has been included in Table

TABLE 4.5.2

Properties of some piezoelectric ceramics; thickness-expander mode at room temperature

Material	Free dielectric constant	Piezoelectric coefficient g_{33}	Volume resistivity	Density	Young's modulus E_{33}	Tensile strength	Curie temperature	Remarks
Units	—	$C N^{-1}$	Ωm	$kg\, m^{-3}$	$N m^{-2}$	$N m^{-2}$	°C	
Multiply by	1	10^{-12}	1	10^{3}	10^{9}	10^{6}	1	
PX4†	1500	265	10^{11}	7·5	79	80	265	High voltage output, high Q
PX5†	1750	356	10^{14}	7·6	59	80	285	High charge output, low Q
PZT-5A‡	1700	374	10^{11}	7·8	53	76	365	High charge output, low Q
PZT-5H‡	3400	593	10^{11}	7·5	48	76	193	Very high charge output, low Q
PZT-7H‡	425	150	10^{9}	7·6	72	76	350	Very high voltage output, high Q
G-2000§	250	80	10^{9}	5·8	47	—	400	Very high Curie temperature, low Q

† Lead zirconate–titanate, made by Mullard Ltd., London, England [3].
‡ Lead zirconate–titanate, made by Vernitron Ltd., Southampton (Brush–Clevite Corp., U.S.A.) [12].
§ Lead metaniobate, made by Gulton Industries, Inc., Metuchen, New Jersey, U.S.A. [13].

4.5.2 for comparison, has the highest Curie point. Since the d_{31}-coefficients of piezoelectric ceramics are negative the (weak) volume expander mode is of little interest (see eqn (4.5.21) above).

The temperature characteristics of charge sensitivity, dielectric constant, and dissipation factor depend on the particular brand of the ceramic material, though in general they follow a tendency shown schematically in Fig. 4.5.9. Most material now available are fairly stable over a wide temperature range. Since the voltage output of a piezoelectric transducer is proportional d/ε, similar variations with temperature of d

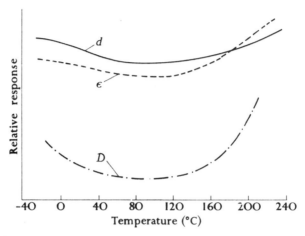

FIG 4.5.9. Schematic temperature characteristics of ferroelectric ceramics (d = piezoelectric coefficient, ε = dielectric constant, D = dissipation factor).

and ε cancel out and the voltage output curve is thus flatter than would appear from the individual d and ε curves.

Piezoelectric ceramics are manufactured by pressing, extrusion, or film-casting. The final heat-treatment is a sintering process during which a shrinkage of up to 20 per cent occurs. Close tolerances of dimensions require grinding. The electrodes are usually fired-on coats of silver or palladium, to which thin connecting wires can be soldered on, prior to polarization. Barium titanate can be polarized by the application of about 2 kV mm^{-1} for a few minutes. Lead zirconate–titanate bodies are somewhat more difficult to polarize because of their higher Curie points, but for this same reason they are less affected by environmental heat and can thus be polarized by the manufacturers.

If two rectangular length-expander elements are cemented together in a sandwich fashion so as to give additive output under bending stress, a new piezoelectric element is created which permits a much larger tip-deflection for a given load than a solid element. In a similar fashion a pair of FS elements produce a sensitive twister sensor. However, the higher output has been obtained at the expense of mechanical stiffness, and the choice between a single element and a so-called 'bimorph' element is therefore governed by the required mechanical input impedance. A particular case for the bimorph is the gramophone pick-up, where a low mechanical stiffness of the needle is an advantage. Bimorphs are less suitable for use in acceleration and pressure transducers which require high mechanical stiffnesses.

A bimorph can be connected internally either in parallel or in series, as shown in Fig. 4.5.10(a) and (b). The parallel connection doubles the

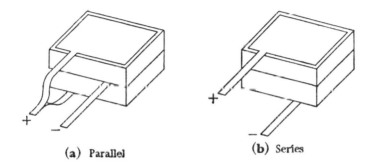

(a) Parallel   (b) Series

FIG. 4.5.10. Parallel and series connection of sandwich-type piezoelectric elements ('bimorphs').

capacitance and in a piezoelectric sender-type transducer admits the full excitation voltage across each plate. Instrument transducers generally employ series connection because of the higher output voltage (halved capacity), provided the low transducer capacitance can be tolerated in the presence of, say, appreciable cable capacitance. The polarization of the two plates with respect to one another differs for series and parallel elements and the intended character must therefore be known at the time of manufacture. Examples of mounting bimorph-type elements are shown in Fig. 4.5.11(a)–(e).

A special type of bimorph, a so-called 'multimorph' [6], is still more flexible than a standard bimorph and may thus be used with advantage in gramophone pick-ups and similar low-stiffness applications, though it has also been employed in a miniature acceleration transducer, the multimorph 'cantilever' representing the (distributed) seismic mass [14]. These elements are solid strips of piezoelectric ceramic with a number of axial holes in the centre plane which can be filled with conductive matter, such as graphite, constituting the centre electrode of a bimorph. Their manufacture is said to be simpler than that of cemented bimorphs.

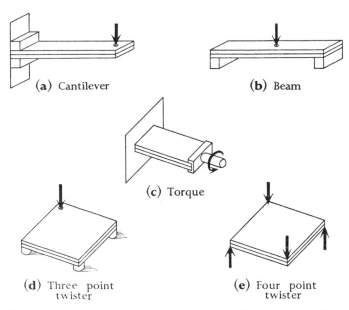

FIG. 4.5.11. Types of mounting of sandwich-type piezoelectric elements ('bimorphs').

4.5.3. Piezoelectric transducers

4.5.3.1. Dynamic performance, general

For a general analysis of the dynamic performance of piezoelectric transducers we must return to section 2.3 on the electromechanical coupling characteristics of bilateral transducers.

Force equilibrium in a piezoelectric transducer, such as a simple thickness-expander type, yields

$$f = Z_m v + Se. \tag{4.5.24}$$

For the notations of force f, mechanical impedance Z_m, velocity v, and voltage e see Fig. 2.2(c) (p. 12). S is a piezoelectric 'sensitivity' of dimension $\mathrm{N\,V^{-1}}$; it depends on the piezoelectric coefficient d $(\mathrm{m\,V^{-1}})$, but also on crystal dimensions. If a $(\mathrm{m^2})$ is the active crystal area and t (m) the thickness,

$$S = \varepsilon\varepsilon_0 a/td, \tag{4.5.25}$$

where ε is the relative dielectric constant of the piezoelectric material, and

$$\varepsilon_0 = 8\cdot854\times10^{-12}\,\mathrm{F\,m^{-1}}, \tag{4.5.26}$$

the permittivity of empty space. Also, since a thickness variation

$\Delta t = x$ (m) produces a charge q_t (C) of magnitude,

$$q_t = -Sx, \qquad (4.5.27a)$$

or, generally, in terms of the time derivatives of q_t and x,

$$i_t = -Sv, \qquad (4.5.27b)$$

the current equation

$$i = -Sv + Y_e e = -Sv + j\omega Ce \qquad (4.5.28)$$

holds, where C (F) is the capacitance of the piezoelectric element, the leakage current being neglected.

Comparison of above eqns (4.5.24) and (4.5.28) with eqns (2.26) and (2.28) (p. 14) shows that, in contrast to the electrodynamic transducer, a current-source representation is here a natural choice, so that

$$N = S \ (\text{N V}^{-1}) \quad \text{and} \quad -N^* = -S \ (\text{A s m}^{-1}). \qquad (4.5.29)$$

Extracting the mechanical impedance Z_m according to eqn (2.10) (p. 10), then the appropriate characteristic transfer matrix equation (2.30) (p. 15) for the piezoelectric transducer becomes (see Fig. 2.3(b), p. 20)

$$\begin{bmatrix} f_t \\ v \end{bmatrix} = \begin{bmatrix} S & 0 \\ j\omega C/S & -1/S \end{bmatrix} \begin{bmatrix} e \\ i \end{bmatrix}. \qquad (4.5.30)$$

The transfer function of the piezoelectric force receiver is obtained from eqn (2.42) (p. 18). The input force may be derived from a pressure, incident on the crystal surface, or from acceleration induced in a seismic mass loading the crystal. Letting $N^* = S$ and $N^2 = S^2$, and assuming also the dynamic properties of the crystal can be represented by discrete values of mass m, stiffness k and damping c,

$$\frac{e}{f} = \frac{S}{S^2 + \{c + j(\omega m - k/\omega)\}(Y_0 + Y + j\omega C)} \quad (\text{V N}^{-1}). \qquad (4.5.31)$$

With $Y_0 = 1/R_0$, where R_0 is the load resistance, and neglecting leakage $(Y = 0)$, eqn (4.5.31) becomes

$$\frac{e}{f_0} = \frac{SR_0/(1 + j\omega CR_0)}{S^2 R_0/(1 + j\omega CR_0) + \{c + j(\omega m - k/\omega)\}}. \qquad (4.5.32)$$

We note that at frequencies above $\omega = 1/CR_0$, but below resonance, i.e. when k/ω dominates ωm, the frequency dependence in the numerator of eqn (4.5.32) cancels that in the denominator and the transducer output is constant in this useful range. Also, since the voltage output is seen to be proportional to $1/C$, piezoelectric transducers require a very high load resistance R_0 in order to secure a high voltage output, as well as a wide useful-frequency range.

The mechanical damping c of piezoelectric materials is very small, and so is the electrical damping S^2R_0, since the electromechanical coupling is relatively weak. According to eqn (4.5.32) a series resonance of the mechanical components occurs for $\omega_r m \simeq k/\omega_r$. At frequencies above ω_r they represent an inductance which resonates, in parallel configuration, with the transducer capacitance C at ω_a ('anti-resonance'). Fig. 4.5.12(a) and (b) illustrate the equivalent circuit and the reactance-frequency characteristic, respectively [2]. (The mechanical components are represented here by their electrical analogues, according to section 2.3.1.)

For completeness we also give the transfer function of the piezoelectric

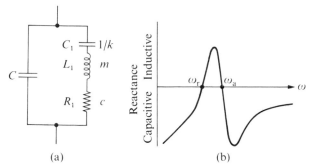

(a)   (b)

FIG. 4.5.12. (a) Equivalent circuit. (b) Reactance variation with frequency of piezoelectric element.

sender. Since it operates at a low source impedance, $R_0 \ll 1/\omega C$, and we have from eqn (2.45) of section 2.3.2,

$$\frac{-v}{i_0} = \frac{-S}{S^2 + \{c + \mathrm{j}(\omega m - k/\omega)\}/R_0} \quad (\mathrm{m\ A^{-1}s^{-1}}). \qquad (4.5.33)$$

Under these conditions, only the mechanical terms matter, and the response is a simple resonance curve. The damping term $(c + S^2 R_0)$ is again insignificant; the resonance of a piezoelectric sender is therefore very sharp.

4.5.3.2. *Sensitivity and frequency response*

At frequencies well below the (first) resonant frequency the equivalent circuit diagram of Fig. 4.5.12(a) can be simplified by lumping the capacitance values of transducer (C_E), cable (C_C), and load (C_L) in one single capacitance C, and the (loss and leakage) shunt resistance of the transducer (R_E) and the load resistance (R_L) into R, as shown in Fig. 4.5.13(a) and (b). Then, at sinusoidal excitation, the primary quantity generated is the electric charge

$$q = Q_0 \sin \omega t, \qquad (4.5.34)$$

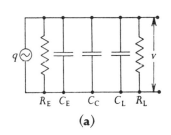 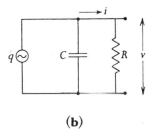

$$R_E \quad C_E \quad C_C \quad C_L \quad R_L$$

(a)   **(b)**

FIG. 4.5.13. Equivalent circuits of piezoelectric transducers well below mechanical resonance.

and hence

$$q = Cv + \int i \, dt = Q_0 \sin \omega t, \tag{4.5.35}$$

giving, with $v = iR$,

$$\frac{di}{dt} + \frac{1}{RC} i - \frac{\omega Q_0}{RC} \cos \omega t. \tag{4.5.36}$$

The complete solution of eqn (4.5.36) yields the output voltage

$$v = V_1 \sin\left(\omega t + \tan^{-1} \frac{1}{\omega RC}\right) - V_2 e^{-t/RC}, \tag{4.5.37}$$

where

$$V_1 = \frac{Q_0/C}{\{1 + (1/\omega RC)^2\}^{\frac{1}{2}}}, \tag{4.5.38a}$$

and

$$V_2 = \frac{Q_0/C}{(1/\omega RC) + \omega RC}. \tag{4.5.38b}$$

Q_0/C is the peak voltage V_0 which occurs across C for $R \to \infty$. From eqns (4.5.37) and (4.5.38) it is seen that the output voltage v, and the phase angle ϕ depend on the time-constant RC and frequency ω. In eqn (4.5.37) there is a sinusoidal term which is proportional to the physical quantity to be measured (e.g. acceleration) and a transient term which disappears for large values of ωRC (eqn (4.5.38b)). The magnitude of the sinusoidal 'sensitivity' term V_1 approaches unity at high values of ωRC. For finite values its response is shown in Fig. 4.5.14, where V_1/V_0 has been plotted against ωRC. Fig. 4.5.14 also shows the phase error

$$\phi = \tan^{-1} \frac{1}{\omega RC} \tag{4.5.39}$$

plotted against ωRC. It is seen that the low-frequency response of a piezoelectric transducer depends upon the time-constant of the circuit consisting of transducer, cable, and load. It is therefore common practice

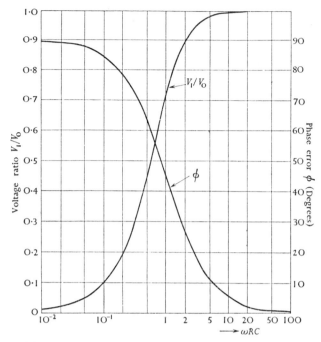

FIG. 4.5.14. Voltage ratio and phase error of piezoelectric transducer circuit.

first to decide upon the minimum time-constant RC required to keep voltage and phase errors within specified limits. This may be done by reference to Fig. 4.5.14. A high value of load resistance is particularly advantageous, though there is no point in choosing one of much higher value than the total leakage resistance of transducer and cable. With the load resistance decided upon a suitable circuit may be selected which is capable of producing the required load resistance as seen by the trans-ducer. The required total capacitance is then determined by the specified time-constant T since $C = T/R$. If this value is higher than the combined transducer and cable capacitance, shunt capacitance may be added which, however, reduces the voltage output according to $V_0 = Q_0/C$.

If, in the other extreme, ωRC is made small compared with unity it can be shown that the output becomes the first time derivative of the input quantities. Such an arrangement can be used for the measurement of rate-of-change of pressure (e.g. rate-of-climb indicator for aircraft), or for the measurement of 'jerk', the time derivative of acceleration (see Table 3.1, p. 34).

While the low-frequency response depends mainly on electrical parameters, the high-frequency response of piezoelectric transducers is a purely mechanical problem. Pressure and acceleration transducers consist, in addition to the piezoelectric element, of a number of components such as mass, loading springs, housing, and electrical leads. All these, as single components or in combination with one another, may give rise to resonances which are usually lower than the lowest element mode. Even small components such as pins or wires used for internal connections may be troublesome in instruments intended for high-frequency applications. If, however, the lowest resonant frequency f_r is known, the response of the

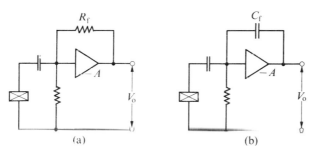

FIG. 4.5.15. (a) Voltage amplifier with resistive feedback. (b) Charge amplifier with capacitive feedback. Schematic diagram.

transducer at a frequency f in the neighbourhood of f_r is

$$\frac{v}{v_0} = \frac{1}{1-(f/f_r)^2},\tag{4.5.40}$$

since piezoelectric transducers are virtually undamped. Above the lowest resonant frequency, resonance and anti-resonance pairs follow each other at small frequency intervals. These intervals get closer with increasing frequency owing to the large number of possible vibratory modes, and it is usually quite impracticable to predict with any accuracy the behaviour of a transducer in this region.

4.5.3.3. Associated circuits [16]

Although here we cannot go into any detail of electronic circuitry, a few comments on amplifier configurations suitable for use with piezoelectric transducers may be useful.

4.5.3.3.1 Voltage amplifiers. Conventional voltage amplifiers are normally unsuitable because of their relatively low input impedances. In the past, electrometer-valve input stages and cathode-follower circuits have been employed. They offer high input impedances—the latter by way of resistive feedback (Fig. 4.5.15(a))—but voltage amplifiers are limited in their frequency response, particularly at the low-frequency end. Since the output voltage also depends on cable capacitance, piezoelectric measuring systems using voltage amplifiers must be re-calibrated when changing the transducer cable.

4.5.3.3.2. Charge amplifiers [17].   The charge amplifier employs capacitance feedback (Fig. 4.5.15(b)). At high amplifier gain A the output voltage of the charge amplifier can be written approximately

$$V_0 \simeq -Q/C_f A, \qquad (4.5.41)$$

where Q (C) is the charge generated at the transducer, and C_f (F) the feedback capacitance. The output voltage remains constant over a wide frequency range and is substantially independent of cable length. Since it is wholly controlled by the feedback capacitance C_f the circuit can be designed to provide long time-constants. The low-frequency response is therefore almost unlimited except for the need to suppress spurious signals from slow temperature fluctuations (pyroelectric effect). Charge amplifiers are now in almost exclusive use with piezoelectric transducers.

4.5.3.4. Environmental effects

4.5.3.4.1. Temperature and humidity.   Temperature variations affect charge sensitivity, permittivity, dissipation factor, and volume resistivity of piezoelectric materials. There is no simple way of predicting the thermal performance characteristics of natural crystals and piezoelectric ceramics with any certainty, though a few guide lines have been given in section 4.5.2, such as safe maximum temperatures and Curie temperatures (Tables 4.5.1 and 4.5.2, pp. 258 and 265). The diversity of performance data can be illustrated by looking at the permittivity–temperature curves of Fig. 4.5.16 [18], which afford a first selection criteria for suitable materials, particularly since it is known from experience that the charge sensitivity curves roughly follow the tendencies of the permittivity curves. Published experimental results often depend on transducer design and on methods of mounting and loading and are therefore of little general value. However, modern lead zirconate–titanate ceramics are quite reliable and some have attained the same high-temperature stability as quartz.

All piezoelectric materials suffer from loss of insulation resistance when water vapour is permitted to condense on their faces. Water-soluble crystals, if unprotected, can stand only a limited range of humidity (Table 4.5.1). They loose their water of crystallization in dry air, and in moist air the crystals dissolve.

4.5.3.4.2. Transverse sensitivity.   The transverse or cross-axis sensitivity of piezoelectric acceleration transducers follows a two-lobed characteristic, as shown schematically in Fig. 4.5.17. Transverse sensitivity can be explained by electrical and mechanical asymmetries in the disc or the

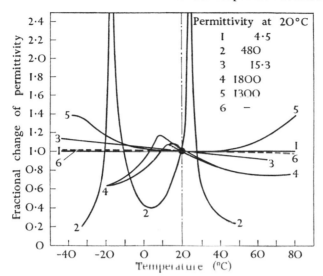

FIG. 4.5.16. Fractional change of permittivity with temperature of (1) quartz, (2) Rochelle salt, (3) ammonium dihydrogen phosphate (ADP), (4) barium titanate A (Brush), (5) barium titanate B (Brush), (6) lead zirconate–titanate (Endevco).

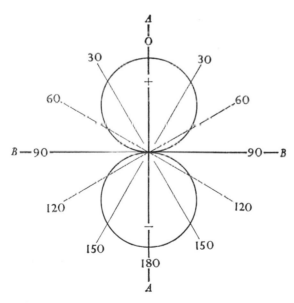

FIG. 4.5.17. Transverse sensitivity pattern of piezoelectric acceleration transducer (barium titanate type). A——A: direction of maximum transverse sensitivity. B——B: direction of minimum transverse sensitivity.

mounting of a practical transducer. Its value is normally expressed as a percentage of the full-scale sensitivity of the transducer in the direction of its main axis. Its maximum value should not be greater than a few per cent [19].

4.5.3.4.3. Sound-pressure sensitivity [20]. Piezoelectric materials having a volume-expander mode produce spurious signals if exposed to high-intensity air-borne noise. Similar effects occur if the noise is admitted to some, but not all, of the crystal faces. In a particular test on a miniature accelerometer the output owing to 'white' noise of intensity 160 dB above threshold was equivalent to an acceleration of ±0·01g to ±0·1g.

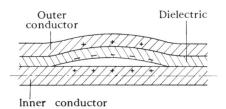

FIG. 4.5.18. Schematic diagram of separation between inner conductor and insulation of coaxial cable.

4.5.3.4.4. Cable noise. Early investigations of the mechanism of the generation of cable noise caused by whipping and twisting the cable during calibration and measurement has led to the design and manufacture of special low-noise cables [21]. Fig. 4.5.18 shows schematically the inner and outer conductor of a concentric cable with a dielectric layer between them, which in places has been separated from the inner conductor. Unless the resulting charges generated by friction and electrostatic induction are conducted away by conductive layers, appreciable spurious voltages can occur [19]. Earlier types of low-noise cable relied on rigidity, thus avoiding separation except under very strong forces. More reliable bonding between inner conductor and insulation has also been tried, but introduction of conductive layers has now become common practice.

Cable noise can also be caused by the electrostrictive properties of Teflon and PVC, though the effect is weaker by 2 or 3 orders as compared with piezoelectricity [22].

4.5.4. Design and construction of piezoelectric transducers

4.5.4.1. Acceleration types

Up to and during the Second World War quartz was almost always used in piezoelectric acceleration transducers, with the probable exception of a few transducers employing Rochelle salt. These instruments are now supplemented by transducers with polarized ceramics of ferroelectric materials such as barium titanate and its derivatives or, in a few cases, by ADP. The quartz types were usually large and heavy and consequently had a fairly restricted upper frequency limit. They were frequently operated in conjunction with d.c. amplifiers, the first stage being an electrometer valve circuit of the highest possible insulation resistance. Rochelle-salt types of earlier design have become less attractive, in spite of their high sensitivity, because they have too low a temperature and humidity limit for the majority of practical applications.

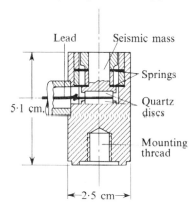

FIG. 4.5.19. Acceleration transducer with two quartz discs loaded by annual springs.

Fig. 4.5.19 shows an early arrangement with two x-cut quartz crystals oriented so that the charges generated at the centre electrode are additive if compression is applied to the crystal stack. The mechanical pre-load by the annular springs must be higher than the expected maximum acceleration force applied in an upward direction. The two outer surfaces of the crystals are in contact with the housing and thus at earth potential. The transducer [43] had an output of 0·62 pC g^{-1} and the lowest resonance occurred at about 15 kHz. The capacitance of the two crystals was only a few pF, and hence negligible compared with that of a cable of any length. The transducer weighed about 275 g and was mounted by means of a $\frac{1}{2}$ in. stud. With an assumed total capacitance, including cable, of 500 pF, the open-circuit output voltage was

$$V = \frac{0·62 \times 10^{-12}}{500 \times 10^{-12}} = 1·25 \text{ mV g}^{-1}. \tag{4.5.42}$$

The transducer was suitable for the measurement of medium and high accelerations only, unless extremely short cables were acceptable.

An instrument [23] built on the same principle, though of much higher sensitivity, is shown in Fig. 4.5.20. The seismic mass has been increased

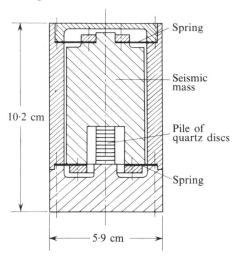

FIG. 4.5.20. Acceleration transducer of high sensitivity employing ten quartz discs.

as far as possible and the crystal pile consisted of ten quartz discs connected in parallel, which produced a charge output of about 80 pC g^{-1}. The lowest resonance occurred at 1500 Hz. Again, with an assumed total capacitance of 500 pF the voltage output was 160 mV g^{-1}, so that accelerations as low as fractions of the earth's gravitation could be measured with ease at frequencies up to, say, 300 Hz. In constructing quartz piles, especially with a large number of discs, care must be taken to attain parallel crystal faces. The transverse sensitivity must be controlled by accurate and sufficiently strong guide springs. In the orientation, if using individual discs, one should take notice of the direction of the y-axes of the discs and attempt to arrange them in such a way that the transverse sensitivities due to inertia forces on the discs are cancelled. However, acceleration transducers of such large sizes and weights are now of little use in air-borne applications, although they have still some attraction for ground vibration work in civil engineering and geology.

An acceleration transducer [24] with theoretical compensation of transverse errors is shown schematically in Fig. 4.5.21. Here two independent

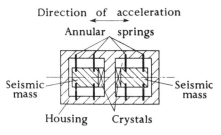

FIG. 4.5.21. Twin-type acceleration transducer, schematic diagram.

seismic systems are arranged so that longitudinal acceleration forces (in the direction of the transducer axis) cause additive outputs, while the outputs from transverse forces are cancelled, if the y-axes of the discs are pointing in the same direction. Output from air-borne sound pressure is also compensated in this arrangement. However, there is danger of beating between the resonant frequencies of the two independent vibratory systems, and since they are similar the beat frequencies are low and sure to fall inside the useful frequency range of the transducer.

Early acceleration transducers [44], made with Rochelle salt, employed twister elements (see Fig. 4.5.11(d)). These so-called three-point twister bimorphs had one free corner for loading. Their dimensions were still

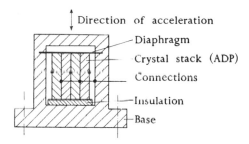

FIG. 4.5.22. Acceleration transducer with ADP crystal stack, schematic diagram.

relatively large but, since they had high sensitivities of about 80 mV g^{-1} at high transducer capacities, they were quite attractive, in spite of the basic shortcomings of Rochelle salt. (Vibration probes of similar design, which were held by hand and in contact with the vibrating body, were made in the U.S.A. and in Germany.) These instruments are now superseded by acceleration transducers with bender bimorphs of ferroelectric ceramics.

The schematic arrangement of an acceleration transducer [45] with ammonium dihydrogen phosphate crystals is shown in Fig. 4.5.22. A stack of length-expander ADP crystals was assembled in a metal housing, any transverse movement being prevented by a diaphragm at the free end of the stack. The resulting transverse sensitivity was about 7 per cent (maximum) of the longitudinal sensitivity of 5·4 pC g^{-1}. Its capacity (with a 0·5 m long cable) was 168 pF, and hence its voltage output was 32 mV g^{-1}. At room temperature the insulation resistance was 70 MΩ, falling to 1 MΩ at the maximum working temperature of 70 °C. The output voltage was fairly stable over a temperature range of −40 °C to +70 °C, since charge sensitivity and capacitance declined roughly at the same rate. The transducer weight was 25 g. Except for its limited temperature range and its relatively low insulation resistance this ADP-type transducer had the advantages of small size, high natural frequency (about 50 kHz), and wide acceleration range (±500 g). ADP has no noticeable pyroelectric effect.

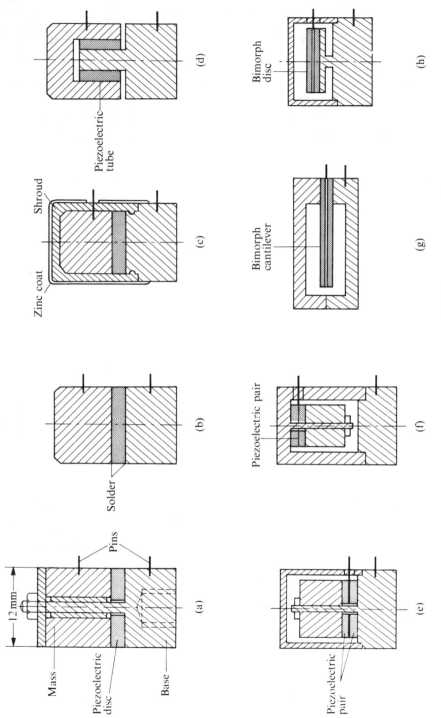

Fig. 4.5.23. Miniature acceleration transducers with ferroelectric ceramic discs: (a) single disc, clamped; (b) single disc, soldered; (c) single disc, soldered and shrouded; (d) shear tube; (e) twin discs, inverted; (f) twin discs, spring loaded; (g) bimorph bender; (h) bimorph umbrella.

With the advent of the polarized ferroelectric ceramics, piezoelectric acceleration transducer design and application acquired momentum. The high permittivity and sensitivity of these materials promoted instruments which are smaller than those containing quartz, whilst the higher mechanical stiffness and strength of these ceramic bodies gave higher natural frequencies and wider acceleration ranges than were possible with Rochelle salt.

The first of a family of miniature acceleration transducer of wide frequency range [25] was developed at the National Bureau of Standards, Washington D.C., U.S.A., using a barium titanate disc sandwiched between transducer base and seismic mass and assembled by clamping (Fig. 4.5.23(a)). In 1953 a soldered version was produced at the Royal Aircraft Establishment, Farnborough, and an epoxy-resin shroud, coated with zinc, was added for humidity protection and electrostatic shielding (Fig. 4.5.23(b) and (c)). However, the hope that the shroud would appreciably increase the damping proved unsuccessful. Other types have been developed through the years [19]. In Fig. 4.5.23(d) a tubular element is employed in a shear mode, and Fig. 4.5.23(f) shows an inverted type which avoids the direct transmission of strains in the substructure to the sensitive elements.

Acceleration transducers with ferroelectric bimorphs are usually of the bender type, employing a cantilever element, as shown in Fig. 4.5.23(g). Acceleration is applied normal to the plane of the bimorph and there is usually no seismic mass except the self-mass of the bimorph [14]. A variation of this design, consisting of a piezoelectric disc bonded to a thin metal disc and supported at its centre by a stud, has been employed successfully. The deflection of the disc under acceleration resembles the shape of an open umbrella (Fig. 4.5.23(h)).

Modern polarized ceramics for use in piezoelectric accelerometers have high temperature stabilities which are comparable with that of quartz. In order to avoid loss of frequency response at the high-frequency end the method of attachment of transducers must be considered carefully [26]. The dynamic characteristics of the bimorphs proper are more complex than those of thickness- and length-expander types and require a more elaborate equivalent circuit if the various loading conditions are to be represented by the circuit [27]. Tables [6] have been computed for a number of common materials and representative cases of loading which may be used with advantage by the designer.

4.5.4.2. *Pressure types*

Through the years the main application of piezoelectric pressure transducers has been the 'engine indicator' for use with internal combustion engines, which employs quartz discs or piles of discs. The dynamic

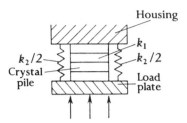

FIG. 4.5.24. Pre-loaded crystal pile.

characteristics of engine indicators have been investigated thoroughly [7], [28] and the basic ideas [29] will be summarized below. Transducers for the measurement of combustion pressures of modern engines need a very wide frequency range, because the 40th, or even higher, harmonics of the engine speed may contribute appreciably to the correct shape of the pressure–time curve. Even at a conventional engine speed of 3000 r.p.m. the highest frequency to be recorded is therefore around 2 kHz, and since piezoelectric transducers are usually undamped the requirement for the lowest resonance at 10–20 kHz is therefore not excessive. Still wider frequency ranges may be wanted for the measurement of transient pressures in detonation, internal ballistics, and in hypersonic work.

The sensitive element of a piezoelectric engine indicator consists essentially of a pair, or a pile of pairs, of quartz discs, as shown schematically in Fig. 4.5.24. The optically flat faces of the quartz crystals are held between similarly flat metal faces of the load plate and the transducer body by way of a pre-loading spring of stiffness k_2, k_1 being the stiffness of the crystal pile. It has been shown that high frequencies are achieved only if residual air between the faces is removed by a high pre-load (say 500–1000 N for discs of 6–10 mm diameter). Bending of the loading plate or the housing would have similar detrimental effects on the natural frequency and may also cause non-linearity of calibration. In earlier designs the pre-load was obtained by a relatively stiff diaphragm which also served as a gas seal (Fig. 4.5.25(a)). The main disadvantages were temperature sensitivity,

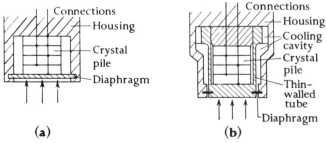

FIG. 4.5.25. Schematic arrangement of piezoelectric pressure transducers: (a) diaphragm-loaded type, (b) tube-loaded type.

owing to variable pre-load with temperature, and non-linearity even at high values of pre-loads. A better arrangement is shown in Fig. 4.5.25(b), where the pre-load is produced by a thin-walled tube under tension, the sealing being provided by a very thin diaphragm of flexible material. In considering Fig. 4.5.24 it is seen that the total force P produced by external pressure is split into the force P_1 in the crystal pile and a shunting force P_2 in the pre-loading member which is either a diaphragm or a thin-walled tube. At a pile compression Δx we have

$$P = P_1 + P_2 = (k_1 + k_2)\,\Delta x; \qquad (4.5.43)$$

hence the useful portion of the total force is

$$\frac{P_1}{P} = \frac{k_1}{k_1 + k_2} = \frac{1}{1 + k_2/k_1}. \qquad (4.5.44)$$

P_1/P, and therefore the sensitivity, increases with decreasing k_2/k_1, i.e. at a given pile stiffness it increases with the flexibility of the loading springs. Linearity is obtained only if k_2/k_1 remains constant over the pressure range, i.e. at high pre-loads when air-cushioning is eliminated and bending stresses are swamped. Other features to be avoided in high-frequency transducer assemblies are loaded screw connections and ball contacts (once very fashionable), since both are likely to introduce non-linearities. In fact, modern assembly techniques employ spot-welding [7] or electron-beam-welding. The tube-loaded design also facilitates the use of an effective water-cooling system around the crystal pile. Piezoelectric engine indicators, designed to these principles and manufactured with great care, had natural frequencies of 30–50 kHz, verified experimentally by exciting the transducer electrically over a wide frequency range by way of the inverse piezoelectric effect [28].

Because of the small masses loading the crystal pile, and the high natural frequencies, the well-designed piezoelectric engine indicator is relatively insensitive to vibration. However, attempts have been made to compensate for vibration by employing two identical systems arranged such that the outputs due to pressure are additive while those due to acceleration cancel. Such a system is an inversion of the compensated acceleration transducer of Fig. 4.5.21, though similar cautionary remarks on the danger of beat frequencies apply here too. A somewhat superior design [30] is shown in Fig. 4.5.26, in which a (slightly rounded) quartz crystal is held between two loading diaphragms. It is easily seen that to a first approximation vibration will cause no output due to opposing forces on the crystal of the two diaphragms. The pressure sensitivity is about half that of a rigidly mounted crystal of similar dimensions.

The resonant frequencies mentioned above are, of course, not the natural frequencies of the crystal pile itself, which may be in the region of

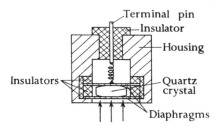

FIG. 4.5.26. Piezoelectric pressure transducer with acceleration compensation, schematic diagram.

500–1000 kHz, depending on size. The measured resonances are due to complex vibratory modes of the total transducer assembly when mounted on a structure. Experience has shown that even minor components, such as terminal pins or electrical connections, can cause resonances at disappointingly low frequencies, and great care must be exercised in the design of all details of a transducer, including the mounting facilities. Screw connections are often the cause of undue flexibility and must therefore be regarded with suspicion.

Techniques and transducers for the measurement of air-blast pressures and of underwater pressure transients are usually classified, but the underlying basic principles are well known. The ideal blast-pressure transducer should be non-directional, which would be achieved with a pressure-sensitive sphere of dimensions small compared with the wavelength of the blast wave. For practical purposes a small cylinder is the second best choice. A small hollow cylinder [31] of piezoelectric material, such as barium titanate, is shown in Fig. 4.5.27. Its outer and inner surfaces are metallized, and the walls are polarized in a radial direction. 'Hydrostatic' air pressure may be applied to the outer surface, and it is assumed that the tube cavity is sealed against external pressure.

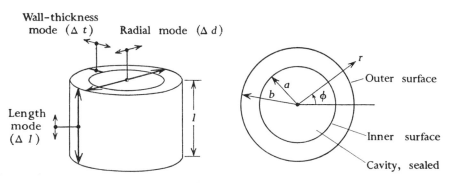

FIG. 4.5.27. Piezoelectric hollow cylinder showing three possible modes of vibration.

The electrical capacity of the transducer element (see section 4.4, Table 4.4.1) is

$$C = \frac{2\pi\varepsilon\varepsilon_0 l}{\ln(b/a)} \quad \text{(F)}, \tag{4.5.45}$$

or for a thin-walled tube,

$$C = \frac{\pi\varepsilon\varepsilon_0 l(b+a)}{(b-a)}, \tag{4.5.46}$$

where

ε = relative dielectric constant of tube material,
$\varepsilon_0 = 8\cdot854\times10^{-12}\,\mathrm{F\,m^{-1}}$,
l = length of cylinder (m),
a = inside radius (m),
b = outside radius (m).

The low-frequency sensitivity can be computed from the tangential and radial stresses set up by the external hydraulic pressure. Let (Fig. 4.5.27)

p_e = external pressure ($\mathrm{N\,m^{-2}}$),
p_i = internal pressure ($\mathrm{N\,m^{-2}}$),
g_{31}, g_{33} = piezoelectric constants of the tube material (section 4.5.2.1),
σ_r = radial stress ($\mathrm{N\,m^{-2}}$),
σ_ϕ = tangential stress ($\mathrm{N\,m^{-2}}$),

then the open-circuit voltages generated by the stresses σ_r and σ_ϕ are

$$V_r = g_{33} \int_{r=a}^{b} \sigma_r \, dr, \tag{4.5.47a}$$

$$V_\phi = g_{11} \int_{r=a}^{b} \sigma_\phi \, dr, \tag{4.5.47b}$$

where [32]

$$\sigma_r = \frac{a^2 b^2 (p_e - p_i)}{r^2(b^2 - a^2)} + \frac{p_i a^2 - p_e b^2}{b^2 - a^2}, \tag{4.5.48a}$$

$$\sigma_\phi = \frac{-a^2 b^2 (p_e - p_i)}{r^2(b^2 - a^2)} + \frac{p_e a^2 - p_i b^2}{b^2 - a^2}. \tag{4.5.48b}$$

From (4.5.47) and (4.5.48), and assuming $p_i = 0$, we have

$$\frac{V}{p_e} = \frac{V_r + V_\phi}{p_e} = g_{33} b \frac{a-b}{a+b} - g_{31} b. \tag{4.5.49}$$

It is seen from eqn (4.5.49) that the output voltage depends upon the radial tube dimensions only and is independent of tube length. In the case of appreciably shunt capacitance in the circuit, however, this is no longer true and the transducer capacitance and the shunt capacitance must be considered. Since g_{31} and g_{33} have opposite signs and since $(a-b)$ is negative, eqn (4.5.49) represents the difference of two opposing effects, which becomes zero at some critical ratio of outside to inside diameter.

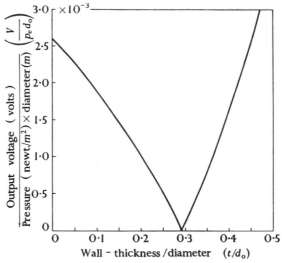

FIG. 4.5.28. Voltage sensitivity to external pressure of barium-titanate tube at various t/d_0 ratios.

In Fig. 4.5.28 the sensitivity curve has been plotted for a barium-titanate material with $g_{31} = -5 \cdot 2 \times 10^{-3}$ and $g_{33} = 12 \cdot 6 \times 10^{-3}$ V m N^{-1}. The independent variable is the wall thickness to outside diameter ratio t/d_0. Rearranging eqn (4.5.49) we have

$$\frac{V}{p_e d_0} = \frac{1}{2}\left(g_{33}\frac{t/d_0}{1-t/d_0} + g_{31}\right), \tag{4.5.50}$$

in which the output is given in volt per unit hydrostatic pressure (N m^{-2}) and per unit diameter of tube (m). It is seen that in the neighbourhood of $t/d_0 = 0 \cdot 29$ the sensitivity is negligible due to cancellation of output from radial and tangential stresses. The minimum is, however, fairly pronounced, and effective transducers can be made for lower pressures with their walls ($t/d_0 \leqslant 0 \cdot 2$) and for higher pressures ($t/d_0 \geqslant 0 \cdot 35$).

The rather complex dynamic characteristics [31], [33] of the tube are concerned with three possible vibratory modes (Fig. 4.5.27):

   (a) length mode (Δl),

   (b) radial mode (Δd),

   (c) wall thickness mode (Δt).

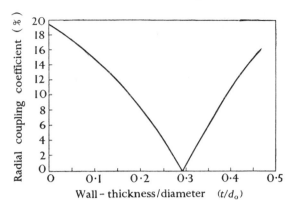

FIG. 4.5.29. Radial coupling coefficient of barium-titanate tubes at various t/d_0 ratios.

Between length mode and radial mode there is considerable coupling, while the wall-thickness mode is relatively independent of the other two. A cross-over point for the lowest resonant frequency occurs at a length–diameter ratio of about 1·5, at which for a given diameter the length mode takes over from the radial mode. The radial coupling coefficient of Fig. 4.5.29 follows a pattern similar to that of Fig. 4.5.28.

Transducers for the measurement of transient pressures are usually calibrated in shock tubes or in gas- or liquid-filled chambers which have been pressurized to a required level. On release of the pressure by a quick-acting solenoid-operated valve the charge generated in the transducer is then measured by a charge amplifier with a suitable time-constant.

Short-duration pressure pulses or step-function pressures with steep fronts have been measured with a travelling-wave arrangement [34], the so-called pressure bar gauge, which is schematically shown in Fig. 4.5.30. A quartz disc of thickness t is bonded between two metal rods of length l_1 and l_2. If a pressure wave impinges on the front f_1 of rod l_1 a travelling wave soon reaches the quartz disc and continues along rod l_2 till it is

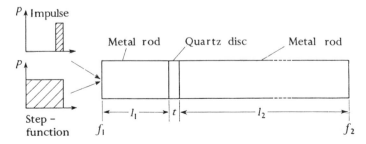

FIG. 4.5.30. Pressure-bar gauge arrangement.

reflected at f_2. During the passage of the pressure wave through the quartz disc a charge is generated, which is proportional to the pressure wave if no reflection takes place at the bonding faces between the quartz and the metal rods. If the duration of the pulse or the wavefront is short compared with twice the travelling time of the wave between quartz and f_2, the record of the reflected wave can be separated from that of the primary wave without difficulty, i.e. if l_2 is sufficiently long at a given sound velocity. In order to record the wavefront faithfully the thickness t should ideally be zero. With a practical disc thickness of $t = 1$ mm a perfect step-function pressure is spread over a time-interval of about $0 \cdot 18 \, \mu s$ and, since the area under the pressure–time curve remains constant, a single rectangular pulse is recorded, with an accordingly lower peak. The length l_1 of the front section of the pressure bar assembly need not be longer than about eight bar diameters in order to ensure an even pressure distribution across the bar cross-section when concentrated forces are applied. With uniform air or hydraulic pressures the front section is not required.

The condition for passage without reflection through the quartz disc requires the reflection coefficient to be zero. If E_1, E_2, A_1, A_2, and c_1, c_2 are the Young's moduli, the areas, and the sound velocities, respectively, of two adjacent materials, we have

$$\alpha = \frac{E_2 A_2/c_2 - E_1 A_1/c_1}{E_2 A_2/c_2 + E_1 A_1/c_1}, \tag{4.5.51}$$

or, for $A_1 = A_2$ (pressure bar of uniform diameter),

$$E_2/c_2 = E_1/c_1. \tag{4.5.52}$$

Materials with E/c ratios close to that of quartz are aluminium and lead, but since the velocity of sound in lead is about a quarter of that of aluminium the lead rod need only be a quarter of length l_2 of an equivalent aluminium rod [35].

Pressure bars are conveniently calibrated by means of a ballistic pendulum [36]. A bouncing ball represents only a short-duration pulse impact, but the force exerted by a rod bouncing on a rigid surface is constant for the period of contact, i.e. the time it takes for sound to travel from one end of the rod to the other and back. Thus the bounce of the rod is equivalent to the application of a positive step function to the gauge face, followed a short time later by a negative step [37].

4.5.4.3. Strain-gauge types

If a thin slab of quartz, or any other piezoelectric material with an active length-expander mode, is cemented to a structure under stress, the strain in the structure is transmitted to the slab and a charge is generated across the crystal. The output voltage can be computed from the h-coefficient of the length-expander mode of the piezoelectric material (see section 4.5.2.1),

$$h = \frac{dE}{\varepsilon\varepsilon_0} \quad \left(\frac{V\ m^{-1}}{m\ per\ m}\right), \tag{4.5.53}$$

where

$d = d$-coefficient of length-expander mode,
$E = $ Young's modulus of the piezoelectric material.

The voltage output is therefore

$$V = het \quad (V), \tag{4.5.54}$$

where

$e = $ strain (m per m),
$t = $ crystal thickness (m).

As an example assume an x-cut quartz slab of thickness $t = 1$ mm and strain $e = 10^{-6}$ m per m. The other crystal dimensions (length and width) are immaterial as long as the shunt capacitance (cable, etc.) is negligible. Using the following data from Table 4.5.1,

$$d_{11} = 2 \cdot 3 \times 10^{-12}\ C\ N^{-1},$$

$$E = 80 \times 10^9\ N\ m^{-2},$$

$$\varepsilon = 4 \cdot 5,$$

and eqns (4.5.53) and (4.5.54), the output voltage becomes

$$V = \frac{d_{11}Eet}{\varepsilon\varepsilon_0} = 4 \cdot 6\ V. \tag{4.5.55a}$$

A slab of similar dimensions and at similar strain, but made of PXE4 lead zirconate–titanate ($d_{31} = 141 \times 10^{-12}\ C\ N^{-1}$, $E = 7 \cdot 7 \times 10^9\ N\ m^{-2}$, $\varepsilon = 1500$), would generate an output of

$$V = 0 \cdot 82\ V. \tag{4.5.55b}$$

The capacitance value of the ceramic gauge, however, would be $1500/4 \cdot 5 = 333$ times higher, and thus of great help in easily obtaining low-frequency response. It would also swamp the capacitance of (short) transducer connections.

Compared with wire resistance strain gauges (see section 4.2.3.1) the output from piezoelectric strain gauges is several orders higher, and piezoelectric strain gauges are therefore of considerable interest. They are, however, suitable only for the measurement of dynamic strains, and calibration requires dynamic equipment. A commercial ceramic strain

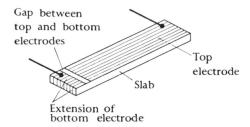

Gap between top and bottom electrodes

Top electrode

Slab

Extension of bottom electrode

FIG. 4.5.31. Piezoelectric strain gauge. The bottom electrode is extended round the one edge onto the small region on the top for ease of terminal attachment.

gauge is shown in Fig. 4.5.31. It has been used in the measurement of dynamic strain in steam-turbine blades [38], [39], giving an output of about $0 \cdot 1$ V for a strain of only 10^{-6} m per m. A piezoelectric strain gauge consisting of a thin slab of x-cut quartz has also been used as a sensing element on a diaphragm of a pressure transducer [40] for the measurement of high dynamic pressures, and internal stresses have been measured by a quartz disc bonded between aluminium cylinders [41]. Direct interaction between piezoelectric material and the channel region of field-effect transistors have been employed in strain sensors of high sensitivity and fast response [42].

References

1. W. G. Cady. *Piezoelectricity*. McGraw-Hill, New York (1946).
2. B. Jaffe, W. R. Cook, and H. Jaffe. *Piezoelectric ceramics*. Academic Press, London (1971).
3. J. van Randeraat (ed.). *Piezoelectric ceramics*. Mullard Ltd., London (1968).
4. A. E. H. Tutton. *Crystallography and practical crystal measurements*, Vol. I, p. 685. Macmillan, London (1922).
5. A. Meissner. Über piezoelektrische Krystalle bei Hochfrequenz, *Z. tech. Phys.* **8,** 74 (1927).
6. A. D. Dobelli. Piezoelectric transducers. *Acustica* **6,** 346–56 (1956).
7. S. Meurer. Beitrag zum Bau piezoelektrischer Indikatoren. *Forschung* **8,** 260 (1937).
8. W. P. Mason. *Piezoelectric crystals and their application to ultrasonics*, p. 259. Van Nostrand, New York (1950).
9. H. Sachse. *Ferroelectrica*. Springer, Berlin (1955).
10. H. D. Megaw. *Ferroelectricity in crystals*. Methuen, London (1957).
11. A. E. Crawford. Lead zirconate piezoelectric ceramics. *Br. Commun. Electron.,* **6,** 516–19 (1959).
12. Anon. *Piezoelectricity*. Vernitron Ltd., Southampton (1966, revised 1969).
13. Anon. *Glennite piezoceramics*. Bulletin H-500. Gulton Industries Inc., Metuchen, New Jersey.
14. H. K. P. Neubert, W. R. Macdonald, and P. W. Cole, Sub-miniature pressure and acceleration transducers. *Control* **4,** 104–6 (1961).
15. H. K. P. Neubert. Bilateral electromechanical transducers; a unified theory. R.A.E. Technical Report No. TR 68248 (1968).
16. R. R. Bouche. Accelerometers for shock and vibration measurements. Technical Paper TP 243. Endevco Corp., Pasadena, California (1967).
17. J. G. Graeme, G. E. Tobey, and L. P. Huelsman (eds). *Operational amplifiers*, pp. 233–5. McGraw-Hill, New York (1971).

18. ANON. *Piezotronic technical data.* Brush Electronics Co., Cleveland, Ohio (1953).
19. D. PENNINGTON. *Piezoelectric accelerometer manual.* Endevco Corp., Pasadena, California (1965).
20. A. D. KAUFMAN. Acoustic sensitivity of accelerometers. *Instrums Control Syst.* **32,** 720 (1959).
21. T. A. PERLS. Electrical noise from instrument cables subjected to shock and vibration. *J. appl. Phys.* **23,** 674 (1952).
22. J. D. ZOOK and S. T. LIU. Piezoelectric effect in Teflon cables. *J. appl. Phys.* **43,** 1304–1306 (1972).
23. P. HACKEMANN. Ein Beschleunigungsmesser nach dem piezoelektrischen Verfahren. *Jb. Dt. Luftf. Forsch.* **1,** 628–9 (1938).
24. G. KLUGE, G. BOCHMANN, and F. BRASACK. Der Bremsvorgang an Kraftfahrzeugen und seine Messung. *Dt. Kraftf. Forsch.* No. 7 (1940).
25. L. T. FLEMING. A ceramic accelerometer with wide frequency range. *Instruments* **24,** 968 (1951).
26. V. MARPLE and N. A. GRAHAM. The dependence of piezoelectric accelerometer response on methods of attachment. *J. Soc. Environmental Engrs.* **12,** 19–20 (1973).
27. P. MASON. *Electromechanical transducers and wave filters,* pp. 209–15. Van Nostrand, New York (1942).
28. W. GOHLKE. *Schwingungseigenschaften von Quarzdruckmessgeräten,* V.D.I. Forschungsheft, Vol. 407, pp. 1–17. V.D.I. Verlag, Berlin (1941).
29. W. GOHLKE. *Mechanisch–elektrische Messtechnik.,* pp. 124–9. C. Hanser Verlag, Munich (1955).
30. S. FAHRENHOLZ, J. KLUGE, and H. E. LINCKH. Über eine Quarzdruckmesskammer für das piezoelektrische Messverfahren. *Phys. Z.* **38,** 73–8 (1937).
31. Brush Development Co., Cleveland, Ohio, Technical Bulletin E 104 (1950).
32. S. TIMOSHENKO. *Theory of elasticity,* p. 56. McGraw-Hill, New York (1934).
33. T. F. HUETER and R. H. BOLT. *Sonics,* p. 147. Wiley, New York (1955).
34. G. TURETSCHEK. *Ein neues Verfahren zur Messung des Druckverlaufes in Stosswellen; Probleme der Detonation.* Schriften der Deutschen Luftfahrtforschung, Berlin (1941).
35. D. H. EDWARDS. A piezoelectric pressure bar gauge. *J. scient. Instrum.* **35,** 346–349 (1958).
36. R. M. DAVIES. A critical study of the Hopkinson pressure bar. *Phil. Trans. R. Soc. A,* **240,** 375 (1948).
37. H. S. CARSLAW and J. C. JAEGER. *Operational methods in applied mathematics.* Clarendon Press, Oxford (1941).
38. R. C. KELL, G. A. LUCK, and L. A. THOMAS. The use of ferro-electric ceramics for vibration analysis. *J. scient. Instrum.* **34,** July issue (1957).
39. G. L. LUCK and R. C. KELL. Measuring turbine blade vibrations *Engineering, Lond.* **182,** 271 3 (1956).
40. P. L. EDWARDS. High speed high-pressure gauges. *Instrums Automn* **30,** 1504–6 (1957).
41. C. A. CALDER and R. G. DRAGNICH. Internal dynamic stress measurement with an embedded quartz crystal transducer. *Rev. scient. Instrum.* **43,** 883 886 (1972).
42. J. R. FIEBINGER and R. S. MULLER. Electrical performance of metal-insular piezoelectric semiconductor transducers. *J. appl. Phys.* **38,** 1948 (1967).
43. Made by the Westinghouse Electric Corp., Baltimore, U.S.A.
44. Made by the Brush Development Co., Cleveland, Ohio, U.S.A.
45. Made by Massa Laboratories, Hingham, Mass., U.S.A.

5. Force-balance transducers

5.1. The force-balance principle

FORCE-BALANCE transducers differ from all other types discussed in Chapter 4 by a feedback loop (section 2.4) which effects comparison between the electrical output quantity and the mechanical input quantity. A state of equilibrium can therefore be attained between the input force caused, for instance, by acceleration or pressure and an opposing force exerted by the output current or voltage, which is thus a measure of the input quantity. In order to achieve equilibrium an electrical pick-off of the resistive, inductive, capacitive, or photoelectric type can be employed to sense the physical movement, the output of which is converted into the required feedback current or voltage by way of a suitable amplifier. The displacement sensor can be supplemented by a velocity-sensing pick-off, such as an electrodynamic generator-type transducer, in a subsidiary feedback loop or by phase-shifting networks, which provide velocity error compensation, in the main loop. Force-balance transducers have a number of advantages which can be summarized as follows.

(a) Instabilities due to environmental factors in the displacement and velocity sensors and in the amplifier do not affect the accuracy of the force-balance transducer. However, the instability in the force generator (servo-actuator) does, since the feedback current or voltage is taken as a measure of the input force, so that the problem of transducer stability has been shifted from sensing element and amplifier to the servo-actuator.

(b) In principle, damping can be obtained by suitable phase-shifting networks in the feedback loop, or by subsidiary loops, and is therefore almost independent of temperature.

(c) Force-balance transducers have better dynamic response characteristics than conventional (open-loop) transducers; they offer a wider useful frequency range at sinusoidal excitation and a more faithful reproduction of transient input–time signatures, particularly at high values of feedback damping.

(d) Their dynamic performance depends in the first instance upon the feedback characteristic (loop gain) and is therefore independent of mechanical restraint (springs) with zero-instabilities, variation of stiffness with temperature, and hysteresis. In principle there need be no mechanical springs in force-balance transducers other than perhaps highly flexible electrical connections to the moving parts. However, the elimination, or even the reduction, of mechanical stiffness, for instance, in guides for linear movement, in pivots or in pressure-sensitive elements, is a difficult problem in many practical transducer designs, and so is the reduction of friction between moving parts.

(e) Since the primary sensing elements (capsules, seismic mass, etc.) travel only a minute distance before force equilibrium is attained, sensitivity to transverse acceleration, for instance, in linear accelerometers of the force-balance type is negligible.

(f) Within limits range, natural frequency, and damping can be adjusted and varied by purely electrical means in the feedback loop. A force-balance transducer is thus more readily adaptable and more flexible than the conventional type requiring mechanical modifications.

(g) From these characteristics it can be expected that force-balance transducers are inherently more stable, and thus more accurate, than conventional transducers, if residual mechanical spring and friction forces can be reduced to well below the levels of input and balancing forces. They are therefore often considered for use as secondary standards.

The disadvantages of force-balance transducers are usually greater complexity and higher cost, and often larger size and weight, since displacement sensor and servo-actuator have to be carried in the same housing, but amplifiers are often simpler than those for use with conventional transducers of comparable performance. The basic theory and the design procedure of force-balance transducers are somewhat more complex since the aspect of stability (i.e. freedom from oscillation within the feedback loop) and of feedback damping must be considered as additional design features.

Most of the more recent books on transducers, such as those by Doebelin, Giles, Rohrbach, and Oliver (see list of references following Chapter 2) contain valuable information on force-balance transducer principles and properties which should satisfy most potential users of instrumentation. However, the designer will probably need a more detailed and—if possible—more unified treatment of the subject. This has been attempted in the present chapter, which derives mathematical models of typical force-balance transducers and discusses their sensitivity and linearity at static loading, followed by a general treatment of their dynamic responses to sinusoidal and transient inputs. The chapter then concludes with a survey of integrated transducer systems for present and future applications.

5.2. Static performance

5.2.1. *Electrodynamic acceleration transducer* ([1], [2])

Consider the schematic arrangement of Fig. 5.1 of a typical electrodynamic force-balance transducer for the measurement of linear acceleration. The transducer housing contains a seismic mass m which is free to

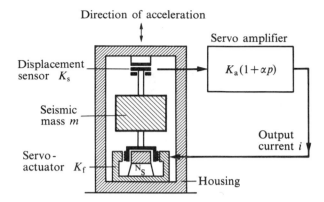

FIG. 5.1. Electrodynamic force-balance transducer for the measurement of linear acceleration, schematic diagram.

move in the axial direction (guides not shown). Acceleration applied along this line causes a relative displacement of the mass which is sensed by a push–pull variable-capacitance detector with a 'sensitivity' K_s (V m^{-1}). Its output voltage is fed to the servo-amplifier of 'gain' K_a (A V^{-1}), comprising in its simplest form a phase-advancing network for velocity compensation with a transfer function $(1+\alpha p)$, where p is the Laplace operator, and providing feedback damping. The output current i (A), in turn, is fed to the electrodynamic servo-actuator with the 'force factor' K_f (N A^{-1}), thus closing the loop by generating the force necessary to balance the input inertia force caused by acceleration.

5.2.1.1. Sensitivity. In an ideal electrodynamic force-balance transducer spring forces (of mechanical guides, etc.) can be ignored in comparison with feedback forces, and damping forces are irrelevant at static loading. The remaining restoring force provided by the electrodynamic servo-actuator, then, is

$$F = Bli \qquad \text{(N)}, \tag{5.1}$$

where B (T or Wb m^{-2}) represents the magnetic flux density in the (annular) air gap of the magnet assembly and l(m) the active wire length of the moving coil coupled with the magnetic field. Equilibrium between the restoring force of eqn (5.1) and the inertia force derived from acceleration A (m s^{-2}) applied to the seismic mass m (kg) occurs for

$$Bli = Am, \tag{5.2}$$

and the transducer sensitivity to static acceleration input becomes

$$S_0 = i/A = m/Bl = m/K_f \qquad \text{(A m}^{-1}\text{ s}^2\text{)}, \tag{5.3}$$

indicating that seismic mass m and force factor $K_f = Bl$ alone determine the static performance characteristics of the electrodynamic force-balance transducer, provided K_s and K_a are not limiting.

5.2.1.2. Linearity. According to eqn (5.3) the output current i of the transducer shown in Fig. 5.1 is a linear function of acceleration A, as long as m, B, and l remain constant over the operational displacement range. Non-linearities in the displacement sensor and/or the servo-amplifier do not affect the linearity of static measurement. Considering force-balance acceleration transducers with a seismic mass represented by a pendulum,

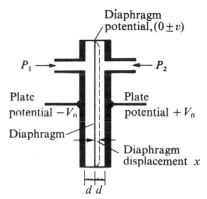

FIG. 5.2. Electrostatic force-balance transducer for the measurement of gas pressures, schematic diagram.

however, note must be taken of the cross-coupling effect [3]. An acceleration component normal to the working axis will cause an error proportional to the product of the normal component and the (small) angular displacement of the pendulum.† A further error specific to pendulous accelerometers also stems from this finite angular displacement which can be regarded as an effective non-linearity with even-order components ([4], [12]) (see Fig. 3.1(b), p. 25), and which, at oblique sinusoidal excitation, produces a d.c. component in the output by 'rectification'. Fortunately, in transducers with high values of feedback stiffness (high loop gain) these errors are usually negligible.

5.2.2. Electrostatic pressure transducer [5]

The functional components of a typical electrostatic force-balance pressure transducer are shown schematically in Fig. 5.2. Two (circular) plates of area a are fixed at a distance of $2d$ apart. A third 'plate', or diaphragm, of negligible thickness is initially positioned in the centre. For simplicity let the diaphragm be displaced in a piston-like manner parallel to itself by a distance $\pm x$ in response to a pressure difference $\pm P = P_1 - P_2$

† Typically a few seconds of arc per g.

across it. According to Fig. 5.2 the two outer plates are at fixed potentials $+V_0$ and $-V_0$, respectively, relative to the diaphragm, which is initially at zero potential.

The diaphragm displacement is sensed by connecting the two condenser halves to a capacitance bridge (see Fig. 4.4.17, p. 241), and the bridge output, after amplification and rectification, is fed to the diaphragm. The diaphragm potential v is thus made to interact with the electrostatic field $E = V_0/d$ between the outer plates in such a way as to oppose the diaphragm displacement by electrostatic forces. However, in contrast to the electrodynamic feedback transducer described above, the electrostatic restoring forces are inherently non-linear functions of displacement; they are also much smaller in magnitude.

5.2.2.1. Sensitivity [6].   Generally, the electrical energy stored in a condenser consisting of two parallel plates of area a (m²) and separated by a distance d (m) is

$$W_e = q^2/2C \qquad \text{(J)}, \tag{5.4}$$

where q(C) is the charge, and

$$C = \varepsilon\varepsilon_0 a/d \qquad \text{(F)} \tag{5.5}$$

the capacity of the condenser. $\varepsilon_0 = 8\cdot854\times10^{-12}$ (F m^{-1}) is the permittivity of empty space, and ε the relative permittivity of the gas. With a potential difference

$$v = q/C \qquad \text{(V)} \tag{5.6}$$

between the plates, eqn (5.4) can be written

$$W_e = \varepsilon\varepsilon_0 a v^2/2d, \tag{5.7}$$

and the electrostatic force acting between the two plates becomes

$$F = -\partial W_e/\partial d = \tfrac{1}{2}\varepsilon\varepsilon_0 a(v/d)^2 \qquad \text{(N)}. \tag{5.8}$$

In the particular case of the arrangement of Fig. 5.2 the two outer plates are at potentials $+V_0$ and $-V_0$, the diaphragm at potential v (which is also the output voltage of the transducer), and the displacement is x (m). Then, the restoring force acting on the diaphragm (of negligible mechanical stiffness) becomes, by analogy to eqn (5.8),

$$F = \tfrac{1}{2}\varepsilon\varepsilon_0 a\left\{\left(\frac{V_0+v}{d+x}\right)^2 - \left(\frac{V_0-v}{d-x}\right)^2\right\} \qquad \text{(N)}. \tag{5.9}$$

The electrostatic restoring force is thus clearly a non-linear function of diaphragm displacement.

Conditions of equilibrium between the restoring force of eqn (5.9) and the input force Pa (N), derived from the pressure difference P (N m^{-2})

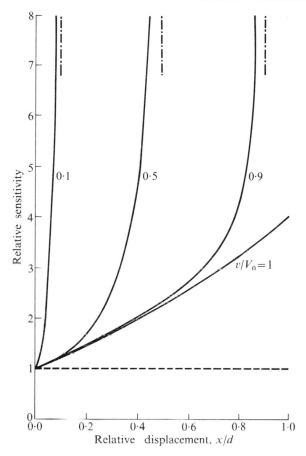

FIG. 5.3. Variation of static sensitivity.

acting on the diaphragm area a, yields the sensitivity to static pressure,

$$S_0 = \frac{v}{P} = \frac{d}{2\varepsilon\varepsilon_0 V_0} \left[\frac{\{1 - (x/d)^2\}^2}{1 - \left(\frac{v}{V_0} + \frac{V_0}{v}\right)\frac{x}{d} + \left(\frac{x}{d}\right)^2} \right] \quad (\text{V N}^{-1} \text{ m}^2). \qquad (5.10)$$

The dimensionless term of eqn (5.10) has been plotted in Fig. 5.3. Sensitivity rises in a non-linear fashion† and becomes infinite for $x/d = v/V_0$, where the restoring force vanishes. At still higher values of x/d the restoring force becomes negative, i.e. the diaphragm flops over to one of the fixed plates and the transducer ceases to function. We also note with

---

† For a linearized expression of sensitivity see eqn (5.20) at the end of the section to follow.

particular interest that according to eqn (5.10) the sensitivity to static pressures is independent of diaphragm area a, and thus of transducer size.

5.2.2.2. Linearity. In dynamic measurements sizeable non-linearity of the feedback system in the operational displacement range may generate troublesome harmonics in the output voltage. For a closer study of non-linearity we therefore return to eqn (5.9), representing the non-linear equation of the servo-actuator, which will be supplemented by the sensor equation

$$v = K_s K_a f(x) = K_r f(x), \tag{5.11}$$

where $f(x)$ is the non-linear sensor displacement function, K_s (V m^{-1}) the (initial) sensitivity of the displacement sensor, K_a (V per V) the amplifier gain, and $K_r = K_s K_a$ (V m^{-1}) the 'total' sensor sensitivity v/x.

Substituting eqn (5.11) into (5.9) yields

$$F = \tfrac{1}{2}\varepsilon\varepsilon_0 a K_r^2 \left[\left\{ \frac{\mu + f(v)}{1+v} \right\}^2 - \left\{ \frac{\mu - f(v)}{1-v} \right\}^2 \right], \tag{5.12}$$

where

$v = x/d =$ relative diaphragm displacement,

$\mu = E/K_r =$ dimensionless design parameter,

$E = V_0/d =$ static electric field strength (V m^{-1}).

The function $f(v)$ now represents the relative capacitance variation $\Delta C/C$ of the displacement sensor which, for the push–pull arrangement of Fig. 5.2, can be written

$$\Delta C/C \propto (v + v^3 + v^5 + \dots) = f(v). \tag{5.13}$$

Substituting eqn (5.13) into (5.12), but ignoring higher than third-order terms in v, then expanding eqn (5.12) into a series for v (where $v < 1$), but again retaining only up to and including third-order terms in v, we obtain the expressions for the restoring force, valid for small displacements,

$$F = 2\varepsilon\varepsilon_0 a E^2 M(v - Nv^3), \tag{5.14}$$

and for the feedback stiffness,

$$K = \frac{F}{x} = \frac{2\varepsilon\varepsilon_0 a E^2}{d} M(1 - Nv^2), \tag{5.15}$$

where

$$M = (1-\mu)/\mu \tag{5.16}$$

is the 'stiffness factor', and

$$N = \frac{1 - 4\mu + 2\mu^2}{\mu(1-\mu)} \tag{5.17}$$

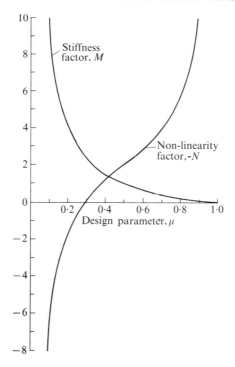

FIG. 5.4. Stiffness and linearity factor of electrostatic force-balance pressure transducer.

is the 'non-linearity factor'. (The constant terms $2\varepsilon\varepsilon_0aE^2$ and $2\varepsilon\varepsilon_0aE^2/d$, respectively, are only scale factors and of no particular interest.)

We see that positive restoring forces, and stiffnesses, occur only for $\mu < 1$ (or $x/d < v/V_0$), and since the operational condition $v < V_0$ requires $v < \mu$, the ranges of validity for v and μ are given by

$$0 < v < \mu < 1. \tag{5.18}$$

The stiffness factor M and the (negative) non-linearity factor N have been plotted in Fig. 5.4 as functions of the dimensionless design para- menter $\mu = E/K_r = (x/d)/(v/V_0)$ over the range $0 \le \mu \le 1$. At low values of μ, i.e. for large K_r/E ratios (high gains), M is high and, consequently, so are feedback force F and stiffness K, as would be expected. For $\mu \to 1$ the M-factor, and thus F and K, vanish and the feedback system becomes statically unstable, as discussed earlier. The non-linearity factor N grows beyond all limits for both $\mu \to 0$ and $\mu \to 1$, but—quite unexpectedly— the restoring force F becomes a linear function of $v = x/d$ for $\mu \simeq 0.3$, independent of diaphragm displacement.

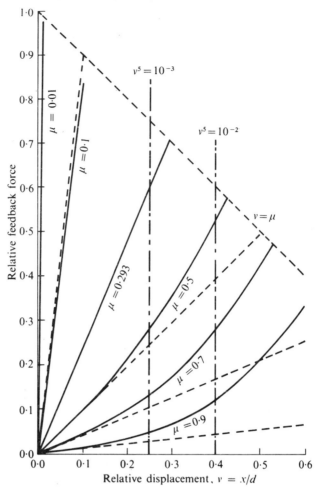

FIG. 5.5. Generalized force–displacement curves of electrostatic force-balance pressure transducer.

These results have practical implications on the design and operation of electrostatic force-balance transducers. Fig. 5.5 shows generalized force–displacement curves of the transducer system, and it is clearly seen that, within the limitations assumed in the derivation given above, linear systems analysis is generally justified for realistic conditions of operation, since the loop gain in practical transducers must be made high in order to swamp the mechanical diaphragm stiffness, i.e. μ is usually small (typically 10^{-2} or lower), and at near-balance conditions the diaphragm displacement is also small, i.e. ν is usually less than 10^{-2}. Even for these rather pessimistic values the maximum deviation from linearity would amount to less than 1 per cent.

In the vast majorities of cases it will therefore be permissible to linearize the expression for the restoring force (eqn (5.9)) and write, for small displacements and high loop gains,

$$F = \frac{2\varepsilon\varepsilon_0 a V_0}{d^2}\left\{1-\left(\frac{v}{V_0}+\frac{V_0}{v}\right)\frac{x}{d}\right\}v \quad (N). \tag{5.19}$$

The linearized transducer sensitivity then becomes

$$S_0 = \frac{v}{P} = \frac{d^2}{2\varepsilon\varepsilon_0 V_0}\left\{1+\left(\frac{v}{V_0}+\frac{V_0}{v}\right)\frac{x}{d}\right\} = \frac{a}{K_f} \quad (\text{V N}^{-1}\,\text{m}^2), \tag{5.20}$$

where $K_f = F/v$ (N V^{-1}) is the force factor of the electrostatic servo-actuator. This result may be compared with the static sensitivity of the

TABLE 5.1

Theoretical design and performance characteristics of electrostatic force-balance transducers for the measurement of small gas pressures

Air gap d (mm)	0·1	0·05	0·1	0·05
Polarizing voltage V_0 (V)	100	100	500	500
Peak electric field strength E (kV mm^{-1})	2	4	10	20
Sensitivity S_0 (V N^{-1} m^2)	5·7	1·43	1·14	0·29
Useful pressure range P_{max} (N m^{-2})†	17·7	71	88	350

† $1\ \text{N m}^{-2} \simeq 0\cdot1$ mm water column.

electrodynamic force-balance acceleration transducer given by eqn (5.3); both expressions are similar and exceedingly simple. Note, however, that the electrostatic pressure transducer provides an output voltage v, whilst the output of the electrodynamic acceleration transducer is a current i. Therefore, the units of measurement of sensitivity S_0 and force factor K_f are not the same for both transducer types.

5.2.2.3. Range of applications. Table 5.1 lists theoretical design and performance characteristics of typical electrostatic force-balance pressure transducers. The values assumed for air gap d, polarizing voltage V_0, and electrical field strength E are realistic and quite safe with regard to electrical breakdown. Sensitivity v/P and useful pressure range P_{max} relate to transducers with small diaphragm displacements and high feedback gains. The table features remarkably high voltage–pressure sensitivities which, unfortunately, are coupled with low values of useful pressure range. Both these properties stem, of course, from the fact that electrostatic forces are weak. Electrostatic force-balance pressure transducers are therefore at their best where small static and dynamic gas-pressure differences, equivalent to a few millimetres of water column, must be measured accurately (see also section 5.4.2.2).

5.3. **Dynamic performance** [6]

The force-balance principle has been in use for a very long time indeed, the earliest application probably being the mechanical balance with a potentiometer pick-up measuring the position of a motor-driven counter-weight sliding on a horizontal arm. But these and similar 'transducers' of old were suitable, and intended, for the measurement of static forces only. Force-balance transducers for dynamic applications are a fairly recent requirement, and only in the last few years have their response characteristics at sinusoidal and transient excitations attracted attention. However, modern force-balance transducers are expected to measure, with high accuracy, dynamic as well as static input quantities,

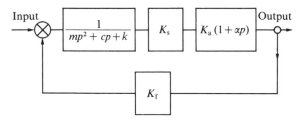

FIG. 5.6. Block diagram of an ideal force-balance transducer.

and it is therefore appropriate to study their dynamic performances in some detail since both the steady-state and the transient responses of closed-loop transducers differ appreciably from those of conventional open-loop transducers discussed in section 3.2.

5.3.1. *Sinusoidal excitation*

Fig. 5.6 shows, in block-diagram form, the generalized feedback system of a force-balance transducer. The mechanical components of the moving parts are mass m (kg) (seismic mass of an accelerometer or diaphragm mass of a pressure transducer), spring stiffness k ($N m^{-1}$) (stiffness of guides or of diaphragm), and (viscous) damping c ($N s m^{-1}$) (inherent or artificial). In the first block of the forward loop they are combined to represent the mechanical transfer function $1/(mp^2+cp+k)$ of the moving parts. In the second block we have the sensitivity K_s ($V m^{-1}$) of the displacement sensor, and in the third the servo-amplifier of gain K_a (measured in ($A V^{-1}$) or (V per V), depending on whether we employ current or voltage feedback), combined with an (ideal) phase-advancing network $(1+\alpha p)$ providing rate compensation and feedback damping. In the feedback loop proper we have the force factor K_f of the servo-actuator, measured in $N A^{-1}$ or $N V^{-1}$, as the case may be. All systems parameters are assumed constant over the working ranges of the trans-ducers and linear systems analysis is employed throughout.

Then, for closed-loop operation (see eqn (2.50), p. 22), the dynamic sensitivity of the force-balance transducer can be written generally [1]

$$\bar{S} = S_0 \frac{K(1+\alpha p)}{mp^2+(c+\alpha K)p+(k+K)},\qquad(5.21)$$

where $K = K_s K_a K_t$ is the feedback, or loop stiffness, and S_0 the appropriate static transducer sensitivity of either the electrodynamic acceleration transducer (eqn (5.3)), or of the electrostatic pressure transducer (eqn (5.20)). In fact, eqn (5.21) holds for all second-order feedback transducer types (p is again the Laplace operator).

Now, if mechanical stiffness k and mechanical damping c are small compared with feedback stiffness and feedback damping, as they should be in a well-designed force-balance system, eqn (5.21) simplifies and Laplace inversion yields the complex sensitivity to sinusoidal excitation,

$$S = S_0 \frac{1+2jh\omega/\omega_0}{1-(\omega/\omega_0)^2+2jh\omega/\omega_0},\qquad(5.22)$$

where

$\omega = $ excitation circular frequency (s^{-1}),

$\omega_0 = (k/m)^{\frac{1}{2}} = $ undamped natural circular frequency (s^{-1}),

$h = \frac{1}{2}\alpha\omega_0 - $ feedback damping ratio.†

The complex response of eqn (5.22) can be split into the amplitude response, or 'sensitivity',

$$|S| = S_0 \left[\frac{1+4h^2(\omega/\omega_0)^2}{\{1-(\omega/\omega_0)^2\}^2+4h^2(\omega/\omega_0)^2}\right]^{\frac{1}{2}},\qquad(5.23a)$$

and the phase response,

$$\phi = \tan^{-1}\frac{-2h\omega/\omega_0}{(\omega/\omega_0)^2+4h^2-1}.\qquad(5.23b)$$

The relative sensitivity (i.e. the square brackets of eqn (5.23a)) and the phase angle (eqn (5.23b)) have been plotted in Fig. 5.7(a) and (b), the frequency ratio ω/ω_0 being the independent variable and the feedback damping ratio h the parameter. These curves should be compared with the response curves of conventional vibratory systems (Fig. 3.6(a) and (b), p. 32). At low frequency ratios the closed-loop response resembles that of a second-order open-loop system (resonance); at higher frequency ratios it behaves similar to a first-order system (6 dB per octave drop and 90° phase shift). From a practical point of view there is a wider useful frequency range above $\omega/\omega_0 = 1$, particularly for high values of feedback damping.

† For the definition of damping ratio see section 3.3.3.

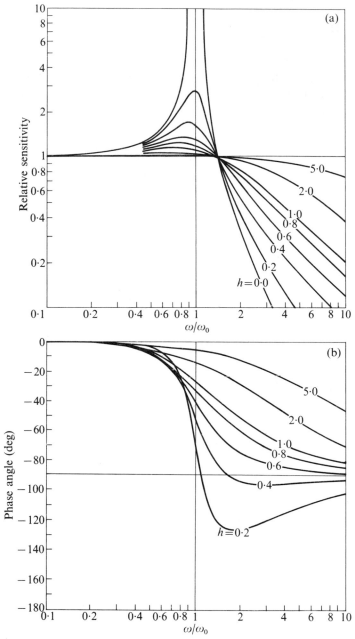

FIG. 5.7. Dynamic response of force-balance transducers to sinusoidal excitation: (a) relative sensitivity; (b) phase angle.

Before leaving the sinusoidal response characteristics attention must be drawn to the problem of dynamic instabilities peculiar to closed-loop transducers generally and to the electrostatic force-balance transducer in particular. Although the ideal feedback system of Fig. 5.6, as it stands, is dynamically stable at all frequencies and damping ratios (except for $h = 0$), additional phase shifts in practical transducer systems may be responsible for the excitation of spurious oscillations within the feedback loop. This can occur—and has occurred in practice—with the electrostatic pressure transducer affected by its 'acoustic' components which consist of subsidiary mass, compliance, and damping of the gas in the pressure chambers and inlet pipes. Since the acoustic system is closely coupled to the electromechanical feedback system, standing acoustic waves may occur, for instance, in the pressure inlet pipes to the transducer which can be easily suppressed, however, by introducing (acoustic) damping material in the pipes [6].

5.3.2. Transient excitation

We now assume that the force-balance transducer system of Fig. 5.6 is excited by a transient input force, and here we shall investigate the response of the feedback transducer to the three most elementary transient force–time signatures: impulse, step, and ramp function.†

Since the Laplace transformation of the unit impulse (Dirac) function is unity, of the unit step function is $1/p$, and of the unit ramp function is $1/p^2$ (p being the Laplace operator), the 'subsidiary' equations [7] to be solved by Laplace inversion can be written (see eqn (5.21) (assuming again $k = 0$ and $c = 0$):

impulse function $\quad \bar{s} = \bar{X} S_0 \dfrac{K(1+\alpha p)}{mp^2 + \alpha K p + K}$;   (5.24a)

step function $\quad \bar{s} = \bar{Y} S_0 \dfrac{K(1+\alpha p)}{p(mp^2 + \alpha K p + K)}$;   (5.24b)

ramp function $\quad \bar{s} = \bar{Z} S_0 \dfrac{K(1+\alpha p)}{p^2(mp^2 + \alpha K p + K)}$,   (5.24c)

where s on the left-hand side of the equations is here the output quantity of the particular transducer (i.e. current $\bar{i}$ for an electrodynamic, and voltage $\bar{v}$ for an electrostatic type, not sensitivity $\bar{S}$ as before) and $\bar{X}$, $\bar{Y}$, and $\bar{Z}$ represent the input quantities (unit impulse, step, and ramp functions, respectively); their magnitudes may be unity, but their dimensions vary. For the pressure transducer, for instance, $\bar{X}$ is measured in $N \, m^{-2} \, s$, i.e. peak pressure P multiplied by the impulse time-interval

---

† For the definition of unit impulse, step, and ramp input functions see section 3.2.2.

$\varepsilon \to 0$; $\bar{Y}$ is simply the height of the pressure step P; and $\bar{Z}$ is in $\mathrm{N\,m^{-2}\,s^{-1}}$, i.e. the time-rate of pressure rise of the ramp function.

The transient response as a function of time $(s = f(t))$ can now be obtained from the appropriate subsidiary equation $(\bar{s} = f(p))$ by applying the rules of inverse Laplace transformation, or—after some rearrangements—by reference to tables of transform pairs.† The solutions fall into three distinct response modes, depending on the magnitude of the feedback damping ratio h:

(a) $h < 1$, 'undercritical' response,

(b) $h = 1$, 'critical' response,

(c) $h > 1$, 'overcritical' response,

which are also well-known features of the transient response curves of the conventional second-order vibratory systems discussed in section 3.2.2.

5.3.2.1. Impulse-function input.

The time-response of a force-balance transducer with undercritical damping $(h < 1)$ to an impulse function is obtained by Laplace inversion of eqn (5.24a), yielding

$$s = X\omega_0 S_0 \frac{\exp(-h\omega_0 t)}{(1-h^2)^{\frac{1}{2}}} [(1-2h^2)\sin\{(1-h^2)^{\frac{1}{2}}\omega_0 t\} +$$

$$+2h(1-h^2)^{\frac{1}{2}}\cos\{(1-h^2)^{\frac{1}{2}}\omega_0 t\}]. \quad (5.25)$$

Similarly, for $h = 1$, we have the critical response

$$s = X\omega_0 S_0 \exp(-\omega_0 t)(2-\omega_0 t), \quad (5.26)$$

and for $h > 1$, the overcritical response

$$s = X\omega_0 S_0 \frac{1}{2(h^2-1)^{\frac{1}{2}}} \left(\{h+(h^2-1)^{\frac{1}{2}}\}^2 \exp[-\{h+(h^2-1)^{\frac{1}{2}}\omega_0 t\}] - \right.$$

$$\left. -\{h-(h^2-1)^{\frac{1}{2}}\}^2 \exp[-\{h-(h^2-1)^{\frac{1}{2}}\}\omega_0 t] \right). \quad (5.27)$$

For a generalized display of the transient response curves only the non-dimensional terms of eqns (5.25) to (5.27) are of interest; the common factor $X\omega_0 S_0$ is ignored. Fig. 5.8 thus shows the generalized transient response of a force-balance transducer to a unit-impulse input, plotted as a function of the non-dimensional 'time' $\omega_0 t$, for undercritical, critical, and overcritical damping ratios h. We note with particular interest that the output quantity (current or voltage) follows the shape of the input pulse more and more closely the higher the feedback damping ratio h becomes. This is in striking contrast to the impulse-response curves of a

---

† A number of books on Laplace transformation techniques can be found in the list of references at the end of Chapter 3.

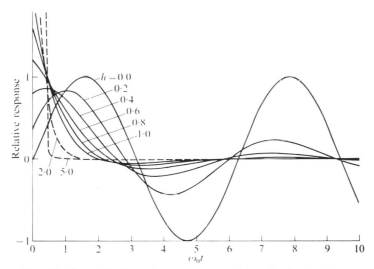

FIG. 5.8. Generalized transient response to unit impulse-function input.

conventional second-order vibratory system with viscous damping (see Fig. 3.8(d), p. 39) which, at higher values of damping, leads to feeble outputs resembling decaying step functions.

Some readers may be worried by the shape of the response curves for finite values of damping in Fig. 5.8, which seem to indicate zero rise-times. This unrealistic feature is explained by the fact that the 'ideal' model of Fig. 5.6 has a phase-advancing term only and therefore does not cater for any time delays which are normally present in a practical transducer system. In order to demonstrate the effect of such time delays, particularly on the initial parts of the impulse response curves, we replace K by $K/(1+\beta p)$ in eqn (5.21), β being a small delay time in the feedback loop, and write

$$\bar{s} = S_0 \frac{K(1+\alpha p)/(1+\beta p)}{mp^2+cp+k+K(1+\alpha p)/(1+\beta p)}. \tag{5.28}$$

For $k = 0$ and $c = 0$ (as before), and introducing the time-ratio $\eta = \beta/\alpha \ll 1$, the solution related to critical damping ($h = 1$) yields

$$s = X\omega_0 S_0 \frac{1}{(1-2\eta)^2}\Big[\{2-(1-2\eta)\omega_0 t\}\exp(-\omega_0 t)-2\exp(-\omega_0 t/2\eta)\Big]. \tag{5.29}$$

Eqn (5.29) shows that for $\eta = 0$ (i.e. for $\beta = 0$) the unit impulse response simplifies to the expression given by eqn (5.26), as it should. It also shows that for $t = 0$ the response is zero, in contrast to the 'ideal' system of Fig. 5.6 which gives $s = 2$ for $t = 0$. Eqn (5.29) has been plotted in Fig. 5.9 for $\eta = 0.05$, 0.1, and 0.25, indicating finite rise-times, as predicted, in

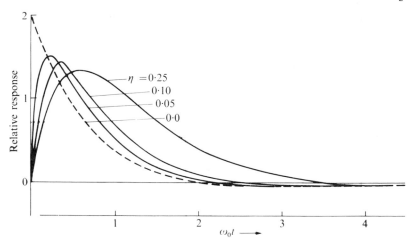

FIG. 5.9. Generalized transient response to a unit-impulse input; critical damping; effect of time-lag.

comparison with the dotted response curve related to the ideal system with $\eta = 0$.

5.3.2.2. Step-function input.   Returning to the system of Fig. 5.6 the transient response to a step function is controlled by the subsidiary equation (5.24b). Its solutions by Laplace inversion are

$$s = YS_0\left(1 + \frac{\exp(-h\omega_0 t)}{(1-h^2)^{\frac{1}{2}}}\left[h\,\sin\{(1-h^2)^{\frac{1}{2}}\omega_0 t\} - \right.\right.$$
$$\left.\left. -(1-h^2)^{\frac{1}{2}}\cos\{(1-h^2)^{\frac{1}{2}}\omega_0 t\}\right]\right), \quad (5.30)$$

valid for undercritical damping ($h < 1$);

$$s = YS_0\{1 - \exp(-\omega_0 t)(1-\omega_0 t)\}, \quad (5.31)$$

valid for critical damping ($h = 1$); and

$$s = YS_0\left\{1 + \frac{1}{2(h^2-1)^{\frac{1}{2}}}\left(\{h-(h^2-1)^{\frac{1}{2}}\}\exp[-\{h-(h^2-1)^{\frac{1}{2}}\}\omega_0 t] - \right.\right.$$
$$\left.\left. -\{h+(h^2-1)^{\frac{1}{2}}\}\exp[-\{h+(h^2-1)^{\frac{1}{2}}\}\omega_0 t]\right)\right\}, \quad (5.32)$$

valid for overcritical damping ($h > 1$).

The generalized transient response curves of a force-balance transducer with different degrees of feedback damping, to a unit step-function input are plotted in Fig. 5.10. Again, these response curves differ from those related to a conventional second-order system mainly by their faithful reproduction of the input signature, particularly at high values of feed-back damping. The introduction of a small delay-time (see previous

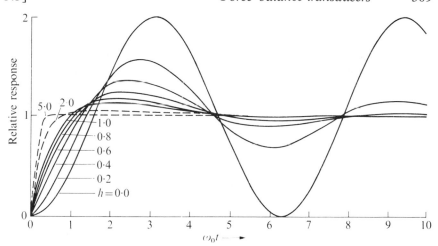

FIG. 5.10. Generalized transient response to unit step-function input.

section) would not affect the character of the response curves to a step-function (or, for that matter, to ramp-function) input; it would only increase the rise-times slightly.

5.3.2.3. Ramp-function input. Similarly, eqn (5.24c) yields the response of the force-balance transducer to a ramp-function input, namely:

$$s = \frac{Z}{\omega_0} S_0 \left[\omega_0 t - \frac{\exp(-h\omega_0 t)}{(1-h^2)^{\frac{1}{2}}} \sin\{(1-h^2)^{\frac{1}{2}} \omega_0 t\} \right] \qquad (5.33)$$

valid for undercritical damping ($h < 1$);

$$s = \frac{Z}{\omega_0} S_0 \{1 - \exp(-\omega_0 t)\} \omega_0 t \qquad (5.34)$$

valid for critical damping ($h = 1$); and

$$s = \frac{Z}{\omega_0} S_0 \left\{ \omega_0 t + \frac{1}{2(h^2-1)^{\frac{1}{2}}} \left(\exp[-\{h+(h^2-1)^{\frac{1}{2}}\}\omega_0 t] - \right.\right.$$
$$\left.\left. -\exp[-\{h-(h^2-1)^{\frac{1}{2}}\}\omega_0 t] \right) \right\} \qquad (5.35)$$

valid for overcritical damping ($h > 1$).

The generalized transducer response curves computed for unit ramp-function input have been displayed in Fig. 5.11, and we notice the same general characteristics mentioned earlier in relation to Figs. 5.8 and 5.10.

5.3.3. Phase-advancing networks

A straightforward computation of the closed-loop characteristics of force-balance transducers is possible only if dealing with ideal systems.

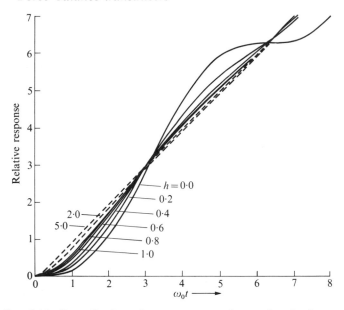

FIG. 5.11. Generalized transient response to unit ramp-function input.

This becomes apparent especially when introducing practical phase-advancing networks. The most commonly used circuit is shown in Fig. 5.12. It consists of a series element of resistance R, with a capacitance C in parallel, and a shunt resistance $R' = \gamma R$, γ being a design parameter. The transfer function of this network can be shown to be

$$\frac{E_2}{E_1} = \frac{\gamma(1+RCp)}{1+\gamma+\gamma RCp}, \tag{5.36}$$

which, for sinusoidal excitation ($p = j\omega$), becomes

$$\frac{E_2}{E_1} = \frac{\gamma\{1+\gamma(1+\omega^2R^2C^2)\}}{(1+\gamma)^2+\gamma^2\omega^2R^2C^2}\left\{1+j\omega\frac{RC}{1+\gamma(1+\omega^2R^2C^2)}\right\} \tag{5.37}$$

$$= K'(1+j\omega\alpha'),$$

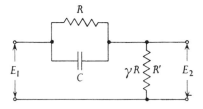

FIG. 5.12. Passive phase-advancing network.

where

$$K' = \frac{\gamma\{1+\gamma(1+\omega^2R^2C^2)\}}{(1+\gamma)^2+\gamma^2\omega^2R^2C^2},$$ (5.38a)

$$\alpha' = RC/\{1+\gamma(1+\omega^2R^2C^2)\},$$ (5.38b)

$$R' = \gamma R.$$ (5.38c)

The phase angle ϕ' is

$$\phi' = \tan^{-1}[\omega RC/\{1+\gamma(1+\omega^2R^2C^2)\}].$$ (5.39)

The transfer factor K' (eqn (5.38a)) and the phase angle ϕ' (eqn (5.39)) have been plotted in Fig. 5.13(a) and (b), respectively, for various values of ωRC, the γ-ratio being a parameter. It is seen immediately that the passive phase-advancing network of Fig. 5.12 is far from providing an ideal (i.e. constant) α'-value, and an ideal (i.e. unity) transmission. A single network of Fig. 5.12 in fact cannot produce a phase shift of $\phi' = 90°$ for finite values of γ. Two such networks in series would cause matching problems and it is mainly for this reason that a combination of a passive network followed by an active phase-shifting network—which comprises a feedback amplifier and provides the necessary buffering—is often used. Even then, acceptable values of transfer factor and phase angle can be achieved only over a limited frequency range.

Another method of obtaining feedback damping involves the use of a secondary sensor, such as an electrodynamic pick-off, which is then employed in a subsidiary feedback loop. Its voltage output is proportional to velocity and thus strictly 90° in advance of the output voltage of the displacement sensor (transducers with velocity feedback will be discussed in the section to follow). However, there may be severe practical problems in fitting a secondary sensor to the moving parts of force-balance transducers; for instance, the electrostatic pressure transducer treated in section 5.2.2 above could hardly accommodate it.

5.3.4. Systems with velocity feedback only

The response curves of Figs. 5.7(a) and (b) for high feedback damping ratios h suggested a possibility of obtaining uniform amplitude and low values of phase shift over a wide frequency range if the velocity term αp is large compared with unity. It is therefore of some practical interest to study the performance of feedback systems which operate with velocity compensation only. Fig. 5.14 shows such a system schematically. The displacement sensor (transfer function K_s (V m^{-1})) of Fig. 5.1 has been replaced by a velocity sensor of transfer function $K_s\alpha$ (V m^{-1} s), which is connected to the servo-actuator (transfer function K_f (N A^{-1})) via an amplifier of transfer function K_a (A V^{-1}), but without introducing a phase-shifting network. The open-loop transfer function of the system

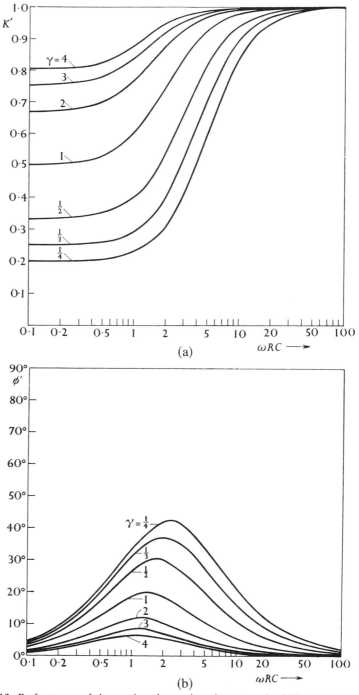

FIG. 5.13. Performance of the passive phase-advancing network of Fig. 5.12: (a) Transfer factor K'; (b) phase angle ϕ'.

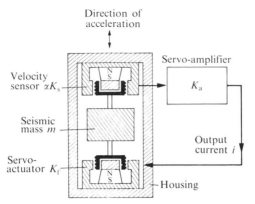

FIG. 5.14. Feedback-type linear-acceleration transducer with velocity compensation only, schematic diagram.

(with negligible mechanical stiffness k and damping c) can be written, for $K - K_s K_a K_f$,

$$K\alpha p/mp^2 = K\alpha/mp, \tag{5.40}$$

and at sinusoidal excitation ($p = j\omega$),

$$1/j\omega T = j/\omega T, \tag{5.41}$$

where the time-constant T (s) is

$$T = m/\alpha K. \tag{5.42}$$

Hence the closed-loop transfer function assumes the form

$$\frac{1/j\omega T}{1+1/j\omega T} = \frac{1}{1+j\omega T}. \tag{5.43}$$

Then, the frequency response, i.e. the (dimensionless) relative sensitivity, becomes

$$1/(1+\omega^2 T^2)^{\frac{1}{2}}, \tag{5.44}$$

and the phase angle

$$\phi = \tan^{-1}(-\omega T). \tag{5.45}$$

Eqns (5.44) and (5.45) indicate that the transducer behaves like a simple first-order system, characterized by its time-constant T, as shown by the plot of Figs. 5.15(a) and (b). In order to obtain uniform response over a wide frequency range the time-constant should be made small. According to eqn (5.42) this calls for a high loop gain K and a small seismic mass m. In practice, their values will be limited by design considerations, except perhaps for K_a which can be adjusted to meet the particular requirements.

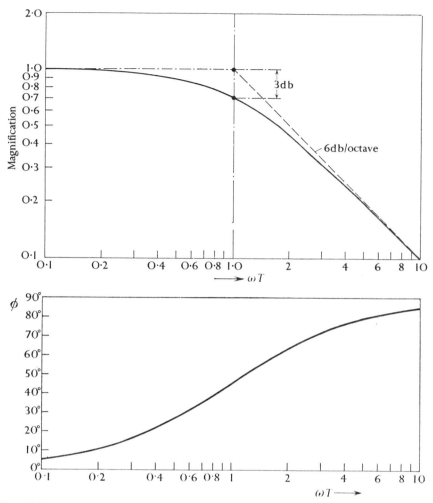

FIG. 5.15. Dynamic performance of feedback-type acceleration transducer to sinusoidal excitation, with velocity compensation only: (a) amplitude response; (b) phase response.

In principle, the response curve of Fig. 5.15(a) includes the static response at $\omega = 0$, but it is obvious that static values of linear acceleration cannot be measured, because balance conditions require a finite and unidirectional velocity

$$v = mA/\alpha K = TA, \tag{5.46}$$

where A (m s^{-2}) is the linear acceleration input and T (s) is the system's time-constant. (At standstill the velocity sensor would have no output and a balancing current would not be available in the actuator.)

However, with an angular acceleration transducer a continuous uni-directional rotation would be feasible. In Fig. 5.16 a rotary mass is coupled to an angular velocity sensor (d.c. generator), the output of which is fed to a d.c. servo-motor, located on the same shaft with the rotary mass and the generator. The servo-motor current is thus a measure of the angular acceleration applied to the (total) rotor mass. This system is at equilibrium at an angular velocity, analogous to the linear velocity of eqn (5.46), which is determined by the angular-acceleration input. At finite amplifier gain the velocity must be finite and greater than zero,

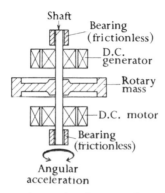

FIG. 5.16 Feedback-type angular-acceleration transducer with velocity compensation only, schematic diagram.

since at standstill the generator output, and thus the feedback current, would vanish. This angular velocity, then, is a direct measure of angular acceleration, but since it also depends on loop gain, stability and accuracy of this transducer type are not in the same class as those of force-balance transducers with displacement (and velocity) compensation. However, with highly stable amplification, or no amplification at all, the system may lead to an angular-acceleration transducer with digital output, obtained from the shaft speed, by using a pulse-generator pick-off and a counter.

In the field of pressure transducers consider a moving-coil microphone, shown schematically in Fig. 5.17. Sound pressure applied to the freely suspended system produces a voltage proportional to the sinusoidal system velocity which is fed back, via an amplifier, to a moving-coil actuator rigidly coupled with the sensor coil. Equilibrium is obtained as described above, and the response curves are similar to those shown in Figs. 5.15(a) and (b). In contrast to a conventional (open-loop) moving-coil microphone, or for that matter to any other known microphone, there is no primary resonance and the frequency range is adjustable by electric means in the amplifier—subject, of course, to any secondary mechanical resonances in the system.

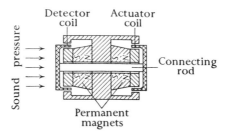

FIG. 5.17. Feedback-type moving-coil microphone, schematic diagram.

5.4. Force-balance transducer systems

This section is to illustrate the scope and diversity of actual force-balance transducer systems devised to measure physical quantities which can be reduced to (small) forces. We shall see that the basic transducer functions—displacement detection, error signal processing, and generation of balancing forces—have been obtained, and combined, in a variety of ways in accordance with specific requirements and the state of the art.

5.4.1. *Force-balance transducers with electrodynamic servo-actuators*

5.4.1.1. Linear acceleration input. The majority of practical force-balance acceleration transducers are pendulous types, i.e. the seismic mass is realized by a physical pendulum measuring linear acceleration applied normal to the axis of the pendulum. An early instrument of this type is based on a conventional d.c. meter movement [8]. The moving-coil frame carries an out-of-balance mass; the angular displacement is measured by a vane constituting a variable-capacitance displacement sensor, and the balancing torque is provided by the feedback current in the moving coil. The variable sensor capacitance is connected to the tank circuit of a super-regeneration r.f. (valve) oscillator which provides short pulses at the quench frequency. Pulse-spacing, and thus average current, depends on sensor capacity, which is proportional to acceleration input. The resulting (averaged) output current is amplified and fed to the servo-coil. In order to obtain positive and negative readings a fixed bias current is supplied from a second d.c. source. A similar instrument also used a pivoted meter movement, but employed a push–pull variable-capacitance sensor in an a.c.-fed bridge circuit [2].

In another pendulous type of force-balance acceleration transducer an inverted pendulum mass is supported by a flexure-type hinge. Force-balance action derives from a differential-transformer-type linear-displacement detector, a high-gain amplifier, and an electrodynamic servo-actuator [9]. A general-purpose instrument of a similar concept, though different in construction and performance, covers acceleration ranges from 0·25 g to 50 g, with good accuracies at a wide range of environmental conditions [10].

Force-balance accelerometers for use in inertial navigation are in a class of their own [3], [11]. Though basically pendulous devices with either inductance- or capacitance-type displacement sensors and electrodynamic servo-actuators, the greatest possible care has been taken in their design, construction, and calibration [4] in order to satisfy the very demanding specification related to instruments for use in inertial navigation systems. Their accuracy is typically 10^{-4} g, and their high feedback stiffness of about $1''$ of arc displacement per g virtually guarantees

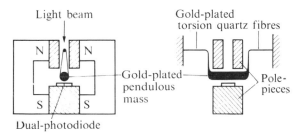

FIG. 5.18. Force-balance-type linear-acceleration transducer with gold-plated pendulous mass and torsion springs, schematic diagram.

freedom from cross-coupling and rectification errors (see section 5.2.1.2). An instrument of comparable performance intended for 'environmental studies' has been described in reference [12]. Flexure elements and a seismic (pendulous) mass are formed from a single piece of specially processed quartz which results in reduced hysteresis, instabilities, and fatigue normally associated with metal flexures. Quartz springs are also in evidence in the accelerometer shown in Fig. 5.18 [13]. The pendulous mass and the torsion fibres are here made from gold-plated quartz rods and fibres, which also carry the servo-current. The position of the shadow of the displaced mass is detected by a dual photo-diode which is illuminated by a small lamp.

Problems arising in the design of multi-axis acceleration transducers are discussed at great length in reference [14]. The solution offered consists of a cluster of single-axis accelerometers which are arranged in such a fashion that sensitivity to angular acceleration and angular velocity about any axis are suppressed. The individual instruments are off-loaded moving-coil types with push–pull capacitance angular-displacement sensors.

The last example in this section concerns a non-pendulous instrument [15]. The (rectilinear) displacement sensor is a conventional push–pull variable-capacitance pick-off in an a.c. bridge circuit. Solid-state oscillator

and servo-amplifier are contained in the transducer housing of about
25 mm diameter and 50 mm length. The servo-actuator coil is suspended
by three pairs of leaf springs which permit (small) coil movements along
the axis of the accelerometer, but act like rigid columns in compression or
tension in other directions (Fig. 5.19).

5.4.1.2. Pressure input.   A number of commercial force-balance pressure
transducers, intended mainly for industrial process-control purposes, are
described by Rohrbach [16] and Oliver [17]. They are usually distin-
guished by a substantial beam pivoted at the centre, and loaded at one end
by a pressure-summing element, such as a capsule or a Bourdon tube, and
on the other by an electrodynamic (or electromagnetic) servo-actuator.

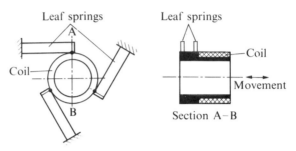

FIG. 5.19. Three-point suspension of servo-coil, schematic diagram.

Here we shall not discuss these instruments any further, since—useful as
they are for static and quasi-static pressure measurements—they are not
suitable, and not intended, for high-frequency or fast-transient inputs.
Their accuracy is generally high, and a specific system [18] may even
qualify for use as a pressure standard.

    A pressure transducer with direct force-balance action was developed
as early as 1953 at the British Iron and Steel Research Association for
the measurement and control of very small pressure differences between
the combustion chamber of an open-hearth furnace and the surrounding
air [19]. Fig. 5.20 shows the familiar elements of a differential-
transformer for displacement sensing of the diaphragm and an electrodynamic servo-actuator. In the early 1950s, of course, oscillator and
carrier amplifier were still valve-operated, but otherwise the offered
solution comprises the essential elements of fast-responding force-balance
transducers, except feedback damping. The author also suggests that, if
the permanent magnet of the actuator is replaced by an electromagnet
energized in series with the actuator coil, the restoring force will be
proportional to the square of the feedback current and, in conjunction
with a (square-law) orifice-type flow-rate sensor, the transducer becomes
a gas flow meter with a linear scale for flow rates.

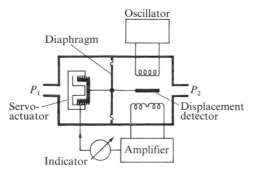

FIG. 5.20. Force-balance-type pressure transducer for small differential gas pressures, schematic diagram.

5.4.1.3. Miscellaneous inputs. In a less well-known type of force-balance transducer the feedback loop comprises a vibrating contact. The restoring force, then, is obtained by time-duration modulation of the (d.c.) feedback current. Fig. 5.21 illustrates the principle, here applied to the measurement of small gas-pressure differences. The moving contact at the bottom of the spindle, normally closed due to gravitation, is broken by an upward force generated by the current flowing through the moving coil in the magnetic field. The de-energized coil, then, allows the contact to be closed again and the cycle restarts. The dwell time of the contact, i.e. the on-off ratio of the current flowing in the circuit, is controlled by an axial force derived from a pressure difference across the (highly flexible) diaphragm. The arrangement of Fig. 5.21 constitutes a discontinuous force-balance system; the input force, generally in a downward direction, is opposed by an electrodynamic upward force acting at closed-contact conditions, i.e. only during part of the cycle of vibration. A closer

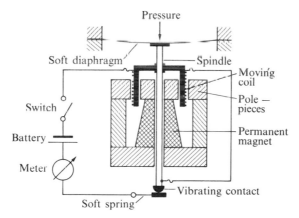

FIG. 5.21. Vibrating-contact-type pressure transducer, schematic diagram.

study of the effect of design parameters on the dynamic performance of the transducer showed [20] that the instrument will be particularly suitable to measure with fair accuracy gas-pressure differences as low as a few millimetres of water column by simply averaging the time-duration modulated feedback current in a 'slow' milliammeter. The advantages of the scheme are, no doubt, simplicity and low cost, but maintaining consistent contact performance over long periods may prove difficult. Because of its sensitivity to environmental acceleration and vibration the instrument is suitable for stationary installation only. The origin of the vibrating-contact principle goes back to the electric bell; its earliest application to the measurement of physical quantities was perhaps the so-called Schmitt accelerometer for use in German missiles [21], [22]. A force-measuring version was suggested by Douce and Yavuz [23] and a pressure sensor by Price and Chapman [24].

The measurement of angular acceleration at practical levels is difficult and leads to unwieldy instruments for the lower ranges. In fact, there seems to be on the market no electrodynamic force-balance transducer for the measurement of angular acceleration, except a complex gyro-based sensor with a servo-motor providing the restoring torque. (An electrostatic force-balance angular accelerometer of high sensitivity will be discussed in section 5.4.2.3 below.)

A feedback angle-of-attack transducer for use in aircraft and missiles is described in reference [25]. A servo-motor automatically rotates an exposed cylindrical probe to face the direction of the local airstream at which the differential pressure between two rows of orifices approaches zero, and an (open-loop) pressure sensor provides an error signal which is fed to the servo-motor via a solid-state amplifier.

5.4.2. Force-balance transducers with electrostatic servo-actuators

5.4.2.1. Linear acceleration input. The need for accurate guidance and control of space vehicles, particularly those with low values of thrust derived from electrical propulsion, has stimulated the development of electrostatic force-balance transducers for the measurement of low-level acceleration, typically 10^{-6} g and lower.

An early single-axis type with a cylindrical mass exhibited a threshold sensitivity of 10^{-5} g [26]. A three-axis instrument aiming at a threshold sensitivity of 10^{-9} g has been developed in France [27]–[29]. The seismic mass is a hollow aluminium sphere of 40 mm diameter suspended by three orthogonal pairs of ring-shaped forcing electrodes (Fig. 5.22). Three separate orthogonal sensing electrode pairs constitute variable capacitances which are used for the measurement of mass displacement. The housing of the transducer is evacuated in order to eliminate excessive air damping, and to avoid electrical breakdown. Since, in the laboratory, the

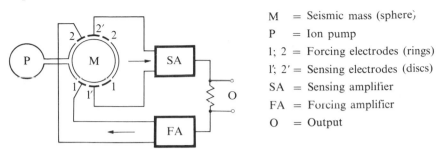

M   = Seismic mass (sphere)
P   = Ion pump
1; 2 = Forcing electrodes (rings)
1'; 2' = Sensing electrodes (discs)
SA   = Sensing amplifier
FA   = Forcing amplifier
O   = Output

FIG. 5.22. Electrostatic force-balance transducer system for low-g acceleration.

environmental random acceleration level is usually at 10^{-5} g or higher, calibration at g-levels below this have to be done under free-fall conditions, i.e. at simulated weightlessness. Reference [28] describes the calibration procedure, and reference [29] describes flight tests of the transducer prototype.

Progress has also been reported recently on highly accurate electrostatically suspended gyroscopes ([11], [30]) using solid or hollow metal spheres spinning at 150 000 r.p.m. and 60 000 r.p.m., respectively, and also on hybrid instruments which would perform the functions of both a gyroscope and a linear accelerometer, for use in inertia navigation systems.

5.4.2.2. Pressure input. Because of the low force levels available from electrostatic servo-actuators, practical force-balance pressure transducers employing electrostatic restoring forces are restricted to low pressure ranges. Based on some earlier work in Holland [31] and Germany [32], Cope [33] developed and analysed an electrostatic force-balance transducer for absolute pressures up to about 30 N m^{-2}. The calibration curve related to a bias voltage of 100 V was linear and had a slope of 1·65 V N^{-1} m^2, or 220 V Torr^{-1}. The sensor capacitances were connected in a resonance-bridge circuit [34], and the feedback voltage was derived from a (valve-operated) phase-sensitive demodulator-cum-amplifier.

The electrostatic force-balance vacuum gauge described by Frank [35] differs in some details from other systems discussed in this section. The capacitance bridge is single-ended and obtains bridge balance by high-stability resistive and capacitive components. After demodulation and amplification the rectified feedback voltage is fed via a diode switch to only one side of the transducer at a time, depending on the sign of the diaphragm displacement. The restoring force thus follows the quadratic law of eqn (5.8), and the results of our discourse in section 5.2.2.2 on the linearity of electrostatic feedback transducers do not apply here. On the other hand, this feature makes the scheme most suitable for use with an orifice-type flow meter, as mentioned in section 5.4.1.2 above.

A low-range differential gas-pressure transducer with electrostatic force-balance action has recently been developed at the Royal Aircraft Establishment, Farnborough [5]. Its high accuracy at static pressure differences of typically ± 40 N m^{-2} should make it suitable as a secondary standard for these very low ranges. Some subsidiary applications of the basic pressure unit will be discussed in the section to follow.

Fig. 5.23 shows the force-balance transducer system for the accurate measurement of small differential gas pressures (see also section 5.2.2).

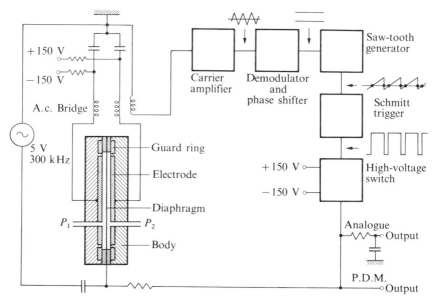

FIG. 5.23. Electrostatic force-balance transducer system for low differential gas pressures.

The displacement of the diaphragm is sensed by push–pull capacitance variations in the air gaps affecting a Blumlein bridge circuit (section 4.4.3.1). The output from the demodulator controls the output level of a saw-tooth generator which, by way of a Schmitt trigger, produces a pulse-duration modulated square-wave. The feedback voltage, then, is obtained from a high-voltage transistor switch controlled by the Schmitt trigger (straightforward analogue high-voltage amplification has proved less attractive). The transducer output is either a pulse-duration modulated voltage or an analogue voltage via a low-pass filter.

In an experimental model the total voltage gain was 10^5, providing a feedback voltage swing of ± 150 V. Calibration of the transducer at the appropriate low pressure ranges was a problem; comparison with a precision water manometer really did not do justice to the inherent high accuracy of the force-balance instrument, but at least showed that the

sensitivity computed from transducer geometry was in close agreement
with, and probably more accurate than, the manometer readings.

5.4.2.3. Miscellaneous inputs.   There are a number of physical quantities
that can be measured with greater accuracy and better dynamic response
by replacing conventional transducers by force-balance types. For instance, the electrostatic differential gas-pressure transducer described
above could be used to design a rate-of-climb sensor with much shorter
than common time-lags for use in aircraft and gliders. These instruments

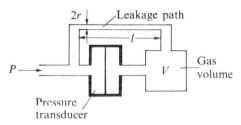

FIG. 5.24. Rate-of-climb transducer, schematic diagram.

operate by employing a controlled leak between the two pressure chambers (Fig. 5.24). The indicated pressure difference ΔP is given by

$$\Delta P = T(dP/dt) \quad (N\ m^{-2}) \tag{5.47}$$

where

$$T \simeq (8\eta lV/\pi\rho c^2 r^4) \quad (s) \tag{5.48}$$

is the time-constant of the leakage path in combination with the gas
volume V (for other symbols see section 3.5). Because of the high
sensitivity of the feedback pressure transducer, time-lags as low as $0\cdot1$ s
would be feasible for a typical climb rate of $1\cdot8$ km (6000 ft) min^{-1} [5].

The high sensitivity of the electrostatic pressure transducer also permits
to measure even the small pressure differences generated by angular
acceleration in a circular gas column. With reference to Fig. 5.25 the
indicated pressure difference can be shown to be [5]

$$\Delta P = \tfrac{1}{2}\pi n\rho D^2 A' \quad (N\ m^{-2}) \tag{5.49}$$

where A' (rad s^{-2}) is the angular acceleration, n the number of turns of
the helix, D (m) its mean diameter, and ρ (kg m^{-3}) the gas density [36].

An experimental model with 13 helical turns and 86 mm mean diameter, filled with dry air at room temperature and atmospheric pressure,
produced an output of about $0\cdot5$ V for an angular acceleration of
1 rad s^{-2}. This high sensitivity can be boosted further, by a factor of at
least 10, if the air is replaced by a heavy chlorofluorhydrocarbon gas
under high pressure. The gas-column concept eliminates suspension problems and, since it forms a closed path, it is not affected by linear

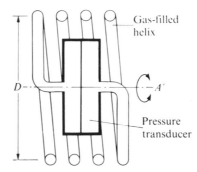

FIG. 5.25. Gas-inertia angular-acceleration transducer, schematic diagram.

acceleration. By virtue of being a sealed system it is also independent of ambient temperature variations. Experimental results and predicted sensitivity were in good agreement.

Finally, we may suggest here a scheme for the (inertial) measurement of extremely low values of angular velocity. Fig. 5.26 shows a 'dumb-bell'-shaped mass floating in a liquid of density ρ_0 (kg m^{-3}); it is guided by weak springs (not shown) to move in a longitudinal direction between two fixed electrodes. If the two masses 1 and 2 are of similar shape and volume V (m^3), and initially at equal distances r (m) from the axis of rotation, but have different densities such that

$$\rho_1 < \rho_0 < \rho_2 \tag{5.50}$$

and, in particular,

$$\rho_0 - \rho_1 = \rho_1 - \rho_0 = \Delta\rho, \tag{5.51}$$

then the dumb-bell floats in the liquid, unless the whole system rotates

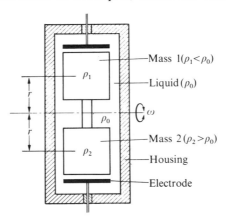

FIG. 5.26. Floating-mass low-range angular-velocity transducer, schematic diagram.

about its axis at an angular velocity ω (s^{-1}). In that case a longitudinal force

$$F = 2Vr\,\Delta\rho\omega^2 \quad \text{(N)} \tag{5.52}$$

is generated which tries to move the dumb-bell in the direction of the heavier mass, 2. This movement can be opposed by an electrostatic balancing force exerted on the float by the two fixed electrodes. The feedback voltage at balance condition, then, is proportional to ω^2. Angular velocities as low as 1 rad s^{-1} should be easily detectable, though the sense of rotation, of course, would not be available. Since the dumb-bell floats, the transducer should be insensitive to linear acceleration and vibration. Also, with the liquid being enclosed in a rigid housing, its density is independent of temperature variations. On the other hand, the zero position of the float can be adjusted by slightly changing the volume available to the liquid, thus changing its pressure and density.

References

1. G. J. SAVANT. *Basic feedback control system design*, Chap. 5. McGraw-Hill, New York (1958).
2. W. R. MACDONALD. Acceleration transducers of the force-balance type. In *Flight test instrumentation* (ed. M. A. Perry), pp. 15–23. Pergamon Press, Oxford (1961).
3. R. H. PARVIN. *Inertial navigation*, Chap. 7. Van Nostrand, Princetown, New Jersey (1962).
4. I. L. THOMAS and R. H. EVANS. *Performance characteristics and methods of testing for force-feedback accelerometers*. Aeronautical Research Council R. & M. No. 3601. H.M.S.O., London (1969).
5. W. R. MACDONALD and C. KING. An electrostatic feedback transducer for measuring very low differential pressures. R.A.E. Technical Report No. TR 71022 (1971)
6. H. K. P. NEUBERT. Notes on the stability and dynamic response of an electrostatic feedback pressure transducer. R.A.E. Technical Report No. TR 70120 (1970).
7. J. C. JAEGER. *An introduction to the Laplace transformation*. Methuen, London (1949).
8. V. B. COREY. Servo accelerometer uses r.f. oscillator. *Electronics* **29**, 151–3 (1956).
9. Kearfott Division, General Precision, Inc., Little Falls, New Jersey.
10. Electro Mechanisms Ltd., Slough, Berks.
11. R. L. GATES. Reviewing the status of inertial sensors. *Control Engng* **18**, 54–8 (1971).
12. E. D. JACOBS. New developments in servo accelerometers. *Proceedings of the 14th Annual Meeting of the Institution of Environmental Sciences*, St. Louis, Missouri (1968).
13. Reeves Instrument Division, Dynamics Corp., Garden City, New York.
14. V. B. COREY. Multi-axis clusters of single-axis accelerometers with coincident centers of angular motion insensitivity. *Proceedings of the 6th International Aerospace Instrumentation Symposium*, Cranfield, Bedford. Peter Peregrinus Ltd., Stevenage, Herts. (1970).
15. Kistler Instruments, Ltd., Farnborough, Hants.
16. C. ROHRBACH. *Handbuch für elektrisches Messen mechanischer Grössen*, pp. 530–2. V.D.I. Verlag, Düsseldorf (1967).
17. F. J. OLIVER. *Practical instrumentation transducers*, pp. 151–3. Hayden Books, New York (1971).
18. Bell & Howell, Ltd., Basingstoke, Hants.
19. B. O. SMITH. A differential pressure meter for open hearths. *Instruments* **26** November issue, 1706–7 (1953).
20. H. K. P. NEUBERT and E. F. PRICE. Vibrating-contact pressure transducer. *Instrums. Control Syst.* **42**, 81–4 (1969).

21. T. M. MOORE. V-2 range control techniques. *Electl. Engng., N.Y.* **65**, 303–5 (1946).
22. T. M. MOORE. German missile accelerometers. *Electl. Engng., N.Y.* **68**, 996–9 (1949).
23. J. L. DOUCE and D. YAVUZ. A novel force-current transducer. *Control* **9**, 624–5 (1965).
24. E. F. PRICE and M. CHAPMAN. A sensitive low-range force-current pressure transducer. *Control* **11**, 128–9 (1967).
25. V. B. COREY. Electrically servoed transducers for flight control. *Proceedings of the 5th International Aerospace Instrumentation Symposium, Cranfield, Bedford.* The College of Aeronautics, Cranfield, Beds. (1968).
26. ANON. Low-g accelerometers await new ground and flight test development. *Missiles Rockets* **15** (23) 32–5 (1964).
27. M. GAY. Recherche sur un principe d'accéléromètre à grande sensibilité. *Rech. Aerospatiale* **110**, January/February issue, 39–44 (1966).
28. M. DELATTRE. The low-g ONERA accelerometer. *Proceedings of the 5th International Aerospace Instrumentation Symposium, Cranfield, Bedford.* The College of Aernnautics, Cranfield, Beds. (1968).
29. M. DELATTRE. Flight-testing the high-sensitivity ONERA accelerometer. *Proceedings of the 6th International Aerospace Instrumentation Symposium, Cranfield, Bedford.* Peter Peregrinus Ltd., Stevenage, Herts. (1970).
30. P. J. KLASS. New gyro nears operational use. *Aviat. Week* **96**, 50–2 (1972).
31. J. J. OBSTELTEN, N. WARMOLTZ, and J. T. ZALLBERG VAN ZELST. A direct-reading double-sided micromanometer. *Appl. scient. Res.* **B6**, 129–36 (1956).
32. H. W. DRAVIN. Über die Verwendbarkeit des Kapazitätsmikromanometers als Absolutdruck-Messinstrument. *Z. InstrumKde* **68**, 1–8 (1960).
33. J. O. COPE, Direct-reading diaphragm-type pressure transducer, *Rev. scient. Instrum.* **33**, 980–4 (1962).
34. J. J. OPSTELTEN and N. WARMOLTZ. A double-sided micromanometer. *Appl. scient. Res.* **B4**, 329–36 (1955).
35. P. M. FRANK. An automatic compensating method for measuring pressures varying within a range of 10^{-6}–10^{-12} torr independent of the type of gas employed. *Microtecnic* **21**, 291–3 and 388–90 (1967).
36. W. R. MACDONALD. A guide to the design and performance of angular acceleration transducers with gas rotors. R.A.E. Technical Report No. TR 74146 (1974).

Appendix 1: The International (SI) system of units and conversion tables[†]

PHYSICAL QUANTITIES	UNITS	SYMBOLS
A. BASIC UNITS		
length	metre	m
mass	kilogram	kg
time	second	s
electric current	ampere	A
temperature	(degree) kelvin	K
luminous intensity	candela	cd
B. SUPPLEMENTARY UNITS[‡]		
plane angle	radian	rad
solid angle	steradian	sr
C. DERIVED UNITS		
area	square metre	m^2
volume	cubic metre	m^3
velocity	metre per second	$m\,s^{-1}$
angular velocity	radian per second	$rad\,s^{-1}$
acceleration	metre per second squared	$m\,s^{-2}$
angular acceleration	radian per second squared	$rad\,s^{-2}$
frequency	hertz	Hz
density	kilogram per cubic metre	$kg\,m^{-3}$
momentum	kilogram metre per second	$kg\,m\,s^{-1}$
angular momentum	kilogram metre squared per second	$kg\,m^2\,s^{-1}$
moment of inertia	kilogram metre squared	$kg\,m^2$
force	newton	$N; kg\,m\,s^{-2}$
moment of force	newton metre	$N\,m$
pressure, stress	newton per square metre	$N\,m^{-2}$
viscosity, dynamic	newton second per square metre	$N\,s\,m^{-2}$
viscosity, kinematic	square metre per second	$m^2\,s^{-1}$
energy, work, heat	joule	$J; N\,m; W\,s$
power, heat flow rate	watt	$W; J\,s^{-1}$
customary temperature	degree (celsius)	$°C = K - 273 \cdot 15$
thermal conductivity	watt per metre degree	$W\,m^{-1}\,K^{-1}$
heat capacity	joule per degree	$J\,K^{-1}$
entropy	joule per (degree) kelvin	$J\,K^{-1}$
electric charge	coulomb	$C; A\,s$
electric potential difference	volt	$V; W\,A^{-1}$
electric resistance	ohm	$\Omega; V\,A^{-1}$
electric resistivity	ohm metre	$\Omega\,m$
electric field strength	volt per metre	$V\,m^{-1}$

[†] ANDERTON, P. and BIGG, P. H. (1967). *Changing to the metric system; conversion factors, symbols and definitions*, Nat. Phys. Lab., H.M.S.O., London.
[‡] These units are dimensionless.

PHYSICAL QUANTITIES	UNITS	SYMBOLS
electric capacity	farad	F; As V^{-1}
magnetic flux	weber	Wb; Vs
magnetic flux density	tesla	T; Wb m^{-2}
magnetic field strength	ampere (turn) per metre	A m^{-1}
magnetomotive force	ampere (turn)	A
inductance	henry	H; Vs A^{-1}
luminous flux	lumen	lm; cd sr
luminance	candela per square metre	cd m^{-2}
illumination	lux	lx; lm m^{-2}
permeability of free space		$\mu_0 = 4\pi \times 10^{-7}$ H m^{-1}
permittivity of free space		$\varepsilon_0 = 8 \cdot 854 \times 10^{-12}$ F m^{-1}

D. MULTIPLES AND FRACTIONS

Multiple	Prefix	Symbol	Fraction	Prefix	Symbol
10^{12}	tera	T	10^{-1}	deci†	d
10^9	giga	G	10^{-2}	centi†	c
10^6	mega	M	10^{-3}	milli	m
10^3	kilo	k	10^{-6}	micro	μ
10^2	hecto†	h	10^{-9}	nano	n
10	deca†	da	10^{-12}	pico	p
			10^{-15}	femto	f
			10^{-18}	atto	a

† The use of these multiples and fractions is not encouraged.

E. CONVERSION TABLES

1. Length

m	in.	ft	yd	mile (U.K.)	nautical mile
1	39·37	3·281	1·093	$6·214 \times 10^{-4}$	$5·397 \times 10^{-4}$
$2·54 \times 10^{-2}$	1	$8·333 \times 10^{-2}$	$2·778 \times 10^{-2}$	$1·578 \times 10^{-5}$	$1·371 \times 10^{-5}$
0·3048	12	1	0·3333	$1·894 \times 10^{-4}$	$1·645 \times 10^{-4}$
0·9144	36	3	1	$5·682 \times 10^{-4}$	$4·935 \times 10^{-4}$
1609	$6·336 \times 10^4$	5280	1760	1	0·8684
1852	$7·291 \times 10^4$	6076	2024	1·151	1

1 Å (ångström) $= 10^{-7}$ mm $= 3·937 \times 10^{-9}$ in.
1 thou (U.K.) $=$ 1 mil (U.S.A.) $= 10^{-3}$ in. $= 2·54 \times 10^{-2}$ mm.

2. Area

m^2	cm^2	in.^{2}	ft^2	yd^2	acre
1	10^4	1550	10·76	1·196	$2·471 \times 10^{-4}$
10^{-4}	1	0·155	$1·076 \times 10^{-3}$	$1·196 \times 10^{-4}$	$2·471 \times 10^{-8}$
$6·452 \times 10^{-4}$	6·452	1	$6·944 \times 10^{-3}$	$7·716 \times 10^{-4}$	$1·594 \times 10^{-7}$
$9·29 \times 10^{-2}$	929	144	1	0·1111	$2·296 \times 10^{-5}$
0·8361	8361	1296	9	1	$2·066 \times 10^{-4}$
4047	$4·047 \times 10^7$	$6·273 \times 10^6$	$4·356 \times 10^4$	4840	1

1 circular mil $= 7·854 \times 10^{-7}$ in.^{2} $= 5·067 \times 10^{-4}$ mm^2.

3. Volume

m³	litre	gal (U.K.)	in.³	ft³	yd³
1	10^3	220	$6\cdot102\times10^4$	35·32	1·308
10^{-3}	1	0·22	61·02	$3\cdot532\times10^{-2}$	$1\cdot308\times10^{-3}$
$4\cdot546\times10^{-3}$	4·546	1	277·4	0·1605	$5\cdot946\times10^{-3}$
$1\cdot639\times10^{-5}$	$1\cdot639\times10^{-2}$	$3\cdot605\times10^{-3}$	1	$5\cdot785\times10^{-4}$	$2\cdot144\times10^{-5}$
$2\cdot832\times10^{-2}$	28·32	6·229	1728	1	$3\cdot704\times10^{-2}$
0·7646	764·6	168·2	$4\cdot666\times10^4$	27	1

1 U.S.A. gal = 0·8326 U.K. gal = 3·785 litre.

4. Angle

radian	degree	minute	second
1	57·3	3438	$2\cdot063\times10^5$
$1\cdot745\times10^{-2}$	1	60	3600
$2\cdot909\times10^{-4}$	$1\cdot667\times10^{-2}$	1	60
$4\cdot848\times10^{-6}$	$2\cdot778\times10^{-4}$	$1\cdot667\times10^{-2}$	1

5. Velocity

m s⁻¹	km h⁻¹	in. s⁻¹	ft min⁻¹	m.p.h.	knot
1	3·6	39·37	196·85	2·237	1·944
0·2778	1	10·93	54·68	0·6214	0·540
$2\cdot54\times10^{-2}$	$9\cdot144\times10^{-2}$	1	5	$5\cdot682\times10^{-2}$	$4\cdot938\times10^{-2}$
$5\cdot08\times10^{-3}$	$1\cdot829\times10^{-2}$	0·2	1	$1\cdot136\times10^{-2}$	$9\cdot876\times10^{-3}$
0·4470	1·609	17·60	88	1	0·869
0·5144	1·852	20·25	101·27	1·151	1

6. Acceleration

m s⁻²	g	in. s⁻²	ft s⁻²
1	0·1019	39·37	3·281
9·81	1	386·2	32·18
$2\cdot54\times10^{-2}$	$2\cdot589\times10^{-3}$	1	$8\cdot33\times10^{-2}$
0·3048	$3\cdot108\times10^{-2}$	12	1

7. Mass

kg	oz	lb	ton	ton (short)	slug
1	35·27	2·205	$9\cdot842\times10^{-4}$	$1\cdot102\times10^{-3}$	$6\cdot852\times10^{-2}$
$2\cdot835\times10^{-2}$	1	$6\cdot25\times10^{-2}$	$2\cdot79\times10^{-5}$	$3\cdot124\times10^{-5}$	$1\cdot942\times10^{-3}$
0·4536	16	1	$4\cdot464\times10^{-4}$	5×10^{-4}	$3\cdot108\times10^{-2}$
1016	$3\cdot583\times10^4$	2240	1	1·12	69·62
907·2	$3\cdot2\times10^4$	2000	0·8929	1	62·16
14·59	514·7	32·18	$1\cdot436\times10^{-2}$	$1\cdot608\times10^{-2}$	1

1 tonne = 10^3 kg.

8. Density

kg m^{-3}	g cm^{-3}†	lb in.$^{-3}$	lb ft^{-3}	ton yd^{-3}	slug ft^{-3}
1	10^{-3}	$3\cdot613\times10^{-5}$	$6\cdot243\times10^{-2}$	$7\cdot524\times10^{-4}$	$1\cdot94\times10^{-3}$
10^3	1	$3\cdot613\times10^{-2}$	62·43	0·7524	1·94
$2\cdot768\times10^4$	27·68	1	1728	20·83	53·7
16·02	$1\cdot602\times10^{-2}$	$5\cdot788\times10^{-4}$	1	$1\cdot205\times10^{-2}$	$3\cdot108\times10^{-2}$
1329	1·329	$4\cdot802\times10^{-2}$	82·97	1	2·579
515·4	0·5154	$1\cdot862\times10^{-2}$	32·18	0·3878	1

† g cm^{-3} = kg l.$^{-3}$ = specific density (relative to water).

9. Moment of inertia

kg m^2	g cm^2	oz in.^{2}	lb in.^{2}	lb ft^2	slug ft^2
1	10^7	$5\cdot467\times10^4$	$3\cdot417\times10^3$	23·73	0·7375
10^{-7}	1	$5\cdot467\times10^{-3}$	$3\cdot417\times10^{-4}$	$2\cdot373\times10^{-6}$	$7\cdot375\times10^{-8}$
$1\cdot829\times10^{-5}$	$1\cdot829\times10^2$	1	$6\cdot25\times10^{-2}$	$4\cdot34\times10^{-4}$	$1\cdot349\times10^{-5}$
$2\cdot926\times10^{-4}$	$2\cdot926\times10^3$	16	1	$6\cdot943\times10^{-3}$	$2\cdot158\times10^{-4}$
$4\cdot214\times10^{-2}$	$4\cdot214\times10^5$	2304	144	1	$3\cdot108\times10^{-2}$
1·356	$1\cdot356\times10^7$	$7\cdot413\times10^4$	4633	32·18	1

10. Force

N	dyne	kgf (kp)	lbf	tonf	pdl
1	10^5	0·102	0·2248	$1\cdot004\times10^{-4}$	7·233
10^{-5}	1	$1\cdot02\times10^{-6}$	$2\cdot248\times10^{-6}$	$1\cdot004\times10^{-9}$	$7\cdot233\times10^{-5}$
9·807	$9\cdot807\times10^5$	1	2·205	$9\cdot846\times10^{-4}$	70·93
4·448	$4\cdot448\times10^5$	0·4537	1	$4\cdot466\times10^{-4}$	32·17
$9\cdot964\times10^3$	$9\cdot964\times10^8$	$1\cdot016\times10^3$	$2\cdot24\times10^3$	1	$7\cdot207\times10^4$
0·1383	$1\cdot383\times10^4$	$1\cdot411\times10^{-2}$	$3\cdot109\times10^{-2}$	$1\cdot389\times10^{-5}$	1

11. Pressure, stress

N m^{-2}	dyne cm^{-2}	kp cm^{-2}	p.s.i.	lbf ft^{-2}	tonf in.$^{-2}$
1	10	$1\cdot02\times10^{-5}$	$1\cdot45\times10^{-4}$	$2\cdot089\times10^{-2}$	$6\cdot477\times10^{-8}$
0·1	1	$1\cdot02\times10^{-6}$	$1\cdot45\times10^{-5}$	$2\cdot089\times10^{-3}$	$6\cdot477\times10^{-9}$
$9\cdot807\times10^4$	$9\cdot807\times10^5$	1	14·22	2049	$6\cdot352\times10^{-3}$
6895	$6\cdot895\times10^4$	$7\cdot033\times10^{-2}$	1	144	$4\cdot466\times10^{-4}$
47·88	478·8	$4\cdot884\times10^{-4}$	$6\cdot943\times10^{-3}$	1	$3\cdot101\times10^{-6}$
$1\cdot544\times10^7$	$1\cdot544\times10^8$	157·5	2239	$3\cdot225\times10^5$	1

1 at (technical) = 1 kp cm^{-2} = 14·223 p.s.i.

12. Pressure (low)

N m^{-2}	m bar	Torr	mm-H$_2$O	in.-Hg	in.-WG
1	10^{-2}	$7\cdot502\times10^{-3}$	0·102	$2\cdot953\times10^{-4}$	$4\cdot014\times10^{-3}$
100	1	0·7502	$1\cdot02\times10^{-3}$	$2\cdot953\times10^{-2}$	0·4014
133·3	1·333	1	13·6	$3\cdot936\times10^{-2}$	0·531
9·807	$9\cdot807\times10^{-2}$	$7\cdot357\times10^{-2}$	1	$2\cdot896\times10^{-3}$	$3\cdot937\times10^{-2}$
3386	33·86	25·40	345·4	1	13·59
249·1	2·491	1·869	25·41	$7\cdot356\times10^{-2}$	1

1 atm (standard) = 760 Torr (mm-Hg) = 14·696 p.s.i.

13. Viscosity, dynamic

Ns m^{-2}	cP	lbf s ft^{-2}
1	10^3	2·089×10^{-2}
10^{-3}	1	2·089×10^{-5}
47·88	4·788×10^4	1

1 P (poise) = 1 dyne s cm^{-2}

14. Viscosity, kinematic

m^2 s^{-1}	cSt	ft^2 s^{-1}
1	10^6	10·76
10^{-6}	1	1·076×10^{-5}
9·29×10^{-2}	9·29×10^4	1

1 St (stoke) = 1 cm^2 s^{-1}.

15. Energy, work, quantity of heat

J	k Wh	kp m	kcal	ft lbf	BTU
1	2·778×10^{-7}	0·102	2·39×10^{-4}	0·7376	9·48×10^{-4}
3·6×10^6	1	3·672×10^5	860·4	2·655×10^6	3413
9·807	2·724×10^{-6}	1	2·344×10^{-3}	7·234	9·297×10^{-3}
4187	1·163×10^{-3}	427·1	1	3088	3·969
1·356	3·767×10^{-7}	0·1383	3·241×10^{-4}	1	1·285×10^{-3}
1·055×10^3	2·931×10^{-4}	107·6	0·2521	778·2	1

1 erg = 1 dyne cm = 10^{-7} N m = 10^{-7} J.
1 eV (electronvolt) = 1·602×10^{-19} J.

16. Power, heat flow rate

W	kp m s^{-1}	ft lbf s^{-1}	kcal h^{-1}	BTU h^{-1}	h.p.
1	0·102	0·7376	0·8598	3·413	1·341×10^{-3}
9·807	1	7·234	8·432	33·47	1·315×10^{-2}
1·356	0·1383	1	1·166	4·628	1 818×10^{-3}
1·163	0·1186	0·8578	1	3·969	1·56×10^{-3}
0·293	2·989×10^{-2}	0·2161	0·2519	1	3·929×10^{-4}
745·7	76·06	550	641·2	2545	1

1 PS = 75 kp m s^{-1} = 0·986 h.p. = 735·5 W.
1 erg s^{-1} = 10^{-7} W.

17. Magnetic units

magnetomotive force: 1 A (turn) = 1·257 Oe cm.
magnetic field strength: 1 A (turn) m^{-1} = 1·257×10^{-2} Oe.
magnetic flux: 1 Wb = 10^8 G cm^2.
magnetic flux density: 1 T = 1 Wb m^{-2} = 10^4 G.
magnetic energy product (B×H): 1 Jm^{-3} = 1 TA m^{-1} = 125·7 G Oe.

Index